Lecture Notes in Physics

Edited by H. Araki, Kyoto, J. Ehlers, München, K. Hepp, Zürich
R. Kippenhahn, München, H. A. Weidenmüller, Heidelberg
and J. Zittartz, Köln

Managing Editor: W. Beiglböck

257

Statistical Mechanics and Field Theory: Mathematical Aspects

Proceedings of the International Conference on the Mathematical Aspects of Statistical Mechanics and Field Theory Held in Groningen, The Netherlands, August 26–30, 1985

Edited by
T. C. Dorlas, N. M. Hugenholtz and M. Winnink

Springer-Verlag
Berlin Heidelberg New York London Paris Tokyo

Editors

T.C. Dorlas
N.M. Hugenholtz
M. Winnink
Institute for Theoretical Physics, University of Groningen
P.O. Box 800, Groningen, The Netherlands

ISBN 3-540-16777-3 Springer-Verlag Berlin Heidelberg New York
ISBN 0-387-16777-3 Springer-Verlag New York Berlin Heidelberg

Printed in Germany

Printing and binding: Beltz Offsetdruck, Hemsbach/Bergstr.
2153/3140-543210

PREFACE

This volume of **LECTURE NOTES IN PHYSICS** contains the text of the lectures presented during the International Conference on the Mathematical Aspects of Statistical Mechanics and Field Theory, held in Groningen, the Netherlands, 26-30 August 1985.

Some of the participants have provided us with an abstract of the poster(s) they presented during the poster sessions. We have incorporated these abstracts as well in this volume. As is unavoidable, in those abstracts claims are presented without proof. Interested readers are encouraged to contact the authors.

The organizers of this conference thought it a good idea to have a conference on mathematical physics in the Netherlands where this subject is receiving more attention nowadays. We hope, among other things, to have given an impetus to mathematical physics in the Netherlands.

We want to express our sincere gratitude to the scientific advisory board, the lecturers, the participants, the Congress Bureau of Groningen University, the secretary of the Institute for Theoretical Physics Ms. M. Boering, and last but not least to the sponsors of this conference, Philips Research Laboratories, Shell Nederland BV, IAMP, Groninger Universiteitsfonds, Koninklijke Nederlandse Akademie van Wetenschappen, College van Bestuur RUG, Fakulteit der Wiskunde en Natuurwetenschappen RUG and Subfakulteit Natuurkunde.

T.C. Dorlas
N.M. Hugenholtz
M. Winnink

Scientifc Advisory Committee: H. Araki, J. Fröhlich, R. Haag,
N.M. Hugenholtz, J.L. Lebowitz

CONTENTS

A Model for Crystallization: A Variation on the Hubbard Model

Tom Kennedy and Elliott H. Lieb
Departments of Mathematics and Physics
Princeton University
Jadwin Hall, P.O. Box 708
Princeton, NJ 08544

Much attention has been paid to the question of proving the existence of long range order in model statistical mechanical systems in which the basic atomic constituents interact with short range forces. An important example is a lattice spin system in which the spin at each site represents the localized spin of an atom located at that site and where the short range, pairwise interaction (Ising or Heisenberg) reputedly comes from an interatomic exchange energy. Another problem - so far unsolved - is the existence of periodic crystals which are supposed to come from short range (e.g. Lennard- Jones) interatomic potentials.

In the real world, however, these interactions are not given a-priori; it is ultimately itinerant electrons and their correlations that give rise to the long range ordering. In other words, a deep unsolved problem is to derive magnetism or crystallization from the Schrödinger equation - or some caricature of it. The construction of a simple model based on itinerant electrons, and the rigorous derivation of ordering from it, is a challenge for mathematical physicists.

A lattice model of itinerant electrons that is believed to display ferro and antiferromagnetism - if it could be solved - is the Gutzwiller-Hubbard-Kanamori model [1-3]. We are also unable to solve it, but we have succeeded in proving that a simplified version of it does display crystallization. It is a toy model but it is, to our knowledge, the first example of this genre. Roughly, it has the same relation to the Hubbard model as the Ising model has to the quantum Heisenberg model. Here we shall give a brief report of our results, the full details of which will appear elsewhere [4].

The Hubbard model, which is the motivation for our model, is defined by the second-quantized Hamiltonian

$$H^H = \sum_{\sigma} \sum_{x,y \in \Lambda} t_{xy} c^{\dagger}_{x\sigma} c_{y\sigma} + 2U \sum_{x \in \Lambda} n_{x\uparrow} n_{x\downarrow} \tag{1}$$

with the following notation: $\sigma = \pm 1$ denotes the 2 spin states of the electrons; Λ is a finite lattice; $c_{x\sigma}$ is a fermion annihilation operator for a spin σ electron at $x \in \Lambda$; $n_{x\sigma} = c^{\dagger}_{x\sigma} c_{x\sigma}$ is the number operator for spin σ at x. Electrons interact only at the same site with an energy $2U$, and $t_{xy} = t_{yx}$ is the hopping amplitude from x to y.

The crucial assumption will be made that Λ is the union of two sublattices $A \cup B$ such that $t_{xy} = 0$ unless $x \in A, y \in B$ or $x \in B, y \in A$. The number of sites in Λ, A, and B are denoted by $|\Lambda|, |A|$ and $|B|$. The two sublattices need not be isomorphic. Thus, for example, a face-centered cubic lattice is allowed with A=face centers and B=cube corners. Λ is said to be *connected* if every $x, y \in \Lambda$ can be joined by a "path" through nonzero t's.

In our model we assume that one kind of electron (say $\sigma = -1$) does not hop. One can say that these electrons are infinitely massive. The Hamiltonian is then

$$H = \sum_{x,y \in \Lambda} t_{xy} c_x^\dagger c_y + 2U \sum_{x \in \Lambda} n_x W(x) \tag{2}$$

with $n_x = c_x^\dagger c_x$ (the subscript σ is omitted since the dynamic electrons have $\sigma = +1$) and with $W(x) = +1$ if a fixed electron ($\sigma = -1$) is at x and $W(x) = 0$ otherwise. It will be recognized immediately that (2) is just a fancy way to say that the movable electrons are independent, with a single particle Hamiltonian

$$\tilde{h} = T + V, \tag{3}$$

with T being the $|\Lambda|$-square matrix t_{xy} and $V_{xy} = 2UW(x)\delta_{xy}$. It is convenient to write $\tilde{h} = h + U$ with

$$h = T + US \tag{4}$$

with $S_{xy} = s_x \delta_{xy}$, and $s_x = 1$ (resp. -1) if x is occupied (resp. unoccupied). The $\{s_x\}$ are like Ising spins.

We shall henceforth call the movable particles "electrons" and the fixed particles "nuclei". This terminology is most appropriate if $U < 0$, for then H does represent a lattice system of electrons and nuclei in which all Coulomb interactions except the on-site electron-nucleus attraction and the on-site infinite nuclear repulsion are regarded as "screened out". This conforms with the spirit of the original Hubbard model. The electron number, the nuclear number and the total particle number, all of which commute with H, are, respectively

$$N_e = \sum_{x \in \Lambda} n_x, \quad N_n = \frac{1}{2} \sum_{x \in \Lambda} [s_x + 1] = \sum_{x \in \Lambda} W(x), \quad \mathcal{N} = N_e + N_n. \tag{5}$$

It is to be emphasized that W does *not* represent a disordered potential. We take the "annealed", not the "quenched" system. The ground state for fixed N_e and N_n is defined by taking the ground state of H (with respect to the electrons) for each W and then minimizing the result with respect to the location of the nuclei. The ground state energy will be denoted by $E(N_e, N_n)$. Likewise, for positive temperature, we take $Tre^{-\beta H}$ with respect to *both* the electron variables and the nuclear locations.

Since $t_{xy} = 0$ for x, y on the same sublattice, the spectra of H and H^H are invariant under $t_{xy} \to -t_{xy}$ (all x, y). There is also a hole-particle symmetry. If $c_x^\dagger \to c_x, c_x \to c_x^\dagger, n_x \to 1 - n_x, t_{xy} \to -t_{xy}$, then $H(U) \to H(-U) + 2UN_n$. If $W(x) \to 1 - W(x)$, then $H(U) \to H(-U) + 2UN_e$. A similar symmetry holds for H^H. Thus, the $U > 0$ and $U < 0$ cases are similar — from the mathematical point of view.

Our results are of two kinds. The first concerns the ground state which we prove always has perfect crystalline ordering and an energy gap (defined later). The second concerns the positive temperature ($1/kT = \beta < \infty$) grand canonical state. For large β and dimension $d \geq 2$, the long range order persists. For small β it disappears and there is exponential clustering of the nuclear correlation functions.

The Ground State

Theorem *1*: (a) *Let $U < 0$. Under the condition $\mathcal{N} \equiv N_e + N_n \leq 2|A|$, the ground state (i.e. we minimize $E(N_e, N_n)$ over the set $N_e + N_n \leq 2|A|$) occurs for $N_e = N_n = |A|$ and a minimizing nuclear configuration is $W(x) = W_A(x) \equiv 1(x \in A), 0(x \notin A)$. Under the condition $\mathcal{N} \leq 2|B|$, the ground state is $N_e = N_n = |B|$ and the B sublattice is occupied ($W = W_B$). If Λ is connected, these are the only groundstates, i.e. if $|A| > |B|$ the ground state is unique; if $|A| = |B|$ it is doubly degenerate. No assumption is made about the sign or magnitude or periodicity of the t_{xy} other than $t_{xy} = 0$ for $x, y \in A$ or $x, y \in B$.*

(b) *Let $U > 0$. Under the condition $\mathcal{N} \geq |A| + |B|$, there are two ground states: $N_e = |A|, N_n = |B|, W = W_B$ and $N_e = |B|, N_n = |A|, W = W_A$. If Λ is connected, these are the only ground states.*

The condition $\mathcal{N} = |\Lambda|$ is called the *half-filled* band. If $|A| = |B| = |\Lambda|/2$, the crystal occurs at the half-filled band. If Λ is a cubic lattice, for example, this means that the ground state is a cubic lattice of period $\sqrt{2}$ oriented at 45° with respect to Λ.

Theorem 1 relies heavily on the fact that the electrons are fermions. The ground state would be completely different if they were bosons. For bosons and for Λ a cubic lattice, the nuclei would all be clumped together in the ground state instead of being spread out into a crystal. By using rearrangement inequalities it is possible to describe this clumping quantitatively.

Next, we define the *energy gap*. Actually two different definitions are of interest. First, let

$$E(\mathcal{N}) \equiv \min\{E(N_e, N_n) | N_e + N_n = \mathcal{N}\}. \tag{6}$$

The *chemical potential* is defined by

$$\mu(\mathcal{N}) \equiv E(\mathcal{N}+1) - E(\mathcal{N}). \tag{7}$$

We say there is a *gap of the first kind* at $\mathcal{N}$ if

$$\mu(\mathcal{N}) - \mu(\mathcal{N}-1) \geq \varepsilon_1 > 0 \tag{8}$$

with ε_1 being independent of the size of the system. We say there is a *gap of the second kind* at N_e, N_n if

$$E(N_e+1, N_n) + E(N_e-1, N_n) - 2E(N_e, N_n) \geq \varepsilon_2 > 0. \tag{9}$$

In other words, the nuclear number is fixed in the second definition.

A gap is one indication that the system is an insulator, for it implies that it costs more energy to put a particle into the system than is gained by removing one. The first kind of gap is relevant if one views our model as an approximation to the Hubbard model; the second is relevant from the "electrons and nuclei" point of view.

Theorem 2: *Assume that Λ is not only connected but that every $x, y \in \Lambda$ can be connected by a chain with $|t_{ab}| \geq \delta$ for every a, b on the chain. Also, assume that $\|T\| \leq \tau$. The following energy gaps exist with $\varepsilon_2 \geq \varepsilon_1 > 0$ and depending only on δ, τ and U:*

$\underline{U<0}$: First kind at $\mathcal{N} = 2|A|$ and at $\mathcal{N} = 2|B|$. Second kind at $N_e = |A|, N_n = |A|$ and at $N_e = |B|, N_n = |B|$.

$\underline{U>0}$: First kind at $\mathcal{N} = |A| + |B|$. Second kind at $N_e = |A|, N_n = |B|$ and at $N_e = |B|, N_n = |A|$.

In order to give the flavor of our methods, the proof of Theorem 1 will be given here. The proof of Theorem 2 is more complicated.

Proof of Theorem 1: Let $\lambda_1 \leq \lambda_2 \leq \ldots$ be the eigenvalues of h in (4). They depend on the nuclei. For N_e electrons the ground state energy, E, of H satisfies

$$E - UN_e = \sum_{j=1}^{N_e} \lambda_j \geq \sum_{\lambda_j<0} \lambda_j = \frac{1}{2}[Trh - Tr|h|]. \tag{10}$$

But $Trh = U\sum s_x = 2UN_n - U|\Lambda|$ and $|h| = \{T^2 + U^2 + UJ\}^{1/2}$ with $J_{xy} = t_{xy}[s_x + s_y]$. Since the function $0 < x \to x^{1/2}$ is concave, $f(y) = Tr\{T^2 + U^2 + yUJ\}^{1/2}$ is concave in $y \in [-1, 1]$. But $f(-1) = f(1)$ (since spec(h) is invariant under $T \to -T$), so $f(1) \leq f(0)$, with equality if and only if $J \equiv 0$. Thus

$$E \geq U\mathcal{N} - \frac{1}{2}U|\Lambda| - \frac{1}{2}Tr(T^2 + U^2)^{1/2}. \tag{11}$$

If Λ is connected, the only ways to have $J \equiv 0$ are either $W = W_A$ or W_B. Consider $U < 0$ and $\mathcal{N} \leq 2|A|$, whence, from (11),

$$E \geq U(2|A| - \frac{1}{2}|\Lambda|) - \frac{1}{2}Tr(T^2 + U^2)^{1/2}. \tag{12}$$

If $W = W_A$ then, as is easily seen, h has precisely $|A|$ negative and $|B|$ positive eigenvalues. Thus, if $W = W_A$ and $N_e = |A|$, then (12) is an equality. The other cases are similar. □

Grand Canonical Ensemble

First, we define the partition function Ξ. A nuclear configuration is denoted by $S = \{s_x\}, s_x = \pm 1$, and the λ_j are the eigenvalues of h in (4). If μ_n, μ_e are the nuclear and electronic chemical potentials,

$$\Xi = \sum_S \exp[\frac{1}{2}\beta\mu_n(\sum_x s_x + |\Lambda|)] \prod_{j=1}^{|\Lambda|} \{1 + \exp[-\beta(\lambda_j + U - \mu_e)]\}. \tag{13}$$

The product in (13) is just the well known Fermi-Dirac grand canonical partition function for the electrons. We want to choose μ_e, μ_n so that $\langle N_e \rangle = \frac{1}{2}|\Lambda|$ and $\langle N_n \rangle = \frac{1}{2}|\Lambda|$, or $\langle \sum s_x \rangle = 0$. From the fact that if $T \to -T$, spec$(h) \to$ spec(h), one has that when $S \to -S$, spec$(h) \to -$spec(h). It is then easy to see that the desired chemical potentials are $\mu_e = \mu_n = U$. Since $\sum \lambda_j = U \sum s_x$, (13) becomes in this case (after dropping an irrelevant factor $2^{|\Lambda|}e^{\beta U|\Lambda|/2}$)

$$\Xi = \sum_S \exp[-\beta F(S)] \tag{14}$$

with

$$\begin{aligned} -\beta F(S) &= \sum_{j=1}^{|\Lambda|} \ell n \cosh(\frac{1}{2}\beta\lambda_j) = Tr \ell n \cosh[\frac{1}{2}\beta h] \\ &= Tr \ell n \cosh[\frac{1}{2}\beta(T^2 + U^2 + UJ)^{1/2}]. \end{aligned} \tag{15}$$

Thus, (14) is like an Ising model partition function but with a complicated, temperature dependent "spin-spin" interaction, $F(S)$, given by (15) in terms of the eigenvalues of h. With respect to this "spin measure" we can talk about the presence or absence of long range nuclear order in the thermodynamic limit.

In order to discuss this limit we henceforth restrict ourselves to a translation invariant nearest neighbor hopping on a cubic lattice in d dimensions.

What we are able to prove is summarized in the schematic figure below and in

Theorem *3: For all U and sufficiently large β there is long range order for $d \geq 2$ (the same kind as in the ground state). For all U and sufficiently small β there is none; indeed there is exponential decay of all nuclear correlation functions.*

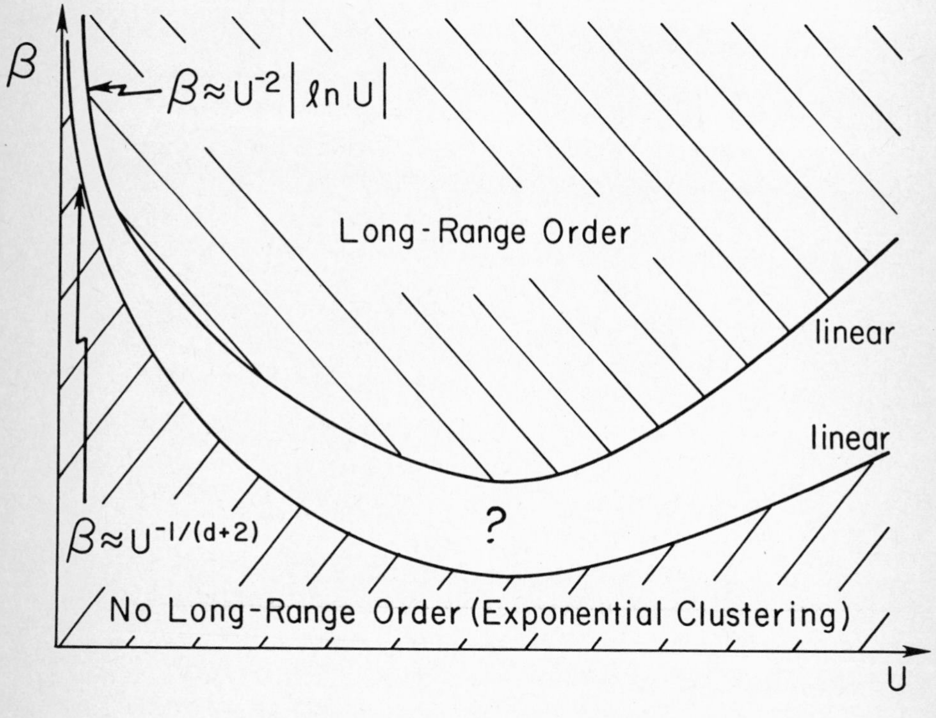

Presumably there is no intermediate phase, but we cannot prove this. For large U, β_c is clearly linear in U. For small U, we have the bound on the lower $\beta_c \sim U^{-1/(2+d)}$ and the bound on the upper $\beta_c \sim |\ell n U| U^{-2}$. Our guess is that the true state of affairs is $\beta_c \sim U^{-2}$.

For large β we use a Peierls argument; for small β we use Dobrushin's uniqueness theorem. A sketch of our proof - omitting many important details - is the following. For simplicity, we here consider only large $U > 0$.

Define, for $x > 0$,

$$P(x) = \ell n \cosh x^{1/2}. \tag{16}$$

[We note in passing that P is concave, and we see from the last expression in (15), using the proof in Theorem 1, that $F(S)$ has its minima at precisely the same values of W (or S) as in Theorem 1.] P is a Pick (or Herglotz) function with the representation

$$P'(x) = \frac{1}{2}x^{-1/2}\tanh x^{1/2} = \sum_{k=0}^{\infty}[(k+\frac{1}{2})^2\pi^2 + x]^{-1}. \tag{17}$$

We are interested in $x = \frac{1}{4}\beta^2(T^2 + U^2 + UJ)$ with $J_{xy} = t_{xy}(s_x + s_y)$.

Long Range Order (large β): Choose S, and then define antiferromagnetic contours in the usual way. If γ is a connected contour component and $\Gamma + \gamma$ is the whole contour, we want to prove that $\Delta \equiv F(\Gamma+\gamma) - F(\Gamma) \geq C|\gamma|$ for a suitable constant $C = C(U)$.

Obviously, $J_{\Gamma+\gamma} = J_\Gamma + J_\gamma$. To remove γ, we change s_x to $-s_x$ inside γ. For $0 \leq t \leq 1$, define $J(t) = J_\Gamma + tJ_\gamma$. Then, assuming for simplicity that Γ lies entirely outside γ, we have, by differentiating (15) and using (17), that

$$-\beta\Delta = \frac{1}{4}\beta^2 U \sum_{k=0}^{\infty}\int_o^1 dt Tr(G_k J_\gamma), \tag{18}$$

with

$$G_k = [(k+\frac{1}{2})^2\pi^2 + \frac{1}{4}\beta^2(T^2 + U^2 + UJ_\Gamma + tUJ_\gamma)]^{-1}. \tag{19}$$

Integrating by parts, this becomes

$$-\beta\Delta = -(\beta^4U^2/16)\sum_k \int_o^1 dt(1-t)TrG_kJ_\gamma G_kJ_\gamma + \ldots \tag{20}$$

[The $\ldots$ terms in (20) come from $t = 0$ in the partial integration. They are small and easily bounded for large U, but they have to be treated more judiciously when U is small.] If A, B, D are matrices with $A \geq B \geq 0$, then $TrAD^\dagger AD \geq TrBD^\dagger BD$. Also, $A \geq B > 0 \Rightarrow 0 < A^{-1} \leq B^{-1}$. Moreover, $T^2 \leq (2d)^2$ and $UJ \leq T^2 + U^2$. Using this, we have the matrix inequality

$$G_k \geq \left[(k+\frac{1}{2})^2\pi^2 + \frac{1}{2}\beta^2\left((2d)^2 + U^2\right)\right]^{-1}. \tag{21}$$

Summing on k,

$$\Delta \geq \text{(const.) } \mathrm{U}^2\left((2\mathrm{d})^2 + \mathrm{U}^2\right)^{-3/2}\mathrm{Tr}(\mathrm{J}_\gamma)^2.$$

Clearly $Tr(J_\gamma)^2 =$ (const.)$|\gamma|$. Thus, for U large, $C(U) \geq$ (const.)U^{-1}, and thus long range order exists in $d \geq 2$ if β/U is large enough.

Absence of Long Range Order (small β): Dobrushin's uniqueness theorem [5-7], together with the modification in [8], gives the following criterion for exponential clustering. We have to bound the change in F when we change the spins at x and y (taking the worst case with respect to the other spins). Call this f_{xy}. The requirement is that for some $m > 0$ and all x,

$$\beta \sum_y f_{xy} \exp[m|x-y|] < 1. \tag{22}$$

By an argument similar to the preceding (large U)

$$\begin{aligned} f_{xy} &\cong (\beta^3 U^2/16) \sum_k T_r G_k J_x G_k J_y \\ &\cong (\beta^3 U^2/16) \sum_k |G_k(x,y)|^2. \end{aligned} \tag{23}$$

Here, $J_x = \delta_x T + T\delta_x$ and $G_k(x,y)$ is the x,y matrix element of (19).

To implement (22) we now require an *upper* bound on $G_k(x,y)$ that has exponential decay. For this purpose the Combes-Thomas argument is ideal. Let Q be the matrix with elements $Q_{xy} = \delta_{xy} e^{n \cdot x}$ and with $|n| = 1$. Then

$$Q G_k^{-1} Q^{-1} = G_k^{-1} + R_k \equiv L_k. \tag{24}$$

The "remainder" R_k can be bounded for large U: $\|R_k\| \leq C\beta^2 U$ for some constant C. Similar to (21), for large U

$$G_k^{-1} \geq [(k+\frac{1}{2})^2\pi^2 + \frac{1}{8}\beta^2 U^2] \equiv \alpha_k. \tag{25}$$

Thus,

$$\|L_k^{-1}\| \leq [\alpha_k - C\beta^2 U]^{-1}. \tag{26}$$

Since $|(L_k^{-1})_{xy}| \leq \|L_k^{-1}\|$, we have from (24), $|(Q G_k Q^{-1})_{xy}| \leq \|L_k^{-1}\|$, and thus

$$|(G_k)_{xy}| \leq \exp[n \cdot (y-x)]\{\alpha_k - C\beta^2 U\}^{-1}. \tag{27}$$

This holds for all n, so summing on k,

$$f_{xy} \leq \text{ (const.) } U^{-1} e^{-2|x-y|}.$$

Hence, (22) holds with $m = 1$ if β/U is small enough.

The support of both authors by the U.S. National Science Foundation, grant PHY 8116101 A03, is gratefully acknowledged.

References

1. M. C. Gutzwiller, Phys. Rev. Letters *10*, 159-162 (1963), and Phys. Rev. *134*, A923-941 (1964), and *137*, A1726-1735 (1965).
2. J. Hubbard, Proc. Roy. Soc. (London), Ser. A*276*, 238-257 (1963), and *277*, 237-259 (1964).
3. J. Kanamori, Prog. Theor. Phys. *30*, 275-289 (1963).
4. T. Kennedy and E. H. Lieb, An itinerant electron model with crystalline or magnetic long range order (in preparation).
5. R. L. Dobrushin, Theory Probab. and Its. Appl. *13*, 197-224 (1968).
6. L. Gross, Commun. Math. Phys. *68*, 9-27 (1979).
7. H. Föllmer, J. Funct. Anal. *46*, 387-395 (1982).
8. B. Simon, Commun. Math. Phys. *68*, 183-185 (1979).
9. J. M. Combes and L. Thomas, Commun. Math. Phys. *34*, 251-270 (1973).

FIRST ORDER PHASE TRANSITIONS AND PERTURBATION THEORY

J. Bricmont and J. Slawny

Institut de Physique Theorique, 2, ch. du Cyclotron
Universite Catholique de Louvain
B - 1348 Louvain-la-Neuve, Belgium
Center for Transport Theory and Mathematical Physics
Virginia Polytechnic Institute and State University
Blacksburg, Virginia 24061, U.S.A.

I. INTRODUCTION

While in most situations the state of a system changes smoothly when external parameters such as pressure, magnetic field, or temperature are varied, there are also sometimes sudden jumps in the density, the magnetization or the energy that occur at some values of these external parameters. One of the basic goals of Statistical Mechanics is to understand these phase transitions starting from a microscopic description.

Since the invention of the Peierls' argument [41, 30], a great variety of models have been shown to exhibit first-order phase transitions [21, 26, 22]. However some basic questions remain unanswered. For example:

- We do not have a proof of the occurrence of first-order phase transitions for a reasonable model of a simple fluid. For a simple liquid-vapour phase transition, we cannot go much beyond van der Waals theory [61, 39].

- We do not have a complete understanding of the Gibbs phase rule, even for classical lattice systems (for partial results see [34, 45]). One would like to show that, typically, given a set of parameters, n phases will coexist on a manifold of codimension n-1 in that space.

- From a more practical point of view, one would like to have reliable techniques to compute the phase diagram of alloys with interactions of interest.

In these lectures, we shall discuss how the Pirogov-Sinai theory [42, 47, 48] and some of its extensions [33, 8, 10, 13, 18, 14, 60] provide at least partial answers to these questions. The main emphasis will be on the use and justification of perturbation theory.

To develop a perturbation theory, we need a reference system which is "completely" known. At low temperatures such a reference system may consist of zero temperature states, or ground states, yielding low temperature expansions.

The first proof of convergence of a low temperature expansion was given by Minlos and Sinai [40] for the Ising model. Minlos and Sinai reformulated this model in terms of a gas of contours separating regions of the lattice where different ground states occur. At low temperatures, activity of this gas is small and therefore one could use methods developed to control low activity expansion to prove convergence here. We review this in Sect. II.

However, it is often useful, and sometimes imperative, to have a perturbation theory around more general objects than ground states, which we call restricted ensembles. Such an ensemble consist of a subset of the phase space, together with a measure (usually a Gibbs state) on it, which gives a better approximation to the true phase than a ground state. A restricted ensemble often consists of small fluctuations around a ground state as in the case of continuous spin models or quantum field theory [26, 33, 18]. However, this notion is useful in other contexts as well, as we shall try to demonstrate on the following examples:

- *Lattice spin systems with a finite number of ground states which are not related by symmetries of the Hamiltonian*. This is the situation considered by Pirogov and Sinai (when the number of ground state is finite). We give an introduction to their theory by discussing a simple example in Sect. III.

- *Lattice spin systems with an infinite number of ground states*. In this situation, the theory is far from complete. The notion of a restricted ensemble is very useful in many cases where an infinite number of ground states occurs because it allows us to reduce this infinite set to the consideration of a finite set of ground states called dominant ground states (Sect. IV).

- *Fluids*. There are very few results on phase transitions for one component fluids (see however [34]), but several results are available for phase separation in mixtures [44, 37, 8]. Restricted ensembles are essential in this case since there is no notion of a ground state for continuum fluids. The reference system here is an ideal or almost ideal gas (Sect. V).

- *Large entropy models*. We discuss an extension of the Pirogov-Sinai theory to the q-state Potts model for q-large and for the temperature where q+1 phases coexist, [8, 14, 60]. q of these phases are small perturbations of the corresponding ground states but there is an additional high-temperature phase which is in equilibrium with the other phases only because of its large entropy. This phase is a small perturbation of the restricted ensemble which is maximally disordered (Sect. VI).

II. THE ISING MODEL: LOW TEMPERATURE EXPANSION

One of the basic open problems in the theory of first-order phase transitions is to understand the phase diagram of simple fluids. There are deep open problems concerning crystallisation but even the liquid-gas transition is understood only in the following lattice-gas approximation (see, for instance, [49, Sect. 5-2]): The intermolecular potential usually consists of a short-range repulsion and of a long-range attractive part. Let us cover the continuum with the cells of the lattice Z^d (d is the dimension of space) and, in place of the repulsive part of the potential, let us require that each cell be occupied by at most one particle. Furthermore, one assumes that the attractive potential is insensitive to the position of the particle in the cell. Then one obtains a lattice gas model, with a variable $\rho(a)$ at each lattice point a, which is equal 1 if the cell is occupied, and 0 if it is empty. A change of variables, $\rho(a) = \frac{1}{2}(1+\sigma(a))$, leads us to a model of ferromagnetism; the attractive potential of the lattice gas yields a ferromagnetic interaction of the magnetic system.

The simplest version of such a ferromagnetic model is the nearest neighbor *Ising model*:

$$H = -\sum_{\langle a,b\rangle} \sigma(a)\sigma(b) - h\sum_{a} \sigma(a) , \quad \sigma(a) = \pm 1 , \tag{2.1}$$

where $\langle a,b\rangle$ means that a and b are nearest neighbors, and the sum is over all nearest neighbor (n.n.) pairs, called *bonds*. Many results are known for this model, especially at low temperatures, but before discussing them, we want to stress the simplifications introduced by this lattice approximation.

- Since the lattice is given a priori, we do not have the problems (massless excitations) associated with the formation of crystals.

- We have also indirectly introduced a *symmetry* into the problem: H in (2.1) is invariant under the change of h into -h and $\sigma(a)$ into $-\sigma(a)$ for all lattice sites a. This symmetry yields a very drastic simplification: because of it, the phase transition, if any, is expected to occur at h=0, i.e. at a *temperature independent* value of h. This is certainly not the case when there is no symmetry between the phases e.g. in a liquid-gas transition, where the pressure for which phase coexistence takes place depends in a non-trivial way on the temperature.

- Finally, there is a well-defined reference system about which one can try to construct a low temperature expansion, namely the ground states. For $h > 0$ (resp $h < 0$) there is a unique ground state with all spins equal to +1 (resp. -1). For h=0 one has both of these ground states.

We shall recall now a derivation of the low temperature expansion for the free energy of the Ising model, [19], an expansion which, in this case, can be proved to converge. It will be convenient to define the free energy with the ground-state energy substracted, like in $\psi(\beta|+)$ below: the $+$ after the vertical bar indicates which ground state is considered. Thus

$$\psi(\beta|+) = -\beta^{-1}\lim_{\Lambda} |\Lambda|^{-1} \log Z(\Lambda|+) ,$$

where $|\Lambda|$ is the number of points in Λ. The limit is taken over an expanding sequence of finite subsets of the lattice with not too large boundaries, like NxN-squares with $N\to\infty$, or, more generally over Van Hove sequences, [43, Chapter 2]. Furhermore,

$$Z(\Lambda|+) = \sum_{\sigma} \exp -\beta H_{\Lambda}(\sigma|+) , \tag{2.2}$$

and

$$H_{\Lambda}(\sigma|+) = \Sigma\ (\sigma(a)\sigma(b)-1) . \tag{2.3}$$

In (2.2) the sum is over all configurations $\sigma=(\sigma(a))_{a\varepsilon\Lambda}$ in Λ; in (2.3) these configurations are extended to all the lattice by setting $\sigma(a)=1$ for a in the complement of Λ, and the summation is over all the bonds intersecting Λ. The Hamiltonian $H_{\Lambda}(\bullet|+)$ is as in (2.1), but with h=0 and the energy of the ground state subtracted: $\sigma(a)\sigma(b)$ replaced by $\sigma(a)\sigma(b)-1$. The expansion will be derived in a manner more involved than necessary for this model. This derivation will, however, be useful when we discuss the Pirogov-Sinai theory in Sect. 3. For simplicity and in order to make the calculations explicit, we shall work in two dimensions.

Let us start by computing the first order term in the expansion of $\psi(\beta|+)$. For β large, the leading term in the partition function (2.2) comes from isolated spin flips i.e. from configurations where all the sites adjacent to a minus-spin are occupied by plus-spins. Restricting the sum in the partition function to these configurations one obtains approximately, by adding the contributions of 0, 1, 2,... spin flips:

$$Z(\Lambda|+)\simeq 1+|\Lambda|e^{-4\beta}+\tfrac{1}{2}|\Lambda|(|\Lambda|-1)(e^{-4\beta})^2+\ldots=(1+e^{-4\beta})^{|\Lambda|}\simeq \exp(|\Lambda|e^{-4\beta}) . \tag{2.4}$$

This yields $\beta\psi(\beta|+) \simeq -\exp-4\beta$, to leading order.

One could obtain in this fashion higher order terms but convergence of this expansion is best proved after an introduction of contours - the basic notion of

Peierls' argument: Consider a configuration $\sigma=(\sigma(a))_{a\in Z^2}$ equal +1 everywhere, with a possible exception of a finite set. Define the *boundary* $B(\sigma)$ of σ by

$$B(\sigma) = \{\langle a,b\rangle : \sigma(a) \neq \sigma(b)\} .$$

Now decompose $B(\sigma)$ into connected components in the usual way: associate to each bond $\langle a,b\rangle$ the bond of the dual lattice which is perpendicular to it, and decompose the resulting set of lines into connected pieces, called *contours*. For each contour γ set

$$[\gamma] = \{a \in Z^d : \text{there exists } b \text{ such that } \langle a,b\rangle \in \gamma\} .$$

We mention that this definition of contours which is convenient for the Ising model needs to be modified for more general models. In the special case of the Ising model this general notion would be defined as follows. Cover the lattice with squares of side R (where R is chosen to be larger than the range of the interaction, which here is 1) and define a square C to be regular if $\sigma|_C$ (the restriction of σ to C) is equal to the restriction to C of one of the ground states (i.e. $\sigma(a)=+1$ for all $a\in C$ or $\sigma(a)=-1$ for all $a\in C$). Then a contour is a pair made of a connected set of irregular squares together with the restriction of σ to these squares. One important property of this general definition is that a contour is no longer a pure geometrical object but rather it is a subset of the lattice together with a configuration on it from which one can read off the bordering ground states.

With any of the two above definitions, contours enjoy the following two important geometric properties:

Property 1. They are closed in the sense that for any contour γ, $Z^d\backslash[\gamma]$ (or the complement of the support of γ in the general case) can be decomposed into several connected components only one of which is infinite. The latter is called the *exterior* of γ while the finite components form the *interior* of γ, Intγ. Moreover, there is a unique ground state (+ or -) along the boundary of each of these components. This property is obvious in the case of the Ising model; it also holds whenever the number of ground states is finite but does not, in general, when the number of ground states is infinite (see Sec. 4).

Property 2. The number of contours of length n (having n bonds) and containing a fixed bond is bounded by c^n where c is a constant (depending only on the lattice and the number R entering the general definition of contours).

Using the first property we can define, for any family of contours, the subfamily of *outer contours*, namely those that do no lie in the interior of any of the contours.

Let us consider the partition function $Z(\gamma)$ of all the configurations having only one outer contour γ and consider the Hamiltonian (2.3). Then

$$Z(\gamma) = \exp(-2\beta|\gamma|) \prod_i Z(\mathrm{Int}_i\gamma|\varepsilon(i)) , \tag{2.5}$$

where $|\gamma|$ is the number of bonds in γ, the product runs over all the connected components $\mathrm{Int}_i\gamma$ of the interior of γ, and $\varepsilon(i) = +$ or $-$ depending on the ground state which, by Property 1, borders $\mathrm{Int}_i\gamma$. Finally, $Z(\Lambda|+)$ is defined by (2.2) and $Z(\Lambda|-)$ is defined similarly, but with $\sigma(b)=-1$ for b in the complement of Λ.

Now for any Λ, we can write

$$Z(\Lambda|+) = \sum_\omega \prod_{\gamma\in\omega} Z(\gamma) , \tag{2.6}$$

where the sum is over *all compatible families* ω *of outer contours* in Λ and the product over contours of the family; compatiblity means here that the supports of the contours are disjoint and that no contour lies in the interior of another contour of the family.

The (important) factorization property implicit in (2.5) derives here from the fact that we have a fixed ground state in the exterior of all outer contours.

Now to get our final expansion we use the symmetry of the Hamiltonian under the global spin-flip relating the two ground states, which implies that for any Λ,

$$Z(\Lambda|+) = Z(\Lambda|-) . \tag{2.7}$$

Using (2.7) and (2.6), one transforms (2.5) into:

$$Z(\Lambda|+) = \sum_\omega \prod_{\gamma\in\omega} \exp(-2\beta|\gamma|)\prod_k Z(\mathrm{Int}_k\gamma|+) , \tag{2.8}$$

with the sum extended over the same set of ω as in (2.6).

This expansion is in a form which is suitable for iteration; iterating it one gets

$$Z(\Lambda|+) = \sum_\omega \prod_{\gamma\in\omega} z(\gamma) , \quad z(\gamma)=\exp-2\beta|\gamma| , \tag{2.9}$$

where the sum now runs over all families ω of contours that are non-intersecting, without the constraint that the contours are outer contours. The identity (2.9) shows that $Z(\Lambda|+)$ is equal to the partition function of a gas of an infinite number of "species of particles" (all contours modulo translations) interacting through a hard-core exclusion (non-intersecting contours), and having activity $z(\gamma)$. This implies immediately that for any contour γ the probability that it is a contour of a configuration is bounded by $\exp-2\beta|\gamma|$ (*Peierls' bound*).

The basic estimate that allows us to control this system is:

$$\sum_{\gamma: b \in \gamma} z(\gamma) \to 0 \text{ (exponentially fast) as } \beta \to \infty , \tag{2.9a}$$

where b *is any fixed bond* (in the case considered here the sum is bounded by exp-4β, for β large enough). This follows easily from the fact that $z(\gamma)$ is decreasing very fast when $|\gamma|$ is increased, here $z(\gamma)=\exp-2|\gamma|$, and that there are not too many contours with given $|\gamma|$ (Property 2).

The convergence of the low temperature expansion follows now from the results on low activity expansion for the gas of contours: Define a *multiplicity function* as a map from the set of contours into non-negative integers, with a finite support. A complex-valued function ϕ defined on the multiplicity functions can be identified with the formal power series

$$\sum_{\vartheta} \phi(\vartheta) z^{\vartheta} , \tag{2.10}$$

where the sum is over all possible multiplicity functions, and

$$z^{\vartheta} = \prod_{\gamma} z_{\gamma}^{\vartheta(\gamma)} ,$$

z_γ being a variable associated to γ.

Define now ϕ as follows: $\phi(\vartheta)=1$ if $\vartheta(\gamma)\leq 1$ for all γ and the contours in the support of ϑ are pairwise disjoint, and $\phi(\vartheta)=0$ otherwise. Then, (2.9) is the value of the formal power series (2.10) for $z_\gamma=\exp(-2\beta|\gamma|)$ if $\gamma\subset\Lambda$, and $z_\gamma=0$ otherwise.

Let ϕ^T be the function corresponding to the logarithm, in the sense of formal power series, of ϕ.

Proposition ([23, 40, 46]). *If* $|z_\gamma| \leq \exp(-2\beta|\gamma|)$ *(all* γ*) then for any contour* γ'

$$\sum_{\vartheta:\vartheta(\gamma')\neq 0} |\phi^T(\vartheta) z^{\vartheta}| \to 0 \quad \text{as} \quad \beta \to \infty . \tag{2.11}$$

Furthermore, setting in

$$\sum \phi^T(\vartheta) z^{\vartheta} \tag{2.12}$$

$z_\gamma=0$ *for* γ *not contained in* Λ*, one obtains for each* Λ *an absolutely convergent series from which* $\log Z(\Lambda|+)$ *is obtained by substitution* $z_\gamma=\exp-2\beta|\gamma|$. *This expansion implies in turn that*

$$\log Z(\Lambda|+) = - |\Lambda|\beta\psi(\beta|+) + \Delta(\Lambda,\beta) ,$$

with $\beta\psi(\beta|+)$ *being an analytic function of* $z=\exp-2\beta$*, while*

$$|\Delta(\Lambda,\beta)| \le \delta(\beta)|\partial\Lambda| \,, \tag{2.13}$$

with $\delta(\beta)=O(\exp(-2\beta))$.

Given the expansion (2.12), the proof of the rest of the Proposition is easy. Using this expansion, one obtains a detailed description of the low temperature phases [40, 23]: There are exactly two pure phases (i.e. extremal translation invariant Gibbs states) related by the spin-flip symmetry, with opposite spontaneous magnetization. In each phase, typical configurations consist of a "sea" of (+)-spins, for example, with small islands of the opposite phase. Correlation functions are exponentially clustering and analytic in $\exp(-2\beta)$. Thus one has quite a complete description of the low-temperature phases. However, in order to obtain these results one has to use many of the simplifications of the Ising model that are not available in more realistic systems.

Not much needs to be changed if we consider higher spins (i.e., $\sigma(a)$ assuming the values $-n,-n+1,\ldots,n$, with the same Hamiltonian or ferromagnetic system with a finite number of ground states. However, if we deal with a continuous distribution of $\sigma(a)$, or with quantum field theories [26], then complications arise.

Let us consider, for example, an Ising model with continuous spin: the Hamiltonian H is as before, (2.1) with h=0, but $\sigma(a)$ is now uniformly distributed in the interval $[-1,+1]$, instead of taking the values +1 and -1 only. As in the Ising model, there are two ground states , the (+)-ground state which is +1 everywhere and the (-)-ground state. The ground states are again related by the spin-flip symmetry. However, one has now the following problem: even in the finite volume, the ground state configurations have probability zero and there are excitations of the ground states with (relative) energy which is arbitrarily small.

To deal with this problem one can introduce two restricted ensembles:

$$X_R^+ = \{\sigma\colon \sigma(a) \ge 0, \text{ all } a \in Z^d\} \,,\ X_R^- = \{\sigma\colon \sigma(a) \le 0, \text{ all } a \in Z^d\} \,.$$

Then one can define contours as connected families of bonds <a,b> for which $\sigma(a)\sigma(b)\le 0$, as in the Ising model. Thus, in each connected component of the complement of a contour, the configuration belongs to one of the restricted ensembles. If one tries to expands the partition function into a sum over outer contours, as in the (2.6), one does not have any more a factorization into partition functions of outer contours. However, it turns out that the restricted ensembles are very dilute at low temperatures. Namely, one has a convergent expansion for these ensembles from which one derives all the properties obtained usually from high-temperature expansion, or from low fugacity expansion of a dilute gas. In particular the factorization in (2.6) holds in the sense that one can construct a low-temperature expansion for the complete model which takes the form of a combined expansion, first into contours and then a "high-temperature expansion" in each restricted ensemble; we refer to [9] for detailes. Such a combined expansion has appeared first in other contexts, particularly in lattice and continuum models of Quantum Field Theory, see [26] and references therein. For other extensions of the Pirogov-Sinai theory to continuous-spin systems we refer to [18] and to Zahradnik's contribution to the present volume.

One can avoid this combined expansion and obtain directly the Peierls' bound on the probability of a contour, as done originally in [55]. This is one of the many examples where chessboard estimates, based on the Reflection Positivity property of the model, [26, 21], proved to be very useful.

To clarify the importance of the symmetry in the preceding arguments consider the Ising model, but with an external field. Now the weight $\exp(-2\beta|\gamma|)$ of a contour is modified slightly, which for a small field is not essential. However, (2.7) would no longer hold and the iteration procedure in (2.8, 2.3) would not converge. Recall that (2.7) comes from the symmetry. Since such symmetry is expected for a realistic liquid-gas model, a first step may be to extend the analysis above to lattice models without symmetry. This is quite non trivial and is the content of the Pirogov-Sinai theory.

III. MODELS WITHOUT SYMMETRY

This section is divided into four subsections. First we introduce a simple model on which the main features of the Pirogov-Sinai theory can be most easily explained, and we discuss its phase diagram from the point of view of perturbation theory. Then we explain some of the ideas of the Pirogov-Sinai theory. In Sect. 3, we state the general result of Pirogov and Sinai for this model and we discuss the perturbation theory for the phase diagram in more details. Finally (Sect. 4) we generalize our model and explain some recent applications to the theory of random surfaces [5].

1. *Phase diagram of the Blume-Capel model.*

This model, [4, 11], is defined by the Hamiltonian

$$H = \sum_{\langle a,b\rangle} (s(a)-s(b))^2 , \qquad s(a) = 0,\pm 1 . \tag{3.1}$$

It has three ground states, (+), (-) and (0), which are equal everywhere to +1, -1 and 0, respectively.

The model has still the $s(a) \to -s(a)$ symmetry but only the (+) and (-) ground states are related by this symmetry; the (0) ground state is in a class by itself. One could expect that at temperatures low enough the model has three pure phases which are small perturbations of the three ground states. We will now give a perturbation-theoretic argument showing that this is not true. More precisely, we will argue that for the original Hamiltonian (3.1) there is no low temperature phase which is a small perturbation of the (+)-ground state, and that for a suitable perturbation of the Hamiltonian such a phase may appear; similarly for the (-)-ground state.

Suppose that in a large but finite region Λ the (+)-boundary conditions produce such a low temperature phase, i.e. that a typical configuration of such a system is +1 almost everywhere, with small excitations distributed rarely but homogeneously in Λ. At very low temperatures the main contribution to the partition function $Z(\Lambda|+)$

```
+  +  +  +  +  +  +  +  +  +  +  +  +  +
+  0  0  0  0  0  0  0  0  0  0  0  0  +
+  0                                0  +
+  0                 0  0  0        0  +
+  0     0  0  0     0  +  0        0  +
+  0     0  -  0     0  0  0        0  +
+  0     0  0  0                    0  +
+  0     0  +  0        0  0  0     0  +
+  0     0  0  0        0  -  0     0  +
+  0                    0  0  0     0  +
+  0           Λ'                   0  +
+  0  0  0  0  0  0  0  0  0  0  0  0  +
+  +  +  +  +  +  +  +  +  +  +  +  +  +
```

Fig. 1. Typical configurations of the low-temperature "(+)-phase" of the model (3.1): tunneling to the (0)-phase.

of this system would come first from the ground state and then from the excitations of lowest energy, for which $s(a)=0$ at a number of isolated points of Λ. Thus, see (2.4),

$$Z(\Lambda|+) \simeq 1 + |\Lambda|e^{-4\beta} + \tfrac{1}{2}|\Lambda|(|\Lambda|-1)(e^{-4\beta})^2 + \dots \simeq \exp(|\Lambda|e^{-4\beta}) \quad (3.2)$$

Similar argument yields for the partition function of the low-temperature (0)-phase

$$Z(\Lambda|0) \simeq 1 + 2|\Lambda|e^{-4\beta} + \tfrac{1}{2}(2|\Lambda|)(2|\Lambda|-1)(e^{-4\beta})^2 + \ldots \simeq \exp(2|\Lambda|e^{-4\beta}) , \tag{3.3}$$

since here lowest-energy excitations can be obtained in two ways: s(a) can be changed either to +1 or to -1.

Now, consider the contribution to $Z(\Lambda|+)$ of configurations of Fig. 1. Here inside Λ one has a slightly smaller region Λ' with the (0)-phase. If Λ is an NxN-square and Λ' is an (N-2)x(N-2)-square, the contribution of such configurations to the partition function is

$$\exp(-4(N-2)4\beta) \bullet Z(\Lambda'|0) \simeq \exp(-4(N-2)4\beta+2(N-2)^2e^{-4\beta}) , \tag{3.4}$$

which for large N is larger than $Z(\Lambda|+)$. This shows that the conjectured (+)-phase is unstable against insertions of the (0)-phase, and that the assumption of the existence of a low-temperature (+)-phase leads to a contradiction.

This perturbation-theoretic argument can be cast into a form which hints at a more systematic development (see the end of Sect. 4). The sum in (3.2) can be interpreted as a partition function of a "gas" of lowest energy excitations of the (+)-ground state. The free energy of this gas is denoted $\psi_R(\beta|+)$ (R stands here for "restricted ensemble", as explained in the introduction). Thus

$$\beta\psi_R(\beta|+) = - e^{-4\beta} + o(e^{-4\beta}) , \tag{3.5}$$

and $\psi_R(\beta|-)=\psi_R(\beta|+)$. Interpreting (3.3) in a similiar manner:

$$\beta\psi_R(\beta|0) = - 2e^{-4\beta} + o(e^{-4\beta}) . \tag{3.6}$$

We stress the point that here, in contradistinction to the Ising model, the partition function depends on the boundary conditions in an essential way.

Since the RHS of (3.4) is essentially $\exp(-\beta(|\partial\Lambda'|+|\Lambda'|\psi_R(\beta|0)))$, it appears that one can infer the low-temperature phases by comparing the thermodynamic potentials $\psi_R(\beta|+)$, $\psi_R(\beta|-)$, and $\psi_R(\beta|0)$ of the corresponding restricted ensembles. Namely, we will say that the (0)-ground state is *dominant* since it has more lowest energy excitations than the other ground states, or equivalently since the leading term of the low-temperature expansion of the free energy of the restricted ensemble of the gas of the low energy excitations is smaller for (0) than for (+) or (-). The above argument suggests that only dominant ground states give rise to low-temperature phases. One can in fact prove (see Appendix):

Theorem. *For* β *large, there is only one translation invariant Gibbs state for the model* (3.1). *In typical configurations of this state the density of sites* a *where* $s(a)=0$ *tends to* 1 *as* $\beta \to \infty$.

This theorem, and its proof, are included here for pedagogical reasons. The method of the proof that we use is applicable in a much broader context (see [10] and Sect. IV). The result itself is a special case of the uniqueness results of Preiss (unpublished) and Zahradnik [54].

Remarks: (1) Using correlation inequalities one can show as in [38] that there is actually a unique Gibbs state for this model. (2) The restriction to large β is inherent in our method. However, in this particular model, we do not expect several phases at higher temperatures.

The perturbation-theoretic argument indicates how to obtain the thermodynamic phases corresponding to the ground states (+) and (-) of (3.1). Namely, perturb βH by replacing it with

$$\beta H_g = \beta H - g\sum_a s(a)^2 \ . \tag{3.7}$$

Now for any $g > 0$ we have only two ground states, (+) and (-). They are related by symmetry of the Hamiltonian and, for temperatures low enough (depending on g) a Peierls argument, similar to the one used for the Ising model, proves the coexistence of two phases. However, one expects that there will be a line $g(\beta)$ on which three phases, (+), (-) and (0) will coexist.

The perturbation-theoretic argument which excluded the (±)-phases for g=0 yields now that $g(\beta)$ is given to order $\exp -4\beta$ by the condition that

$$\beta\psi_R(\beta,g|+) - g = \beta\psi_R(\beta,g|0) \ ,$$

-g on the LHS being the energy of the (+)-ground state. The free energies are (see (3.5), (3.6)):

$$\beta\psi_R(\beta,g|0) = -(e^{g}+e^{-g})e^{-4\beta} + o(e^{-4\beta}) \ , \qquad \beta\psi_R(\beta,g|+) = -e^{-g}e^{-4\beta} + o(e^{-4\beta}) \ .$$

Thus one expects the line of coexistence to be given by: $g(\beta)=\exp -4\beta+$higher order terms. One can obtain a better approximations for $g(\beta)$ by considering the restricted ensembles which include higher-order excitations of the ground states, excitations with energies not exceeding some cutoff energy E (two adjacent 0 spins in the + ground state or one - spin in this ground state, etc.). However this would give only the asymptotics of the line $g(\beta)$ of coexistence; to prove the existence of $g(\beta)$, and the fact that the obove computations yield a curve asymptotic to $g(\beta)$ we need more sophisticated arguments, since there is no evidence that the procedure outlined above will converge when E is increased. The existence of $g(\beta)$ is one of the main results of the Pirogov-Sinai theory, specialized to the present model:

2. *The Pirogov-Sinai theory*

We will sketch now the proof, according to Pirogov and Sinai [42,47], that there exists a line $g(\beta)$ on which the three low-temperature phases coexist. Define:

$$Z(\Lambda|+) = \sum_s \exp{-\beta H_\Lambda(s|+)} \; , \tag{3.8}$$

with

$$H = \sum_{<a,b>} (s(a)-s(b))^2 - g\sum_a s(a)^2 \; . \tag{3.8a}$$

We let e_o, e_+ and e_- denote the energy per lattice site of the corresponding ground states:

$$e_o(g) = 0 \; , \quad e_\pm(g) = - g \; .$$

We introduce also the partition functions with the same boundary conditions as before, but with no substruction of the ground state energy:

$$Z^\varepsilon(\Lambda) = (\exp{-|\Lambda|\beta e_\varepsilon})Z(\Lambda|\varepsilon) \; , \quad \varepsilon = 0, \pm \; .$$

Define contours, almost as in the Ising model, as subsets of the lattice composed of connected families of bonds $<a,b>$ for which $s(a)\neq s(b)$, together with a configuration on it. Unlike in the case of the Ising model, here the energy of a contour,

$$E(\gamma) = \sum_{<a,b>} (s(a)-s(b))^2 \; ,$$

where the summation is over all the bonds of γ, depends not only on its length but also on the neighboring configuration. With these definitions one has the expansion:

$$Z(\Lambda|+) = \sum_\omega \prod_{\gamma\varepsilon\omega} \exp(-\beta E(\gamma)) \prod_i (\exp|\mathrm{Int}_i\gamma|\beta e_+) \; Z^{\varepsilon(i)}(\mathrm{Int}_i\gamma) \; , \tag{3.9}$$

where the sum is over all families of outer contours in Λ, and the second product runs over all the connected components of the interior of the contour γ, indexed by i. $\varepsilon(i)=0,+$ or $-$ according to the value of the spins along the boundary of $\mathrm{Int}_i\gamma$.

Now, we would like to iterate this expansion, as we did in the case of the Ising model, in order to obtain eventually a sum over families of pairwise nonintersecting, but not necessarily outer, contours. This would yield a convergent polymer expansion for $Z(\Lambda|+)$, provided the resulting activities of the contours are small enough. Such an expansion would allow us to derive all the desired properties of the (+)-phase, like Peierls' bound on probability of contours, and clustering properties of pure phases. We would like also to have a similar expansion for the (0)- and the (-)-phases, and the condition of coexistence of such a three phases should determine $g(\beta)$.

To obtain this, we write

$$Z^{\varepsilon(i)}(\mathrm{Int}_i\gamma) = (Z^{\varepsilon(i)}(\mathrm{Int}_i\gamma)/Z^{+}(\mathrm{Int}_i\gamma))\cdot Z^{+}(\mathrm{Int}_i\gamma) \ , \tag{3.10}$$

as we did in the Ising model, except that the ratio was equal to 1 there, by symmetry. Inserting (3.10) into (3.9) and iterating, we obtain the expansion

$$Z(\Lambda|+) = \sum_{\omega} \prod_{\gamma\varepsilon\omega} z^{+}(\gamma) \ , \tag{3.11}$$

where the sum is over all families ω of non-overlapping contours in Λ, not necessarily outer ones, and the activities are

$$z^{+}(\gamma) = \exp(-\beta E(\gamma)) \prod_{i} (Z^{\varepsilon(i)}(\mathrm{Int}_i\gamma)/Z^{+}(\mathrm{Int}_i\gamma)) \ . \tag{3.12}$$

Note that the families of contours over which we sum in (3.11) are not, in general, families of contours of configurations of our model (unlike in the Ising case). Indeed, because of our induction method, all contours have the (+)-ground state in their exterior, even those that may be in the (-)- or (0)-interior of another contour.

The convergence of the polymer expansion depends on an estimate like (2.3a) and would clearly hold here if, in (3.12), we had only the factor $\exp{-\beta E(\gamma)}$, because $E(\gamma)\geq c|\gamma|$, where $|\gamma|$ is the number of bonds in γ. So we would like to show that the factors $Z^{0}(\mathrm{Int}_i\gamma)/Z^{+}(\mathrm{Int}_i\gamma)$ are not too large.

From general arguments we know only that

$$|\log Z^{0}(\Lambda) - \log Z^{+}(\Lambda)| \leq \text{const.}\,|\partial\Lambda| \ . \tag{3.13}$$

Since the boundary of $\mathrm{Int}\gamma$ is just γ we have to find out how does the const. in (3.13) depend on β. From the intuitive picture described in Sect. 1, we expect that it should grow like β if we are *not* on the coexistence line. Indeed, if the (0)-phase is the dominant one, for example, then, in a system with the (+)-boundary conditions, a contour "tunnelling" to the (0)-ground state will develop itself and its length will be approximately equal to $|\partial\Lambda|$. Its weight will be of the order $\exp(-\beta|\partial\Lambda|)$ which will give a constant approximately equal to β. And vice versa, if the const. in (3.13) grows like β then the polymer expansions (2.2a) about the ground states are out of control and there is no reason to expect that we are on the coexistence line.

Now, what is the expected behaviour of the const. of (3.13) *on the coexistence line*? From the results obtained for the Ising model (the convergent polymer expansion of the Proposition), we expect that

$$\log Z^{\varepsilon}(\Lambda) = -|\Lambda|\beta\psi(\beta) + \Delta^{\varepsilon}(\Lambda,\beta), \qquad \varepsilon=+,0,-, \tag{3.14}$$

where in the first (volume) term ψ is the bulk free energy (independent of the boundary conditions); $\Delta^{\varepsilon}(\Lambda,\beta)$ is the boundary term which we expect to satisfy the bound (2.13):

$$|\Delta^{\varepsilon}(\Lambda,\beta)| \le \delta(\beta)|\partial\Lambda|, \tag{3.15}$$

where $\delta(\beta)\to 0$ exponentially fast as $\beta\to\infty$. Now, the common volume terms cancel out in $(\log Z^{0}(\Lambda)-\log Z^{+}(\Lambda))$, and therefore the bound (3.15) implies that the const. in (3.13) is of order $\delta(\beta)$.

Thus, we have two extreme situations: either the const. in (3.13) grows like β for large enough Λ (away from the coexistence line), or it tends exponentially to zero with β (if we are on the coexistence line). Therefore we can expect to be able to construct $g(\beta)$ by imposing the condition that (3.15) holds for $\varepsilon=+$, $-$ and 0 simultaneously. How does one show that such a line $g(\beta)$ exists? If we had only to deal with the contours of lowest energy (isolated zeros in the + ground state) then, as our calculation above shows, one could set $g(\beta)=\exp{-4\beta}+$higher order terms, and one would get equality of the bulk free energies for all the ground states, and also the bound (3.15) to order $\exp{-4\beta}$. It is fairly obvious that we could still find a line $g(\beta)$ if we had only a finite number of contours (up to lattice translations). But we have an infinite number of them. However, $g(\beta)$ can be constructed via an infinite sequence of approximations for the following reasons: when the size $|\gamma|$ of the contour γ increases, $E(\gamma)$ grows proportionally to it: $E(\gamma)\ge c|\gamma|$ (the *Peierls Condition* of [24] and [42]). So, roughly speaking, if we have a line $g_n(\beta)$ such that we would have coexistence if we considered only contours of size less than or equal to n, then, to include contours of size n+1 we shall only have to modify g_n by an amount like $\exp(-\beta(n+1))$. This gives a convergent sequence of approximations to the true phase coexistence line $g(\beta)$.

We stress the importance of the Peierls Condition in the above arguments. Once a notion of a contour can be introduced in a model in such a way that the Peierls Condition holds the most essential elements of the Pirogov-Sinai theory can be taken over (for instance, see [57, 59] for an extension of the theory to some models with an infinite number of ground states, and the model of Sect. 4). When the Peierls Condition does not hold, one can sometimes substitute for it its "averaged" version - the Peierls' bound, for a suitably defined contours, as done in [8] and the work described in Sect. IV.

3. *General phase diagrams and perturbation theory*

We start with the general result of Pirogov-Sinai, specialized to this model. Let us introduce an additional perturbation of the Hamiltonian (3.7):

$$\beta H(g,h) = \beta H - g\sum_a s(a)^2 - h\sum_a s(a) \ . \qquad (3.16)$$

Then we have the following zero-temperature phase diagram in the (g,h)-plane. The ground state is unique except along three half-lines: there are two ground states, (+) and (-), on the half-line $g>0$ $h=0$; two ground states, (+) and (0), for $g<0$ $h=-g$; and (-) and (0) for $g<0$ $h=g$. Of course, for $g=h=0$ one has three ground states (see Fig. 1 in [48]).

A set of perturbations such as $g\Sigma s(a)^2+h\Sigma s(a)$ above "removes completely the degeneracy of the ground states" since every possible combination of the original ground states appear on this zero-temperature phase diagram on manifolds of the correct dimension (n ground states coexist on manifolds of codimension n-1). The Pirogov-Sinai theory then says that for small $|g|$ and $|h|$ (here in fact for all (g,h)), and for low temperatures, the phase diagram is a small deformation of the one at T=0. Thus we have one point, $g(\beta)$, with three phases, three lines emanating from that point with two phases, etc..

Moreover, this theory allows one to justify the perturbation expansion, which was our starting point. Let us explain how does one obtain the perturbation expansion of the phase diagram in a more general situation [48]:

Define an *excitation* of a ground state as a configuration which coincides with this ground state everywhere, with a possible exception of a finite set of sites. Decompose every excitation into connected components, called *elementary excitations*, in such a way that the (relative) energy of an excitation with respect to the ground state is just the sum of the (relative) energies of the components. Note that this decomposition may differ from that into contours. For example, in the Ising model, a large bubble of minus-spins in the (+)-ground state with a plus-spin in its middle is an elementary excitation giving rise to *two* contours: one separating the bubble from the surrounding plus-spins, and another one separating the central spin from the bubble.

To compute the asymptotics of the lines and surfaces of phase coexistence up to an error of order exp-βE, we consider for each ground state the restricted ensemble consisting of its excitations with elementary components of energy not exceeding E. This we call the *gas of excitations* of the ground state G with energy not exceeding E. For example, in the Blume-Capel model considered above, the gas of first-order excitations of the (+)-ground state, i.e. excitations with energy not exceeding 4, consists of all configurations with isolated sites where s(a)=0, surrounded by sites b where s(b)=+1. The free energy $\psi_R(\beta|G)$ of this gas (denoted P_E^G in [48]) has a convergent (polymer) expansion, similar to the one in the Proposition, if β is large enough (depending on E). Indeed, convergence of this expansion rests on an estimate of the form (2.9a) which clearly holds here: we have a finite number of elementary excitations (modulo translations) and therefore in a suitable analogon of (2.9a) one has to sum over a finite set.

The definition of ψ_R, and the construction of the phase diagram which follows, makes sense for a much larger class of systems, than those to which the Pirogov-Sinai theory applies, many of them with an infinite number of ground states (see Sect. IV). Namely, it is enough that the Hamiltonian satisfies the following (regularity) condition: *The energy of an excitation tends to infinity with its support*. Note that this condition is not satisfied by the one-dimensional Ising model (see Sect. 4 for more interesting examples).

To obtain the line of coexistence of all the phases, we set equal the free energies of the restricted ensembles corresponding to all the ground states, and solve for g, h (in our example h will always be zero, by symmetry). This is what we did in our first-order calculations. To obtain the surface of the coexistence of the (+)- and (0)-phases, for example, we again set the corresponding free energies equal, and impose the condition that the restricted (-)-ensemble has a higher free energy.

This yields an asymptotic expansion for the line and for the surfaces of phase coexistence [48]. We do not expect this expansion to be convergent. However the coefficients of the terms in this expansion are easy to compute. Also, it often happens that knowing the first few terms in the expansion gives the topological structure of the phase diagram (i.e. the incidence properties of the lines and surfaces of the diagram). This structure is not affected by higher order terms which only give a better numerical approximation to the true phase diagram.

4. *Applications to surface models*

Let us consider the same Hamiltonian (3.1) as in the Blume-Capel model, but let $s(a)$ take $2n+1$ values: $s(a)=-n,-n+1,\dots,n$. Then the same analysis as in the Blume-Capel model implies that there will be a unique phase at low temperatures (see below). Also, we can add to βH $2n$ perturbations which lift the degeneracy of the ground states, for example,

$$\sum_{k:\ k\neq 0,\ |k|\leq n} \mu(k) \sum \delta(s(a)-k) \ ,$$

and obtain a complete phase diagram for small temperatures and small μ. Considering the lowest energy excitations, we deduce that the coexistence between the phase corresponding to $s(a)=n$ and the one for $s(a)=n-1$ will occur for $\mu(n)-\mu(n-1)=\exp(-4\beta)+o(\exp-4\beta)$, while coexistence between the (0)- and the ($\pm$)-phases will occur for

$$\mu(\pm 1) = \exp(-4n^2\beta) + \text{higher order terms} \ . \tag{3.17}$$

Indeed the (n-1)-ground state has more excitations of energy equal to 4 than the (n)-ground state (like in the Blume-Capel model where $n=1$) while the 0 and ± 1 ground states are identical in their low energy excitations up to energy $4n^2$ which are given, in the 0 ground state, by $s(a)=\pm n$, $s(b)=0$ for $|a-b|=1$.

It turns out that this discussion has an interesting application to the theory of random surfaces (see [5]). First, observe that if we replace $(s(a)-s(b))^2$ by $|s(a)-s(b)|$ in the Hamiltonian (3.1) there are no dramatic changes: n^2 in (3.17) is replaced by n but the phase diagram has the same qualitative properties, in particular there is a unique phase at low temperatures. However, this model is isomorphic to the SOS model of a surface which fluctuates between two hard walls ($s(a)$ is the height of the surface above site a and the walls are at height $-n$ and $+n$).

Now, consider a different model, where the interface is constrained to be above one fixed wall, say at height 0: the model is defined by the Hamiltonian $H=\sum|s(a)-s(b)|$ and the constraint $s(a)\geq 0$ (all a). Let us tie the surface to the wall ($s(a)=0$) on the boundary of a box Λ and ask for the value of

$$\lim_{\Lambda\to\infty} \langle s(0) \rangle_\Lambda \ . \tag{3.18}$$

It is well known [16, 23, 56] that if there is no wall this limit is finite at low temperatures if the dimension of the lattice is larger than or equal to 2 (actually, for at temperatures if $d\geq 3$, [6, 29]).

However, the presence of the wall induces an effective repulsion of the surface so that the limit (3.18) is infinite for all dimensions at low temperatures (and, presumably, at all temperatures). This fact is an easy consequence of the Pirogov-Sinai theory in the form discussed above.

Indeed,

$$< s(0) >_\Lambda \geq < s(0) >_{\Lambda,2n} , \tag{3.19}$$

where on the RHS we have added the constraint $s(a) \leq 2n$ for all a. This is fairly intuitive and actually follows from FKG-inequalities. Now, change the variables on the RHS of (3.19) to $s(a)' = s(a) + n$. The $s(a)'$ variables are exactly like in the model discussed above ($|s(a)'| \leq n$). Since there is a unique phase (actually, a unique Gibbs state) for this model at low temperatures which moreover is concentrated on small perturbations of the (0)-ground state, $< s(0)' >_\Lambda$ ($|s(a)'| \leq n$) will go to zero, as $\Lambda \to \infty$, independently of the boundary conditions. Inserting this into (3.19) we get

$$\lim_{\Lambda \to \infty} < s(0) >_\Lambda \geq n .$$

But since n is arbitrary, our claim is proved. For more details, we refer the reader to [5].

IV. INFINITE NUMBER OF GROUND STATES; OPEN PROBLEMS

We will now discuss on a simple example some of the existing extensions of the Pirogov-Sinai theory, and some of the open problems regarding low-temperature behavior of classical lattice systems. The model is a version of the familiar spin-½ antiferromagnet on a square lattice. It is chosen here because it illustrates most of the points we want to make. At the expense of complicating the notation we could have discussed models of alloys which are of some importance for applications (see Fig. 5 of [48]).

Let $\sigma(b) = \pm 1$ be a spin-½ variable at a point b of the simple square lattice Z^2 and let

$$H = H_{nn} - \alpha H_{nnn} - hM \quad (+ H_V) , \tag{4.1}$$

where

$$H_{nn} = \Sigma\, \sigma(a)\sigma(b) , \quad H_{nnn} = \Sigma\, \sigma(a)\sigma(b) , \quad M = \Sigma\, \sigma(a) ,$$

with the sum extended over all pairs of nearest neighbors of the lattice in H_{nn}, over all pairs of next nearest neighbors in H_{nnn}, and over all points of the lattice in M; the term H_V, see (4.3), will be included later, when we will consider a three-dimensional "stabilization" of this model.

The antiferromagnetic Hamiltonian H_{nn} is well known to have two chessboard-like ground states. To obtain the zero-temperature phase diagram of H one can group its terms as follows:

$$H = \sum_{\square} \left[\tfrac{1}{2} \sum_{nn} \sigma(a)\sigma(b) - \alpha \sum_{nnn} \sigma(a)\sigma(b) - \tfrac{1}{4}h \sum \sigma(a) \right] = \sum_{\square} \Phi_{\square} ,$$

where the first sum is over all elementary squares □ ("plaquettes") of the lattice, the first sum in the square bracket over (all four) pairs of n.n.'s within the plaquette; the second over the two pairs of n.n.n.'s within the plaquette, and the last over the four points of the plaquette.

Now, minimization of $\Phi_{\square}$ yields Fig. 2:

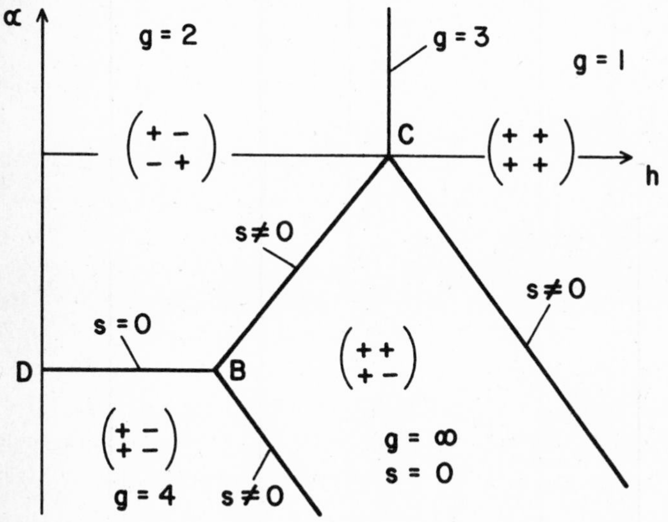

Fig. 2. Configurations of the square minimizing $\Phi_{\square}$ are obtained by rotating those indicated on the figure. g is the number of (periodic) ground states (obtained by patching the indicated configurations of squares). If g is infinite s indicates the entropy per lattice site of the set of all ground states. The Pirogov-Sinai theory applies whenever g is finite.

For points of the lines of this phase diagram minimizing configurations of the plaquettes are those of each of the adjacent open regions. For example, this results in three ground states for the points of the half-line h=4, α>0 and a non-zero entropy for the half-line $h+4\alpha=4$, $\alpha\leq 0$. In the later case, any configuration obtained from the (+)-ground states by flipping spins on a set of points no two elements of which are n.n.'s or n.n.n. s is a ground state again.

In each of the (three) open regions with a finite g the ground states are *equivalent* in the sense that they are related by translations and rotations which are symmetries of the Hamiltonian. For the (g=2)- and (g=4)-regions the Peierls' argument of Dobrushin [15] and Gerzik [24] yields two and four low-temperature pure phases, respectively. The Pirogov-Sinai theory applies both in these regions and in the neighborhood of each point of the half-line $h=4$, $\alpha>0$. In the last region the phase diagram is best computed using the perturbation expansion, which is justified with the help of the Pirogov-Sinai theory: Let $\psi_R(\beta,\alpha,h|G^2nnn)$ and $\psi_R(\beta,\alpha,h|G^0)$ be the free energies of the restricted ensembles of excitations of the corresponding ground states, with energies not exceeding E. For h=4 the lowest order excitations are obtained by flipping one of the spins in such a way that the only non ground states plaquettes are those with P^1-configurations. This can be done in $n_1(G^0)=1$ ways per lattice site in the case of the G^0 ground state, and in $n_1(G^2nnn)=\frac{1}{2}$ ways per lattice site for G^2nnn. Hence G^0 dominates in order 1 and only the G^0-phase is present at low temperatures. Indeed, the equation $\psi_R(\beta,\alpha,h(\beta)|G^0) = \psi_R(\beta,\alpha,h(\beta)|G^2nnn)$ for the line $h(\beta)$ of coexistence of the G^0-phase and the G^2nnn-phases, yields $h(\beta)=4-(1/2\beta)\exp{-8\alpha\beta}+o(\exp{-8\alpha\beta})$, in accordance with the argument based on domination.

The rest of the plane: the interval DB and the strip with base BC, including the points of its boundary, is not covered by the Pirogov-Sinai theory. In fact there are only a few rigorous results about the system in this region, not even a formal perturbation theory which would lead to plausible conjectures on low-temperature behavior (see, however, below, the discussion of a three-dimensional stabilization of this model). For, for the values of the parameters on the boundary of the strip the system admits local changes of the ground states without a change in the energy. Hence, the number of ground states in a square of edge-length L is of order c^{L^2}, and therefore the ground-state entropy is non-zero. Consider, for instance, for $h+4\alpha=0$ and $\alpha<-\frac{1}{2}$ (i.e. on the half-line emanating from the point B of Fig. 2) the ground state G^2nn consisting of alternating rows of plus- and minus-spins. Flipping any number of minus-spins which are not nearest neighbors produces P^1-plaquettes only, which are of the same energy as those occurring in G^2nn. Hence one will have an infinite number of kinds of excitations with the same energy and the standard perturbation expansion does not make sense. Following [63], points of the zero-temperature phase diagram for which the ground-state entropy is non-zero will be called *super-degenerate*.

This perturbation expansion makes also no sense for (α,h) on the interval DB or in the strip, though here local change of a ground state does not yield a ground state any more: now some of the *excited* configurations can be changed locally without changing their energy. We will discuss this in detail for the interval DB, after describing the ground states and low-energy excitations.

All the periodic ground states of the system are among those which can be obtained from G^2nn by flipping all spins on a number of lines parallel to one of the coordinate axes. For instance flipping spins on every second line yields a G^2nnn-ground state.

The plaquette P^1 obtained from the ground state plaquettes P^2 by flipping one of the (-)-spins has the lowest energy

$$\varepsilon = \Phi_\square(P^1) - \Phi_\square(P^2) .$$

Hence the first excitation energy is 4ε. An excitation with this energy is obtained when one (-)-spin is flipped in any of the ground states. However flipping any interval of spins which begins and ends with a minus, like -+- or -+-+-+-, yields again an excitation with four P^1-plaquettes and energy 4ε. Thus if Λ is a square of side L, the coefficient of $\exp-\beta 4\varepsilon$ of the expansion of $(1/|\Lambda|)\log Z(\Lambda|G^2)$ is of order at least L, as $L \to \infty$, and hence has no thermodynamic limit. However if we ignore the fact that the free energy of the gas of low energy excitations does not exist, but just compare the coefficients of the low-temperature expansion of $\log Z(\Lambda|G^2nn)$ and $\log Z(\Lambda|G^2nnn)$ we would say that the ground state G^2nnn is dominant since G^2_{nnn} admits twice as many excitation with, say, (-+-)-intervals flipped (both horizontal and vertical) than G^2nn (only vertical or only horizontal); in a similar way G^2nnn wins not only over G^2nn but over any other periodic ground state.

It is not clear at all whether these considerations lead to right conclusions about low-temperature phases of the system (see, however, below, for results concerning a three-dimensional stabilization of the model). For, most likely, the low-temperature expansion of the free energy of the system starts not with $\exp-4\beta\varepsilon$ but with $\exp-2\beta\varepsilon$, due to presence of nonlocalized excitations: flipping a half-line of alternating spins of a ground state which starts with a minus-spin one gets a configuration with only two P^1-plaquettes. Indeed such non-localized excitations do contribute to low-temperature expansion in one-dimensional models, or ferromagnetic models in any dimension, [32, 48]. Though some models with similiar features have been treated in a number of works (see [50, 51], for example), *the problem of computation of the low-temperature expansion for general models of this type seems to be open*.

We return now to a discussion of the super-degenerate points (the boundary of the strip; non-zero ground-state entropy). Although the standard low-temperature expansion does not make sense at these points, the zero-temperature limits of equilibrium states - zero order of a perturbation expansion, in a sense, are of interest here. A discussion of these limits clarifies the situation conceptually and yields non-trivial information on the low-temperature phase diagram [62, 64, 17, 63]. We explain first the situation on the example of the antiferromagnet in external field [62, 64, 17].

Consider the model (4.1) with $\alpha=0$. We let $h\to 0$ as $T\to 0$ (see Fig. 3). More precisely we approach the point $h=0$, $T=0$ along one of the inclines, i.e. we set $h(T)=4+h'T$, where h' is the inverse slope of the line; for $h'=0$ one has the usual zero-temperature limit. Substituting this into (4.1), with $\beta=T^{-1}$, we obtain

$$\beta H = \beta(H_{nn} - h(T)M) = \beta H_o - h'M , \tag{4.2}$$

where $H_o=H_{nn}-4M$.

Now one would expect that as $\beta\to\infty$ the term βH_o will suppress the contribution to equilibrium states of all but the ground state configurations of H_o [62]. Thus in the limit one obtains a state which is supported by the ground state configurations of H_o with interaction defined by $-h'M$. This is indeed the case (Slawny, 1980, unpublished; [64, 17]).

The ground states of H_o are supported by configurations minimizing $\Phi_\Box$ for each plaquette $\Box$. Inspection of Fig. 2 shows that these are exactly the configurations in which no minus-spin is a nearest neighbor of another minus-spin.

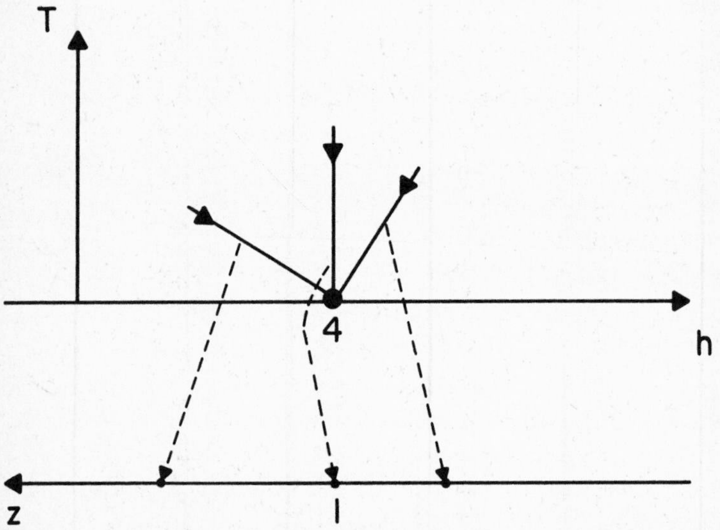

Fig. 3. Blowing-up the super-degenerate points: the limiting states obtained by approaching the point (T=0, h=4) from different directions are identified with equilibrium states of a lattice gas with different values of the activity z.

In an isomorphic lattice gas language in which plus-spins are identified with empty sites and minus-spins with occupied, one obtains a model in which no two particles are allowed to occupy the same or neighboring sites and the activity is $z=\exp -2h'$. This picture suggests that close to a super-degenerate point one should use a perturbation expansion about the limiting state of the hard core lattice gas and to interpret the properties of the system under consideration in terms of the later model. We refer to [62, 63, 64, 17] for an implementation of these ideas.

More generally, for any H_o with non-zero ground-state entropy limits of equilibrium states of $\beta H_o + H'$, as $\beta \to \infty$, are equilibrium states, in the sense of [65], of the Hamiltonian H' restricted to the set of ground states of H_o, and the limits of the pressure and entropy are equal to those of the limiting system (Slawny, 1980, unpublished). The reader may want to work out the limiting systems for the super-degenerate points of Fig. 2 different from C, and the two-parameter families of interactions obtained by approaching these points from different directions.

We now pass to a three-dimensional "stabilization" of the model for which we are able to develop an extension of the Pirogov-Sinai theory describing its low-temperature behavior.

The model is on a simple cubic lattice Z^3, with the Hamiltonian (4.1), but with the term H_v included ("stacking" of [12]): the two-dimensional layers, with intralayer interaction $H^2 = H_{nn} - \alpha H_{nnn} - hM$, are coupled by the term

$$H_v = -J \sum_a \sigma(a)\sigma(a+v) \, , \qquad J > 0 \, , \tag{4.3}$$

where v is the unit vector in the vertical direction. Since the coupling constant J is positive, the ground states of the full Hamiltonian are obtained by repeating in each horizontal plane the same ground state of H^2. Hence the zero-temperature phase diagram remains as on Fig. 2, with the change that the ground-state entropy is now zero everywhere: the number of ground states in a cube of edge-length L is of order c^L (as $L \to \infty$) for the points of the strip and the interval DB, and of order c^{L^2} on the lines where the entropy was non-zero previously.

Now, in the later case, there is still no perturbation theory since there is an infinite number of kinds of excitations with given energy, as in the ($g=\infty$, $s=0$)-case in two dimensions. For, the two-dimensional model admits a *local* change of the ground state without a change in the energy. Repeating this change in a finite number of neighboring planes one obtaines a configuration in which only the vertical bonds going up from the uppermost plane and down from the lowest plane are "excited", independently of the number of planes affected by the change. Thus, as the "c^L-case" in two dimensions, the "c^{L^2}-case" in three dimensions admits no standard perturbation expansion and *the problem of the nature of the low-temperature phase (phases?) is open* (see [12] for a discussion of a similiar situation). However for all other (α,h) the standard perturbation expansion makes sense and conclusions derived from it are justified by an extension of the Pirogov-Sinai theory of [10]. We will now discuss in some detail the example of the interval DB.

First of all we note that for the points of the interior of the interval the regularity condition of Sect. III.3 is satisfied. Furthermore, the two G^2nnn-ground states are dominant there. For, as in the two-dimensional version, the lowest order excitations are obtained by flipping one minus-spin. This has now energy $4\varepsilon+2J$, and $n_1(G)=\frac{1}{2}$ for any ground state G. If J is not too large, the next order is obtained by flipping a (-+-)-interval (energy $4\varepsilon+6J$), with $n_2(G^2nnn)=1$ and $n_2(G^2nn)=\frac{1}{2}$, and intermediate values of n_2 for other periodic ground states. Hence, G^2nnn is dominant in second order, and therefore one would expect two low-temperature phases.

That this conclusions are correct follows from [10]. The main points of the proof ("Peierls' bound for large-scale contours") are explained in the Appendix, especially in the proof of Lemma 1.

V. CONTINUUM FLUIDS

The fluid models for which most results are known about phase transitions, and the Widom-Rowlison model [52] in particular, concern phase separation in mixtures. In the simplest version of this model, one considers two species of particles A and B, with no interaction between particles of the same species and with a hard-core exclusion between A- and B-particles. Then Ruelle proved, [44], that the A-rich and the B-rich phases coexist if the activities z(A) and z(B) of the two species are equal and large enough. This result holds also if there is a suitable interaction among the A particles and among the B-particles provided that they are identical, i.e. that the A-B symmetry of the model still holds [44, 39, 58]. Comparing this to the Ising model, we could say that we have kept the symmetry between the two phases but we do not have a lattice any more, nor do we have ground states to perturb around. In fact this is an example where restricted ensembles are important. A natural problem is an extension of this result to a non-symmetric situation (the interaction between the A-particles is different from the interaction between the B-particles). This is done in [8] but under rather restrictive assumptions on these interactions (see below).

The restricted ensembles considered by Ruelle (in a somewhat different terminology) are defined as follows: Cover the space with non-overlapping cubes so that if we have an A particle in some cube then the hard-core forbids B particles to be in any of the adjacent cubes. Define one restricted ensemble by requiring that there be at least one A-particle in each cell, and another with at least one B-particle in each cell. Now, if we have a configuration of one of the restricted ensembles in one region of space and of the other in another region, the hard-core condition implies that some of the cells must be empty. Define then *a contour* γ as a connected set of empty cells and $|\gamma|$ as the number of cells in γ. For large activities empty cells are unlikely, and we can apply the Peierls argument. Using the A-B symmetry, one shows that for large $z(A)=z(B)$ the probability of a contour γ is bounded by $\exp(-O(z(A))|\gamma|)$. Going from this estimate to the proof of phase separation is standard. The new feature of this proof is the introduction of the restricted ensembles in which outside of a contour we have a gas and not just one configuration. Note that in the A-restricted ensemble only A particles occur while in the actual A-rich phase there will be some (small) "islands" of the B particles.

Using an extension of the Pirogov-Sinai theory in which the ground states are replaced by restricted ensembles, [8] prove an A-B phase separation, even if the A-A and the B-B interactions break the A-B symmetry. However, one must then require that each of the restricted ensembles be *dilute* in the sense that they are both within the region of convergence of their Mayer expansion. Such a limitation is not necessary in the A-B symmetric case. So, at present, we need either symmetry or diluteness which prevents us from dealing with a liquid-gas transition in a one-component fluid: one could invent restricted ensembles for the liquid and the gas phases by locally fixing the fluid density at appropriate values but there is no symmetry between these ensembles and it is not clear what kind of diluteness would hold for the liquid restricted ensemble. Of course, one can obtain a one-component fluid model from the Widom-Rowlison A-B model by "integrating out" the B particles. Then the A-B phase separation translates into a liquid-gas transition. The effective interactions between the A-particles induced by the B-particles can be computed explicitly when there is no interaction among the B-particles since, for a fixed configuration of the A-particles, the B-particles form a free gas [52]. Unfortunately these interactions are quite unrealistic for a model of a one component fluid.

VI. THE POTTS MODEL

At each lattice site we have a variable $s(a)$ taking values $1,\ldots,q$, where q is an integer; the Hamiltonian is

$$H = - \sum_{\langle a,b\rangle} \delta(s(a),s(b)) \ ,$$

where δ is the Kronecker delta-function and the summation is over pairs of nearest neighbors of the lattice.

We shall consider q large and $d \geq 2$. The phase diagram is expected to be as follows [53]: There are q ground states $s(a)=m$ (all $a \in Z^d$), $m=1,\ldots,q$, and, for temperatures low enough, there are q phases which are small perturbations of these q ground states. As the temperature is raised, we do not reach a critical point, as one does in the Ising model (which corresponds to $q=2$), but rather a first-order transition. At this transition temperature $T(q)=\beta(q)$ the mean energy $\langle \delta(s(a),s(b)) \rangle$ (a,b nearest neighbors) and the order parameter $\langle \delta(s(a),m) \rangle$ are discontinuous. For temperatures above $T(q)$, there is only one phase; at $T(q)$, there are $q+1$ phases, one high-temperature phase and q low-temperature phases. These latter phases are referred to as the "ordered phases" because, in each of them, the mean energy is close to 1 (for large q) i.e. typical configurations have most of their bonds "unbroken" ($s(x)=s(y)$ if x and y are nearest neighbors). The high-temperature phase is called "disordered" because its typical configurations have most of their bonds "broken" ($s(x) \neq s(y)$ if x and y are nearest neighbors), and the average energy tends to zero as $q \to \infty$. There are many configurations with broken bonds for large q, and thus the disordered phase has large entropy which compensates for its smaller energy than that of the ordered phases, which explains why all these phases can coexist.

This model has been analyzed using the reflection positivity property of the interaction by Kotecky and Shlosman [35]. Here we want to consider it from the point of view of the Pirogov-Sinai theory because it illustrates nicely the use of restricted ensembles [8, 14, 60]. Also, there is a vague analogy between this model and a liquid-gas transition: the ordered phases have a lot of energy (like the liquid phase) while the disordered one is favored by entropy (like the gaseous phase).

There is a more precise analogy between this model and the Blume-Capel model (3.7) with small positive g: at low temperatures, there are two (ordered) phases and, at some higher temperature $T(g)$ we reach a first-order transition point where three phases coexist. The third phase corresponding to the ground state G^0 of the $g=0$ Hamiltonian, has larger entropy (of low energy excitations) than the two ordered phase and becomes the unique phase for temperatures above $T(g)$.

Returning to the Potts model, we define $q+1$ restricted ensembles: q of them (the ordered ones) are trivial; they consist of the corresponding ground state configuration: $s(x)=m$ (all x), $m=1,\ldots,q$. The last ensemble is made of all the configurations for which all bonds are broken:

$$X_R^D = \{s: s(x) \neq s(y) \text{ if } x \text{ and } y \text{ are n.n.'s}\} ,$$

together with the Gibbs state induced by H on it: the conditional probability distributions of this Gibbs state in finite volumes are just the normalized counting measures, since H=0 on X_R^D.

The free energy of these ensembles is easy to compute. For the ordered ensembles, it is just the ground state energy, equal to -d (d bonds per lattice site in d dimensions). For the disordered ensemble, the free energy equals, for q large enough h,

$$-\beta^{-1}(\log q + f(q^{-1})) , \qquad (6.1)$$

where f is an analytic function, and f(0)=0 (see [8, proof of Theorem 5]).

One expects therefore phase coexistence between the ordered phases to occur for $\beta(q) \simeq (\log q)/d$ (to leading order). Let us write $\beta = \beta' \log q$. Now we should think of log q as playing the role of β in the Blume-Capel model while β' plays the role of g. For $\log q \to \infty$ we expect the following: if $\beta' < 1/d$ then the disordered restricted ensemble is similar to the unique for g<0 ground state G^0 of the Blume-Capel model. For $\beta' > 1/d$, the ordered ensembles replace the ground states G^+, G^-. And for $\beta' = 1/d$ we have coexistence of the ordered and the disordered ensembles (just like at g=0 we had three ground states). Actually, the analogy is more precise if one restricts oneself to quantities which are invariant under the symmetry of the Hamiltonians, i.e. under the global spin flip for the Blume-Capel model or under global permutations of the spins in the Potts model. Then one has only one ground state for g=0 and one restricted ensemble for $\beta' \neq 1/d$.

For the Blume-Capel model, the Pirogov-Sinai theory proves the existence of a line $g(\beta)$ on which all three phases coexist. Moreover, one can compute $g(\beta)$ perturbatively (see Sect. 3). An extension of the Pirogov-Sinai theory [8, 14, 60] gives similar results for the Potts model: for any q large enough there $\beta'(q)$, $\beta'(q) \to 1/d$ as $q \to \infty$, for which q+1 phases coexist.

To find the perturbation expansion for $\beta'(q)$ one has to compute the free energies of the gases of excitations of the restricted ensembles (which replace the ground states), then set them equal and solve for $\beta'(q)$. The lowest order excitations of the disordered ensemble are given by one unbroken bond $\langle x,y \rangle$ (i.e. s(x)=s(y)), surrounded by broken bonds. There are d such excitations per lattice site. One looses a factor of q (to leading order) in the sum over configurations (entropy factor) but one gains a factor exp β on the energy side. Combining these two contributions, one obtains the free energy of this new ensemble where all bonds, adjacent to unbroken bonds are broken; this is similar to the gas of isolated spin flips in Sect. 2. To leading order, it is equal

$$- \beta^{-1}(\log q + f(q^{-1})) - (d \cdot \exp\beta)/\beta q , \qquad (6.2)$$

where

$$|f(q^{-1})| \le O(1/q) \ , \tag{6.3}$$

since f is analytic and zero at the origin.

For the ordered ensemble, the lowest order excitations are equal to the corresponding ground state everywhere apart form one point. This yields

$$- d - (q-1)/\beta \bullet \exp(-2d\beta) \tag{6.4}$$

for the leading order terms of the free energy of the restricted ensemble. Setting equal (6.2) and (6.4), writing $\beta=\beta' \log q$ and solving for β'one obtains 1/d as the term of order zero. Inserting $\beta'=1/d$ in the second term of (6.2) and using (6.3) one obtains

$$-d - O((q \log q)^{-1}) = - \beta'^{-1} - d/(\beta' \log q)q^{(1-1/d)} + O((q \log q)^{-1}) \ ,$$

which, to leading order, yields

$$\beta'(q) = 1/d + 1/(q^{(1-1/d)} \log q) \ . \tag{6.5}$$

This coincides with the leading order terms of the exact answer [53] for d=2:

$$\beta(q) = \beta'(q) \log q = \tfrac{1}{2} \log q + \log (1+(1/\sqrt{q})) = \log(1+\sqrt{q}) \ .$$

However, (6.5) holds for all dimensions, and higher order corrections can be obtained by including higher order excitations (see also [25] and [28]).

Acknowledgements. J. B. thanks A. El Mellouki, J. Frohlich, K. Kuroda and J. L. Lebowitz for discussions and collaboration on some of the topics of the lectures, and Center for Transport Theory and Mathematical Physics at Virginia Tech for hospitality. J. S. thanks Joel Lebowitz and Daniel Styer for discussions, and Insitut de Physique Theorique de Universite Catholique de Louvain for hospitality; his work on this paper has been supported in part by NSF grant MCS 8301709.

APPENDIX: PROOF OF THE THEOREM OF SECT. III.1

We explain first the main ideas of the proof. We want to show that, at low temperatures, there is a unique phase which is a small perturbation of the ground state G^0. If G^0 was the unique ground state, we would define the contours of a configuration s as connected regions where s does not coincide with G^0. Then we would try to prove a Peierls' bound for such contours in the Gibbs state with (0)-boundary conditions. From this estimate one would deduce, using an idea of Gallavotti and Miracle-Sole [23] which is recalled below (see Lemmas 2 and 3), that typical configurations of any translation invariant Gibbs state are small perturbations of the ground state G^0 and that the low-temperature phase is unique.

However, such a simple scheme of proof has to be modified here, because there are three ground states, G^+, G^-, G^0, not one. Clearly, there is no damping factor in the Hamiltonian, and hence no Peierls' bound, for the regions where a configuration coincides with either G^+ or G^- and which would be regarded as part of a contour according to the scheme outlined above.

We know, however, from the discussion of Sect. III that if we associate with each ground state the restricted ensemble consisting of its lowest energy excitations then the (0)-restricted ensemble has the lowest free energy among these three restricted ensemble. This suggests that we use our first idea but with restricted ensembles replacing ground states: the contours of a configuration s are now connected subsets of the lattice on which the restriction of s does not belong to the (0)-restricted ensemble. If we try to estimate the probability of these contours, there will be some damping factor coming from the inclusion of the low energy excitations. The probability that a configuration s restricted to a region Λ belongs to the (+)-restricted ensemble, is bounded by the ratio of the partition functions, in Λ, of the gas of low energy excitations of (+) to the partition function of the gas of low-energy excitations of (0). This ratio is approximately equal

$$\exp(-|\Lambda|\exp -4\beta) , \tag{A.1}$$

(see (3.5), (3.6)).

However, at low temperatures this damping factor becomes very small and it is not obvious how to control the sum over all the contours. To deal with this problem, we shall define contours on a large (temperature-dependent) scale. Cover the lattice with boxes B(i) whose size is chosen so that $|B(i)|\exp(-4\beta)\to\infty$ as $\beta\to\infty$. Then, from (A.1) we can deduce that the probability that the configuration restricted a given box B(i) belongs to the (+)-restricted ensemble goes to zero as $\beta\to\infty$.

This coarse-grained description causes an obvious problem: there are other excitations than the lowest order ones and, if we define our contours on such a large scale, we loose some localization of these excitations. Namely an excitation will have to be in a box B(i) but it may be anywhere in that box. Moreover, we may have many different excitations in a given box. So it seems that we are back to the problem of controlling the sum over the contours. Observe however that the energy of the lowest excitations above the lowest ones (whose energy is equal to 4) is 6 (corresponding e.g. to two adjacent sites with s(a)=0 in the ground state G^+). The probability that such an excitation (or any higher energy excitation) is found in a region Λ can be bounded, by the usual Peierls' argument, by

$$|\Lambda|\exp(-6\beta) \ .$$

Therefore, if we choose the size of the boxes in such away that $|B(i)| \simeq \exp(c\beta)$ with c between 4 and 6, say 5, then, for any box B(i), the following two events will be very unlikely when $\beta \to \infty$:

- the configuration restricted to B(i) belongs to the (+)- or (-)-restricted ensemble.
- there an excitation (of any of the three ground states) of energy greater that 4 in B(i).

From this one deduces rather easily that a typical configuration belong to the (0)-restricted ensemble in most of these large scale boxes, and from this the uniqueness of the phase is deduced using the ideas of Gallavotti and Miracle-Sole mentioned at the beginning of this appendix.

Of course, the numbers 4 and 6 used here are accidental; what matters is the discreteness of the set of excitation energies. We may summarize the above discussion as follows: one associates with each energy E of elementary excitations a distance scale $\ell \simeq \exp(\frac{1}{2}\beta E)$ in the following way: consider a temperature dependent family of boxes $\Lambda(\beta)$ of volume $\exp(\beta E')$. If $E' > E$, $|\Lambda(\beta)| \gg \ell^2$ and there will be many excitations of energy E in $\Lambda(\beta)$, for large β. "Many" means that we are almost in the thermodynamic limit, i.e. the free energy of the gas of excitations of energy $\leq E$ is equal to its bulk contribution plus a (relatively) small boundary term. On the other hand, if $E' < E$, then, as $\beta \to \infty$, it becomes very unlikely to find *even one* excitation of energy E in $\Lambda(\beta)$. It is the presence of these different distance scales associated with the different excitation energies that makes the proof outlined above work.

Now we define the restricted ensembles more precisely and discuss their main properties. Let

$$X = \{-1,\ 0,\ +1\}^{Z^2}$$

be the configuration space of the model (we set d=2 for simplicity) and let

$$X_R^+ = \{s \varepsilon X\text{: } s(a)=0,+1 \text{ for all a, and if } s(a)=0 \text{ then } s(b)=+1 \text{ if } |b-a|=1\}$$

be the restricted ensemble of configurations of the gas of lowest energy excitations of the (+)-ground state. X_R^- is defined similarly, and

$$X_R^o = \{s \in X: \text{ if } s(a)=-1 \text{ or } +1 \text{ then } s(b)=0 \text{ if } |b-a|=1\} .$$

The subscript R at X ("restricted ensemble") indicates that only elementary excitations up to some order (here up to first order) are considered.

For $\Lambda \subset Z^2$, $X_{R,\Lambda}^{\varepsilon}$ $(\varepsilon=+,0,-)$ is the set of restrictions to Λ of the configurations of X_R^{ε}. We define the partition function of the restricted ensemble with boundary conditions $s'\varepsilon X_{R,\Lambda^c}^+$ as

$$Z_R(\Lambda|s') = \Sigma \exp-\beta H_\Lambda(s) ,$$

where the sum is over $s\varepsilon X_R^+$ which are equal to s' on Λ^c, and

$$H_\Lambda(s) = \Sigma (s(a)-s(b))^2 ,$$

where the sum is over pairs <a,b> of n.n.'s with at least one point in Λ. These partition functions depend on the restriction of s' to $\partial\Lambda$ only, and therefore the same notation will be used for any $s'\varepsilon X_{R,M}$ with $\partial\Lambda \subset M$. It is easy to realize that $Z_R(\Lambda|+)$ is just the partition function of a hard-square lattice gas: at each lattice site we can have a particle (corresponding to spin zero) and no two particles can be adjacent. The activity of a particle is equal to the Boltzmann weight of a zero spin, $z=\exp-4\beta$. This identification helps to realize that, for β large, i.e. for z small, we have a convergent low activity expansion for this restricted ensemble. The free energy is analytic in z and we have good clustering properties of the correlation functions. In particular, we may write, for all Λ

$$\log Z_R(\Lambda|s) = - |\Lambda|\beta\psi_R(\beta|+) + o(e^{-4\beta})|\partial\Lambda| ,$$

where

$$\beta\psi_R(\beta|+) = - e^{-4\beta} + O(e^{-8\beta}) ; \tag{A.2}$$

see (3.5). Similar formulas hold for $Z_R(\Lambda|-)$ and for $Z_R(\Lambda|0)$, except that in the latter case we have

$$\beta\psi_R(\beta|0) = - 2e^{-4\beta} + O(e^{-8\beta}) . \tag{A.3}$$

Now we introduce the contours. They will be defined on two different scales. On the first scale, we just have the ordinary contours with energy large enough. The second, large, scale will depend on β and will be used to take advantage of the difference in the number of low energy excitations of the (0)- and the (+)- and (-)-ground states.

Let s be an excitation (Sect. III, especially III.3) of the (+)-ground state. Define its *retouch*, ret(s), by setting ret(s)(a)=+1 if s(a)=0 and s(b)=+1 for b such that |b-a|=1, and ret(s)(a)=s(a) otherwise. Thus, in ret(s) we remove all the lowest energy excitations of s, but not the excitations of higher order. For example, if $s \varepsilon X_R^+$ then ret(s) is the (+)-ground state. The retouch of excitations of the other ground states is defined similarly.

A configuration s is *retouched* if s=ret(s). Let s be such a configuration. Its *boundary* is the set of pairs <a,b> where s(a)≠s(b). A *small scale contour* γ of a configuration s is a pair $\gamma=([\gamma],s_{[\gamma]})$ where $[\gamma]$ (the *support* of γ), is a maximal connected subset of the boundary of ret(s) and $s_{[\gamma]}$ is the restriction of s to $[\gamma]$; this definition of $[\gamma]$ is different from that of Sect. II, but closely related to it. The energy of γ is

$$E(\gamma) = \Sigma\ (s(a)-s(b))^2 \ , \tag{A.4}$$

where the sum is over pairs of n.n.'s contained in $[\gamma]$. Note that since we defined contours by starting with a retouched configuration the lowest energy excitations (s(a)=0, s(b)=+1, for |a-b|=1) appear in the restricted ensembles but do not give rise to contours, as they would with the usual definitions. In particular, we have

$$\min_\gamma E(\gamma) = 6 \ , \tag{A.5}$$

not 4, and it follows immediately from (A.1) that $E(\gamma) \geq |\gamma|$ = number of bonds in $[\gamma]$.

The large scale contours are defined as follows: Let L(β)=exp(5β/2) (the reason for this choice will become clear later). Cover Z^2 with *large boxes*

$$B(i) = B(0) + \tfrac{1}{2}Li \ , \qquad i \ \varepsilon \ Z^2 \ ,$$

where B(0) is a square of side L(β) centered at the origin; |B(i)|=exp(5β). B(i) is a *regular box* of a configuration s if $s|_{B(i)} \ \varepsilon \ X^o_{R,B(i)}$, and it is *irregular* otherwise. Thus we may distinguish between two types of irregular boxes B(i) of a configuration s:

type 1: $s|_{B(i)} \ \varepsilon \ X^+_{R,B(i)} \cup X^-_{R,B(i)}$

type 2: The support of a small scale contour of s intersects B(i).

A *large scale contour* Γ is a connected family of irregular boxes. We set

$||\Gamma||$ = number of boxes in Γ, $|\Gamma|$ = number of sites in Γ.

Thus, in our case, $|\Gamma|$ is bounded from above and from below by a constant times $\exp(5\beta)||\Gamma||$.

Now we state three Lemmas the last two of which are fairly standard, given Lemma 1. The proof of the theorem follows also in a by now standard way from these lemmas, [23], so that we shall concentrate on the proof of the new ingredient (Lemma 1) and sketch only the rest of the proofs.

Lemma 1. *Assume that* β *is large enough. There exists a constant* c *such that, for all finite* $\Lambda \subset Z^2$, *all boundary conditions* $s \in X^o_{R,\Lambda^c}$ *and all contours* $\Gamma \subset \Lambda$,

$$P_\Lambda(\Gamma|s) \le \exp(-c\beta||\Gamma||) ,$$

where $P_\Lambda(\bullet|s)$ *is the Gibbs measure in* Λ *with boundary conditions* s.

If f is a function on X_M, $M \subset Z^2$, we identify it with the corresponding function on X which depends on the restriction of the configuration to M only, and we write $f \in C(M)$. For Λ contained in M and $s \in X$ we define

$$\langle f\rangle_\Lambda(s) = \Sigma f(s')P_\Lambda(s'|s) ,$$

where the sum is over configurations s' in Λ.

Lemma 2. *For any finite subset* M *of the lattice, any* $f \in C(M)$ *and any* $s \in X^o_R$, *the thermodynamic limit*

$$\langle f\rangle^0 \equiv \lim_\Lambda \langle f\rangle_\Lambda(s)$$

exists and does not depend on $s \in X^o_R$. *Moreover,*

$$|\langle f\rangle_\Lambda(s) - \langle f\rangle^0| \le ||f||_\infty \exp(-c\beta d(M,Z^2\setminus\Lambda)) ,$$

where c *is a positive constant and* d(,) *is the distance between sets in large-scale units:* $d(A,B)=L(\beta)^{-1}\inf\{|a-b|: a\in A, b\in B\}$, *where* $|a-b|$ *is the Euclidean distance of* a *and* b, *for example.*

For any configuration $s \in X$ and finite subset Λ of the lattice consider the set of the large-scale contours of s which intersect both Λ and its complement. Let $\Gamma_\Lambda(s)$ be the union of intersections of these contours with Λ - an analogon of the "open contours" of the IM of [23].

Lemma 3 [23]. *There are two positive constants,* c *and* c', *such that*

$$P_\Lambda(\, ||\Gamma_\Lambda|| > c|\partial\Lambda| \;|s) \le \exp(-c'\beta|\partial\Lambda|)$$

for any finite Λ, $s \in X$, *and* β *large enough.*

Given the lemmas, we can prove the theorem: As in [23], we have to show that for any finite subset M of the lattice, any $f \in \mathbf{C}(M)$, and any $s \in X$,

$$\lim_\Lambda |\Lambda|^{-1} \sum_a \langle\tau(a)f\rangle_\Lambda(s) = \langle f\rangle^o \,, \tag{A.6}$$

where $\tau(a)$ is the translation by a (acting on functions) and the sum is over all a such that $M + a \subset \Lambda$. Now, conditioning on Γ_Λ:

$$\langle\bullet\rangle_\Lambda(s) = \sum P_\Lambda(\Gamma_\Lambda|s)\langle\bullet\rangle(\Gamma_\Lambda) \,.$$

By Lemma 3 we may assume that Γ_Λ covers a small part of Λ; indeed,

$$|\Gamma_\Lambda| \le O(\exp 5\beta)||\Gamma_\Lambda|| \le O(\exp 5\beta)|\partial\Lambda| \,,$$

which is much smaller than $|\Lambda|$, for $|\Lambda|$ large.

Decompose the complement of Γ_Λ in Λ into connected components: $\Lambda\backslash\Gamma_\Lambda = \cup_i \Lambda(i)$. For most a of the LHS of (A.6), $\tau(a)f$ will be in $\mathbf{C}(\Lambda(i))$ for some i. But by our definition of contours, the configuration on the boundary of $\Lambda(i)$ belongs to the restricted ensemble X^o_R. Thus, for most a, $\langle\tau(a)f\rangle(\Gamma_\Lambda)$ will be equal to $\langle\tau(a)f\rangle_{\Lambda(i)}(s)$ for some $\Lambda(i)$ and some $s \in X^o_R$. Now we can use Lemma 2, which says that $\langle\tau(a)f\rangle_{\Lambda(i)}(s)$ is almost equal to $\langle f\rangle^o$ provided $a+\Lambda_o$ is far from the boundary of $\Lambda(i)$. But again, since $|\Gamma_\Lambda|$ is much smaller than $|\Lambda|$ the fraction of Λ that is covered by the $\Lambda(i)$'s with $|\Lambda(i)| \ge |\Lambda|^{1-\varepsilon}$, $\varepsilon > 0$, goes to 1 as $\Lambda \to \infty$. So we can assume $\Lambda(i)$ to be large and to have a small boundary (since the latter is contained in Γ_Λ); thus, for most a, we are far from the boundary of $\Lambda(i)$ and this concludes the proof.

Now, given Lemma 1, it is rather easy to prove Lemmas 2 and 3. For Lemma 2, we observe that in the complement of the contours, we have the (0)-restricted-ensemble. However, for this ensemble we have a convergent low-fugacity expansion. Now, combine this expansion with a (large-scale) contour expansion. Convergence of the combined expansion then follows from Lemma 1, which gives a suitable damping factor for the contours, and the following remarks.

- When we sum over all contours containing a given box, the number of contours having a given (large scale) size $||\Gamma||=n$ grows like c^n where c *does not* depend on $L(\beta)$.

- When we consider the combined expansion, we have objects defined on two different scales: the contours and the graphs of the low fugacity expansion in the restricted ensemble. However, one shows that the sum over all graphs connected to a given box B in a contour is at most $\exp(O(\exp{-4\beta})|\partial B|)$, where the boundary ∂B of B has a size $L(\beta)=\exp(5\beta/2)$ which is much smaller than $\exp(4\beta)$ and, therefore, this contribution is a small correction to the weight of the contour.

Once we have a convergent expansion for the Gibbs state with boundary conditions in the (0)-restricted-ensemble, Lemma 2 is easy to prove.

To prove Lemma 3 we follow [23]: we first change boundary conditions from the arbitrary s in the Lemma to (0)-boundary-conditions. Direct estimates on partition functions yield:

$$P_\Lambda(\bullet|s) \leq \exp(4\beta|\partial\Lambda|)P_\Lambda(\bullet|0) \ .$$

Now from Lemma 1 and simple combinatorics (we connect together the possibly disconnected parts of Γ_Λ through $\partial\Lambda$, which, for $||\Gamma_\Lambda||\geq c|\partial\Lambda|$, gives only a small error) we obtain

$$P_\Lambda(\ ||\Gamma_\Lambda||> c'|\partial\Lambda|\ \ |0) \leq \exp(-c''\beta\ |\partial\Lambda|) \ ,$$

from which Lemma 3 follows.

Proof of Lemma 1. We will use the following notation. We write $Z(\Lambda|\bullet)$ for a partition function in a subset $\Lambda\subset Z^2$, where we indicate after the vertical bar the ensemble defining the partition function and the boundary conditions. For example,

$$P_\Lambda(\Gamma|s) = Z(\Lambda|\Gamma,s)/Z(\Lambda|s) \ , \tag{A.7}$$

where in the numerator we sum only over the configurations for which Γ is a (large-scale) contour.

For a contour Γ its *support* is

$$[\Gamma] = \bigcup_{B\in\Gamma} B \ ,$$

and, for $\Lambda \subset Z^2$ its *boundary*, $\partial\Lambda$, is the set of points of its complement which have nearest neighbors in Λ.

Now, we condition on the values of the spins in $\partial[\Gamma]$:

$$P_\Lambda(\Gamma|s) = \sum_{s'} P_\Lambda(\Gamma|s,s')P_\Lambda(\Gamma,s'|s) \ , \tag{A.8}$$

where the sum runs over all $s' \varepsilon X^o_{R,\partial[\Gamma]}$, since, by definition of Γ, configurations in $\partial[\Gamma]$ belong to the restricted ensemble $X^o_{R,\partial[\Gamma]}$.

Conditioning on the spins in $\partial[\Gamma]$ decouples the contour Γ from $\Lambda\backslash[\Gamma]$ (the interaction has a range equal to one). We can therefore write

$$P_\Lambda(\Gamma|s,s') = P_{[\Gamma]}(\Gamma|s,s') = Z([\Gamma]|\Gamma,s,s')/Z([\Gamma]|s,s') \ , \tag{A.9}$$

where s, s' define the boundary conditions for $[\Gamma]$ (s outside Λ, s' inside Λ). Now we shall prove that the RHS here is bounded by $\exp(-c\beta||\Gamma||)$, with c independent of s and s'; by (A.8), this will obviously prove the lemma.

We can write

$$Z([\Gamma]|\Gamma,s,s') = \sum_{\Gamma^2} \sum_{\omega}{}^2 Z([\Gamma]|\Gamma^2,\omega,s,s') \ , \tag{A.10}$$

where the first summation is over all possible families Γ^2 of type-2 boxes of Γ, and the second over families ω of small scale contours in $[\Gamma]$ such that for each box of Γ^2 there is a contour of ω with support intersecting the box.

Let

$$[\Gamma]\backslash[\omega] = \bigcup_i M_i, \text{ where } [\omega] = \bigcup_{\gamma\varepsilon\omega} [\gamma] \ , \tag{A.11}$$

be the decomposition of $[\Gamma]\backslash[\omega]$ into connected components; with a fixed family ω of small-scale contours, $Z([\Gamma]|\Gamma^2,\omega,s,s')$ is a sum over configurations belonging to a restricted ensemble in each M_i, an ensemble determined by ω and the boundary conditions s, s'. We may write:

$$Z([\Gamma]|\Gamma^2,\omega,s,s') = e^{-\beta E(\omega)} \prod_i Z_R(M_i|s_i) \ , \tag{A.12}$$

where

$$E(\omega) = \sum_{\gamma\varepsilon\omega} E(\gamma) \ ,$$

and s_i is the configuration on ∂M_i determined by s, s' and ω. Moreover, by the definition of Γ^2,

$$[\Gamma\backslash\Gamma^2] \subset \bigcup_{\varepsilon(i)\neq 0} M_i \ . \tag{A.13}$$

We insert (A.12) into (A.10), (A.10) into (A.9) and estimate the ratio

$$(\prod_i Z_R(M_i|s_i))/Z([\Gamma]|s,s') \le (\prod_i Z_R(M_i|s_i))/Z_R([\Gamma]|s,s') \ , \tag{A.14}$$

where in the denominator we used the obvious lower bound:

$$Z([\Gamma]|s,s') \ge Z_R([\Gamma]|s,s') \ .$$

We can now use the convergent low-fugacity expansion to estimate the ratio of the partition functions of the restricted ensembles in (A.14).

Using a simple modification of the Proposition of Sec. II, we can write log $Z_R(\Lambda|s)$ for the boundary conditions s belong to the ε-restricted ensemble, as a sum of a volume term and a boundary term:

$$\log Z_R(\Lambda|s) = -|\Lambda|\beta\psi_R(\beta|\varepsilon) + \Delta^{\varepsilon}(\Lambda,\beta,s) \ , \tag{A.15}$$

where $\psi_R(\beta|\varepsilon)$ is as at the beginning of this Appendix. The boundary term is

$$\Delta^{\varepsilon}(\Lambda,\beta,s) = \sum_{\vartheta:\vartheta\ \partial\Lambda\neq\emptyset} \phi'(\vartheta|s)\exp{-4\beta|\vartheta|} \ , \tag{A.16}$$

where the sum is over all multiplicity functions ϑ defined on the "particles" of the "gas" of elementary excitations of the restricted ensemble, $\phi\ \ \partial\Lambda\neq\emptyset$ means that the support of one of the "particles" of ϑ intersects $\partial\Lambda$, and $|\vartheta|$ is the total number of "particles" (counting multiplicities) in the support of ϑ. Finally, $\phi'(\vartheta|s)$ is a sum of $\phi^T(\vartheta)$ (see Sect. 2) which depends only on s through its restriction to $\partial\vartheta$ - the boundary of the support of ϑ. The estimate (2.11) holds if we replace in it $\phi^T(\vartheta)$ by $\phi'(\vartheta|s)$. Thus we have:

$$\sup_X \sum_{\vartheta:X\varepsilon\mathrm{supp}\vartheta} |\phi'(\vartheta|s)|e^{-4\beta|\vartheta|} \le 0(e^{-4\beta}) \ , \tag{A.17}$$

where the sum runs over all ϑ whose support contains a fixed excitation X.

We write the volume terms of (A.14) as:

$$\exp\beta(\sum_i |M_i|\psi_R(\beta|\varepsilon(i))-|\Gamma|\psi_R(\beta|0)) \le \exp(-(e^{-4\beta}+0(e^{-8\beta}))\sum_{i:\varepsilon(i)\neq 0}|M_i|)$$
$$\le \exp(-\tfrac{1}{2}e^{-4\beta}|\Gamma^1|) \le \exp(-ce^{-\beta}||\Gamma^1||) \ , \tag{A.18}$$

where $\Gamma^1 = \Gamma\backslash\Gamma^2$, and where to obtain the first equality we used (A.2) and (A.3), and the last one (A.11); c is a β-independent constant.

Now, in (A.14) we have also to estimate the boundary terms. The family of the boundaries ∂M_i consists of two subfamilies: one contained in $[\omega]$ (coming from the small scale contours), and another one contained in $\partial[\Gamma]$, on which we have the same boundary conditions, s and s', in the numerator and the denominator of (A.14). Since these boundary conditions are the same, the contributions to the boundary term (A.16) from this part of the boundary, cancel each other in $\Sigma_i \log Z_R(M_i|s_i)$ and in $\log Z_R([\Gamma]|s,s')$.

On the other hand, for the first part of the boundary we obtain, using (A.17), a term bounded in absolute value by

$$c'e^{-4\beta}|\omega| , \tag{A.19}$$

where c' is independent of β, and $|\omega|=|[\omega]|$. Hence, we obtain finally that (A.14) is bounded by

$$\exp\,(-ce^{\beta}||\Gamma^1|| + c'e^{-4\beta}|\omega|) .$$

By $E(\gamma)\geq|\gamma|$, this inserted in (A.8) yields

$$Z([\Gamma]|\Gamma^2,\omega,s,s')/Z_R([\Gamma]|s,s') \leq \exp(-\beta' E(\omega)-ce^{\beta}||\Gamma^1||) , \tag{A.20}$$

where

$$\beta' = \beta'(\beta) \equiv \beta - c'e^{-4\beta} , \tag{A.21}$$

with c' from (A.19); for β large, β' differs little from β.

Now, inserting (A.20) into (A.10), and (A.10) into (A.9), gives:

$$P_\Lambda(\Gamma|s,s') \leq \exp(-ce^{\beta}||\Gamma^1||)\sum_\omega{}^2 \exp-\beta' E(\omega) . \tag{A.22}$$

The sum over the families ω of small-scale contours is restricted by the condition that, for each $B\varepsilon\Gamma^2$ there exists at least one contour $\gamma\varepsilon\omega$ with $[\gamma]\cap B\neq\emptyset$. Therefore,

$$\sum_\omega{}^2 \exp\,(-\beta' E(\omega)) \leq \prod_{B\varepsilon\Gamma^2} \Big(\sum_{m\geq1} (1/m!) \sum_{\gamma_1,\dots,\gamma_m}^{B} \exp(-\beta' \sum_j E(\gamma_j))$$

$$\leq \prod_{B\varepsilon\Gamma^2} \Big(\sum_{m\geq1} (1/m!)\Big(\sum_\gamma^B \exp(-\beta' E(\gamma))\Big)^m\Big) , \tag{A.23}$$

where the superscript B indicates summation over contours γ with $[\gamma]\cap B \neq\emptyset$.

Now, since $E(\gamma)\geq|\gamma|$ and, since $E(\gamma)\geq 6$ (see (A.5)), the usual Peierls' argument gives for β large

$$\sum_{\gamma}^{B} \exp(-\beta' E(\gamma)) \leq c'|B| \exp(-6\beta') = c' e^{-\frac{1}{2}\beta} , \qquad \text{(A.24)}$$

where we use $|B|=\exp 5\beta$ and (A.17); c' does not depend on β. (A.24) inserted into (A.23) gives for (A.22) the bound:

$$(\exp(c' e^{-\frac{1}{2}\beta})-1)^{||\Gamma^2||} \leq (c'' e^{-\frac{1}{2}\beta})^{||\Gamma^2||} , \qquad \text{(A.25)}$$

since $\exp(y)-1=0(y)$ as $y\to 0$.

Now, inserting this last estimate into (A.18) and going back to (A.8) concludes the proof.

REFERENCES

1. Aizenman, M. and Lieb E. H.: J. Stat. Phys. **24**, 279 (1981)
2. Baxter, R. J.: J. Phys. **C6**, L 445 (1983)
3. Binder, K., Lebowitz, J. L., Phani, M. K. and Kalos, M. H.: Acta Metall. **29**, 1655 (1981)
4. Blume, M.: Phys. Rev. **141**, 517 (1966)
5. Bricmont, J., El Mellouki, A. and Frohlich, J.: "Random Surfaces in Statistical Mechanics: Roughening, Rounding, Wetting ...". Preprint
6. Bricmont, J., Fontaine, J. R. and Lebowitz, J. L.: J. Stat. Phys. **29**, 193 (1982)
7. Bricmont, J., Kuroda, K. and Lebowitz, J. L.: J. Stat. Phys. **33**, 59 (1983)
8. Bricmont, J., Kuroda, K. and Lebowitz, J. L.: "First Order Phase Transitions in Lattice and Continuous Systems: Extension of the Pirogov-Sinai Theory", to appear in Commun. Math. Phys..
9. Bricmont, J., Lebowitz, J. L. and Pfister. C. E.: Commun. Math. Phys. **78**, 117 (1980)
10. Bricmont, J. and Slawny, J.: "Phase Transitions in Systems with a Finite Number of Dominant Ground States", in preparation.
11. Capel, H. W.: Physica **32**, 966 (1966)
12. Coppersmith, S. N.: "The Low Temperature Phase of Stacked Triangular Ising Antiferromagnet", preprint, Brookhaven National Laboratory
13. Dinaburg, E. I. and Sinai, Ya. G.: Commun. Math. Phys. **98**, 119 (1985)
14. Dinaburg, E. I. and Sinai, Ya. G.: work on the Potts Model, in preparation
15. Dobrushin, R. L.: Func. Anal. and Appl. **2**, 31, 44 (1968)
16. Dobrushin, R. L.: Theory Prob. and Appl. **17**, 582 (1972)

17. Dobrushin, R. L., Kolafa, J. and Shlosman S. B.: "Phase Diagram of the Two-dimensional Ising Antiferromagnet (Computer-assisted Proof)", preprint
18. Dobrushin, R. L. and Zahradnik, M.: "Phase Diagrams for the Continuous Spin Models. Extension of Pirogov-Sinai Theory", preprint
19. Domb, C.: Adv. Phys. **9**, 149 (1960)
20. Fisher, M. E. and Selke, W.: Phil. Trans. Roy. Soc. Lond. **A302**, 1 (1981)
21. Frohlich, J., Israel, R. B., Lieb, E. H. and Simon, B.: Commun. Math. Phys. **62**, 1 (1978); J. Stat. Phys. **22**, 297 (1980)
22. Frohlich, J., Simon, B., and Spencer, T.: Commun. Math. Phys. **50**, 77 (1976)
23. Gallavotti, G., Martin-Lof, A. and Miracle-Sole, S.: In "Statistical Mechanics and Methematical Problems" (A. Lenard, ed.), pp. 162-204, Springer-Verlag, Berlin, 1973
24. Gerzik, V. M.: Izv. Akad. Nauk SSSR Ser. Mat. **40**, 448 (1976)
25. Ginsparg, P., Goldschmidt, Y. Y. and Zuber, J.-B.: Nucl. Phys. **B170**, 409 (1980)
26. Glimm, J. and Jaffe, A.: "Quantum Physics", Springer-Verlag, New York, 1981
27. Glimm, J., Jaffe, A. and Spencer, T.: Commun. Math. Phys. **45**, 203 (1975); Ann. Phys. **101**, 610 (1976); ibid., 631
28. Goldschmidt, Y. Y.: Phys. Rev. **B24**, 1374 (1981)
29. Gopfert, N. and Mack, G.: Commun. Math. Phys. **82**, 545 (1982)
30. Griffiths, R. B.: In "Statistical Mechanics and Field Theory" (C. De Witt and R. Stora, eds.), Gordon and Breach, New York, 1971
31. Griffiths, R. B.: In "Phase Transitions and Critical Phenomena", Vol. 1, (C. Domb and M. S. Green, eds.), Academic Press, London and New York, 1972
32. Holsztynski, W. and Slawny, J.: Commun, Math. Phys. **66**, 147 (1979)
33. Imbrie, J. Z.: Commun. Math. Phys. **82**, 261 (1981); Commun. Math Phys. **83**, 305 (1982)
34. Israel, R. B.: Apendix B to "Convexity in the theory of Lattice Gases", Princeton University Press, Princeton, New Jersey, 1979
35. Kotecky, R. and Shlosman, S. B.: Commun. Math. Phys. **83**, 493 (1982)
36. Lanford, O. E.: In "Statistical Mechanics and Mathematical Problems" (A. Lenard, ed.), pp. 1-113, Springer-Verlag, Berlin, 1973
37. Lebowitz, J. L. and Lieb, E. H.: Phys. Lett. **32A**, 2 (1972)
38. Lebowitz, J. L. and Martin-Lof, A.: Commun. Math. Phys. **25**, 278 (1972)
39. Lebowitz, J. L. and Penrose, O.: J. Math. Phys. **7**, 98 (1966)
40. Minlos, R. A. and Sinai, Ya. G.: Tr. Mosk. Mat. Obshch. **17**, 213 (1967); Mat. Sb. **73**, 375 (1967); Tr. Mosk. Mat. Obshch. **19**, 113 (1968)
41. Peierls, R.: Proc. Cambridge Philos. Soc. **32**, 477 (1936)

42. Pirogov, S. A. and Sinai, Ya. G.: Teor. Mat. Fiz. **25**, 358 (1975); Teor. Mat. Fiz. **26**, 61 (1976)
43. Ruelle, D.: "Statistical Mechanics: Rigorous Results", Benjamin, New York, 1969
44. Ruelle, D.: Phys. Rev. Lett. **27**, 1040 (1971)
45. Ruelle, D.: Commun. Math. Phys. **53**, 195 (1977)
46. Seiler, E.: "Gauge Theories as a Problem in Constructive Quantum Field Theory and Statistical Mechanics", Lecture Notes in Physics, vol. 159, Springer-Verlag, Berlin-Heidelberg-New York, 1982
47. Sinai, Ya. G.: "Theory of Phase Transitions: Rigorous Results", Pergamon Press, Oxford, 1982
48. Slawny, J.: "Low Temperature Properties of Classical Lattice Systems: Phase Transitions and Phase Diagrams", to be published in *Phase Transitions and Critical Phenomena*, C. Domb and J. L. Lebowitz, eds., vol. 10, Academic Press, London, 1985.
49. Thompson, C. J.: "Mathematical Statistical Mechanics", Princeton University Press, Princeton, N. J., 1979.
50. Villain, J. and Bak, P.: J. Physique **42**, 657 (1981).
51. Villain, J., Bidaux, R., Carton, J. P. and Conte, R.: J. Physique **41**, 1263 (1980).
52. Widom, B. and Rowlison, J. S.: J. Chem. Phys. **52**, 1670 (1970)
53. Wu, F. J.: Rev. Mod. Phys. **54**, 235 (1982)
54. Zahradnik, M.: Commun. Math. Phys. **93**, 559 (1984)
55. Bortz, A. and Griffiths, R. B.: Commun. Math. Phys. **26**, 102 (1972)
56. Gallavotti, G.: Commun. Math. Phys. **22**, 103 (1972)
57. Basuev, A. G.: Theor. Math. Phys. **58**, 171 (1984)
58. Cassandro, M. and Da Fano, A.: Commun. Math. Phys. **36**, 277 (1974)
59. Galeb, F. F.: Tr. Mosk. Mat. Obshch. **44**, (1982) Engl. Transl.: Trans. Moscow Math. Soc. **2**, 111 (1983).
60. Kotecky, R. and Preiss, D.: in preparation
61. van der Waals volume: Physica **73** (1974)
62. Baxter, R. J., Enting, I. G. and Tsang, S. K.: J. Stat. Phys. **22**, 465 (1980)
63. Lebowitz, J. L., Phani, M. K. and Styer, D. F.: J. Stat. Phys. **38**, 413 (1985)
64. Dobrushin, R. L. and Shlosman, S. B.: "The Problem of Translation Invariance of Gibbs States at Low Temaperatures", Sov. Sc. Rev. Ser. C, vol. 5 (1985)
65. Ruelle, D.: "Thermodynamic Formalism", Addison-Wesley, Reading, Massachusetts, 1979

Low Temperature Continuous Spin Gibbs States on a Lattice and the Interfaces between Them - a Pirogov Sinai Type Approach

Miloš Zahradník

Faculty of Mathematics and Physics, Charles University,
Prague, Sokolovská 83
186 00 Czechoslovakia

§ 0 .Introduction. The Main Result and the Basic Example.

The Pirogov Sinai theory, which is one of the very general expressions of the so called Peierls approach, is often referred to as a complicated theory one should use only if all the "more common" methods fail to work.
In this note, we aim to contribute in favor of the opposite point of view: namely that this theory is really a convenient and general means for the description of a low temperature behaviour of many lattice spin models of statistical physics.

The theory is not only a "method of construction of the phase diagram". It gives, too, some rather precise information about the structure of coexisting phases.
Even in the uniqueness region, the theory gives some results hardly obtained by another methods (e.g. about the behaviour of the "nonstable" phases) but the region of a coexistence of phases is the more characteristic realm of applications of the theory.

Here, we will formulate one general result, illustrating a possible use of the theory. The investigation was motivated by the example below.

MAIN THEOREM

ASSUMPTIONS.

1) General conditions.

We consider a lattice model with a configuration space

$$X = (\mathbb{R}^k)^{\mathbb{Z}^\nu}, \quad k \geq 1, \quad \nu \geqslant 2$$

and with a finite range hamiltonian of the type

$$H(x_\Lambda \mid x_{\Lambda^c}) = \sum_{A \cap \Lambda \neq \emptyset,\ \mathrm{diam}\, A \leq \kappa} \Phi_A(x_A) \tag{0}$$

where Φ_A are translation invariant, bounded from below, continuous interactions on $X(A) = (\mathbb{R}^k)^A$, with values in $\mathbb{R} \cup \{+\infty\}$, and κ is the "interaction range".

2) Existence of ground states.

We assume that there are two translation invariant ground states of H, denoted by x^+, x^- in the following.
We also assume that there are two "horizontally translation invariant" ground states of H, denoted by $x^\pm$, $\tilde{x}^\pm$ in the following, such that $x^\pm = \tilde{x}^\pm$ for no "vertical shift" of $x^\pm$; both $(x^\pm)_t$, $(\tilde{x}^\pm)_t$ converging to $(x^+)_t$ (resp. $(x^-)_t$) if $t_\nu \to \infty$ (resp. $t_\nu \to -\infty$).

Note. Horizontal (resp. vertical) shift is a shift $\{t \rightsquigarrow t+s\}$ where $s = (s_1, \dots, s_\nu)$ and $s_\nu = 0$ (resp. $s_1 = s_2 = \dots s_{\nu-1} = 0$).

3) Second differentiability of all Φ_A & the positive definiteness of the Taylor approximations.

We assume that the interactions Φ_A are twice differentiable at any $(x+)_A$, $(x-)_A$, $(x\pm)_A$, $(\tilde{x}\pm)_A$. We also assume that there is some $C > 0$ such that for any $y \in X$ such that $|y|^2 = \sum_{t \in \mathbb{Z}^\nu} |y_t|^2$ is sufficiently small, the inequality

$$H(x^+ + y) - H(x^+) \geq C|y|^2 \tag{1}$$

holds, and analogous inequalities hold also for x^-, $x\pm$, $\tilde{x}\pm$.

4) A Peierls (Gertzik, Pirogov, Sinai) type condition.

We assume that for each $\delta > 0$ there is some $C = C(\delta) > 0$ such that for any $y \in X$ satisfying the condition $y_t \equiv 0$ for all except of a finite number of $t \in \mathbb{Z}^\nu$,

$$H(x^+ + y) - H(x^+) \geq C|B(y)| \tag{2}$$

where

$$B(y) = B_\delta(y) = \{t \in \mathbb{Z}^\nu : |y_t| \geq \delta\}$$

and analogously for x^-, $x\pm$, $\tilde{x}\pm$.

Note. We do not assume the smallness of y there. Actually, the condition (2) must be supplemented by some entropy bounds for the "energy shells" of the model, guaranteeing the existence of a Gibbs state at least for low temperatures - see [1].

Until now, the interactions Φ_A were fixed. We assume further that there is a two parameter family of hamiltonians of the type

$$H^{\lambda,\lambda'}(x) = \sum_{A:\,\mathrm{diam}\,A \leq \varkappa} \Phi_A^{\lambda,\lambda'}(x_A), \quad \Phi_A^{0,0} \equiv \Phi_A .$$

We assume that the functions $\Phi_A^{\lambda,\lambda'}(x_A)$ are real analytical both in λ and λ' (having analytical extensions to some neighborhood of zero in $\mathbb{C} \times \mathbb{C}$ not depending on x_A) and also "sufficiently regular" as functions of the triple (λ, λ', x_A) (see [1]).

5) Degeneracy removing condition.

We define the quantities (formally–it is easy to give a precise definition)

$$\begin{aligned} e_{\lambda,\lambda'} &= |\mathbb{Z}^\nu|^{-1}\left(H^{\lambda,\lambda'}(x+) - H^{\lambda,\lambda'}(x-)\right) \\ f_{\lambda,\lambda'} &= |\mathbb{Z}^{\nu-1}|^{-1}\left(H^{\lambda,\lambda'}(x\pm) - H^{\lambda,\lambda'}(\tilde{x}\pm)\right) \end{aligned} \tag{3}$$

Note. $e_{\lambda,\lambda'}$ is the difference in the mean energies of X^+, X^- whereas $f_{\lambda,\lambda'}$ is the difference in the mean interface energies of $X^\pm$, $\tilde{X}^\pm$.

We assume that the mapping

$$\{ (\lambda, \lambda') \to (e_{\lambda,\lambda'}, f_{\lambda,\lambda'}) \} \tag{4}$$

has an invertible differential in (0,0).

RESULT.

Then, under all these assumptions, the following is true:

There is an analytical function

$$\{ T \rightsquigarrow (\lambda(T), \lambda'(T)) \} : (0, T_0) \to \mathbb{R}^2 \tag{5}$$

defined for some (small) $T_0 > 0$ and such that:

I (for $\nu \geq 2$)

There are two translation invariant Gibbs states P_T^+, P_T^- on X corresponding to the hamiltonian $H^{\lambda(T),\lambda'(T)}$ and the temperature T such that P_T^+ is a perturbation of the gaussian field μ^+ arising if each Φ_A is replaced by its second order Taylor expansion around X_A^+. (Analogously for P_T^-).

II (for $\nu \geq 3$)

There are two horizontally translation invariant Gibbs states $P_T^\pm$, $\tilde{P}_T^\pm$ on X corresponding to the hamiltonian $H^{\lambda(T),\lambda'(T)}$, and the temperature T such that e.g. $P_T^\pm$ is a perturbation of the gaussian field $\mu^\pm$ arising if each Φ_A is replaced by its second order Taylor expansion aroud $(X^\pm)_A$.

Notes.

1) A generalization to the case of periodical ground states is possible.

2) In the case I there is even a one dimensional manifold of all possible $(\lambda(T), \lambda'(T))$.

3) Both the Gibbs fields $P_T^\pm$, $\tilde{P}_T^\pm$ converge ("exponentially fast") to P_T^+ in the "vertical" direction $t_\nu \to \infty$ and correspondingly, they converge to P_T^- in the direction $t_\nu \to -\infty$.

References.

The statement I is proven in the paper [1] (there are no $X^\pm$, $\tilde{X}^\pm$, λ' in [1]).

The statement II will be a subject of a paper which is now in preparation [2]. A short, discrete spin version of the method used in II is explained in [3].

A standard reference on the Pirogov Sinai Theory is [4]. Some complements to this theory are given in [5] (for the discrete spin case). The line of study used in II was originated by the paper [6].

The other relevant references are [7] (which contain some results comparable to [1]) and also the contributions of J. Bricmont and J. Slawny in this volume. See also [1] for further references.

Technically, the proof of Main Theorem is based on some cluster expansion techniques for perturbed gaussian fields. Some of these techniques we learned from [8] .

<u>Basic example</u>

Take $X = (\mathbb{R})^{\mathbb{Z}^\nu}$. Consider a hamiltonian

$$H(x) = \sum_{\text{nearest neighbors}} \Phi(x_s, x_t) + \sum \Psi(x_t)$$

where

$$\Phi(x_s, x_t) = |x_s - x_t|^2 + \text{small perturbation}$$

and where Ψ is given as follows:
("double well potential")

Figure 1

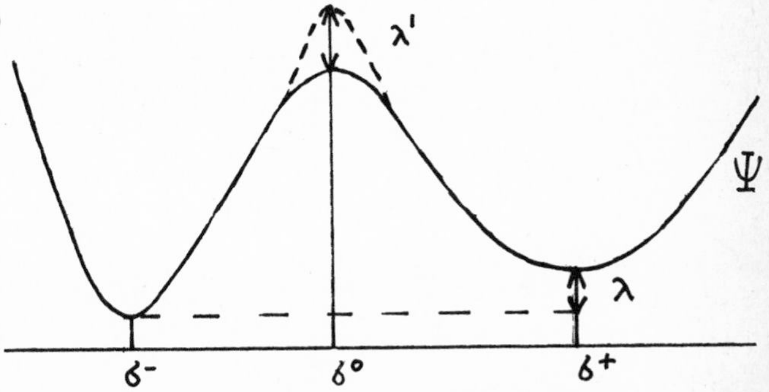

Intuitively, it is understood that the "energy - entropy fighting" principle governs the behaviour of the translation invariant phases in this example. For a properly "balanced" value of λ one expects two translation invariant phases - one centered around σ^+ and the other one around σ^- .
If we further change the height of the "hill above σ^0 " (by varying λ') this has a comparatively small effect on the coexistence of the preceding phases (assuming that the hill remains "high").

The change of λ' may effect, however, very substantially, the structure of interfaces between the phases centered around σ^+ , σ^- . For a "properly balanced λ' "one may even expect two equally good possibilities: either "tunelling" directly from one well to the other or "going over the hill". Graphically, one expects two competing kinds of interfaces: either

```
++++++++++++++++++++
++++++++++++++++++++
--------------------
--------------------
```

or

```
++++++++++++++++++++
++++++++++++++++++++
00000000000000000000
--------------------
--------------------
```

(Hence the notions of $X^\pm, \tilde{X}^\pm$).

Note. One can easily imagine more wells and also more competing interfaces. In fact, if one studies the coexistence of a maximal possible number of phases (and interfaces) then no additional difficulties arise in the general case.
Some modifications are needed, however, if one aims to construct the full phase & interface diagram. This was not studied in detail yet but it seems that a combination of a technique presented there with the technique of [5] will suffice to describe this more general situation.

x x x

In the rest of this contribution we will sketch the main notions and ideas used in the proof of Main Theorem (mostly of its I part; the subject of interfaces being only touched).

I. COEXISTENCE OF PHASES

§ 1. Basic concepts of the Pirogov Sinai Theory. Abstract P.S. setting.

We will present a generalized form of the theory, in the spirit of [1]. See also the contributions of J. Bricmont and J. Slawny in this volume.
Notations. We consider the situation of Main Theorem if not specified otherwise. (Sometimes we will sketch a more general approach, or illustrate our considerations on the Example above).
Traditionally, the theory is considered as a low temperature one. It is, nevertheless clearer to explain it in the case $T = 1$ where T denotes the temperature. By this we mean that the rescaled quantities

$$x^* = T^{-\frac{1}{2}} x \,, \quad H^*(x^*) = T^{-1} H(x) \tag{6}$$

are used (another scaling of x is needed in the "nongaussian" case).

The models studied by the theory must have a structure described by the following two complementary concepts: first, there is a notion of
1) almost ground configuration
(ground state in the original P.S. approach - for the discrete spin systems).
A second notion is that of
2) energy barrier (composed of contours), separating the regions with an almost ground configuration.

The usual choice is as follows.
1) Write the constant ground states x^+ and x^- as

$$x^+ = \{ x_t^+ \equiv \delta^+ \} , \; x^- = \{ x_t^- \equiv \delta^- \} .$$

Take some sufficiently large δ (to be specified later) and denote by $\mathcal{U}^{\pm}$ the δ neighborhood of $\delta^{\pm}$. Take

$$X^+ = (\mathcal{U}^+)^{\mathbb{Z}^\nu}, \quad X^- = (\mathcal{U}^-)^{\mathbb{Z}^\nu} \tag{7}$$

for the sets of almost ground configurations. See Figure 2.
(In the discrete case, we take $\mathcal{U}^{\pm} = \mathfrak{z}^{\pm}$).
(Clearly, δ must not be choosen too large. Actually, δ is small before the scaling (6)).

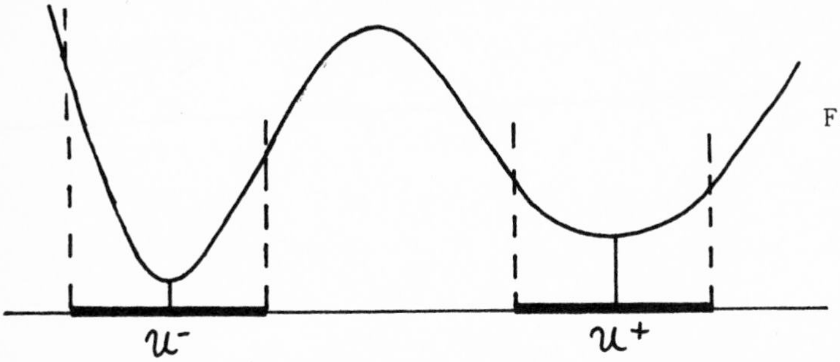

Figure 2

2) The definition of a contour starts with a notion of a correct point: $t \in \mathbb{Z}^{\nu}$ is said to be a + correct point of $x \in X$ (more precisely, $(+, \kappa, \delta)$ correct if $x_s \in \mathcal{U}^+$ for all $s \in \mathbb{Z}^{\nu}$ such that $|s-t| \leq \kappa$.
The set of all noncorrect points of x will be denoted by $B(x)$ (= $B_{\delta}^{\kappa}(x)$).
In the discrete spin case, one then defines a contour simply as a restriction of x to some finite component of $B(x)$ (which is called the support of a contour).
In the continuous spin case, such a definition works only at the "conceptional level". In order to prevent some complications otherwise arising in the technical parts of the theory it is, however, reasonable to introduce a more refined notion of a contour. The following modification is used in [1] : (one can skip this and read only the Note below).

"Wrapped" contours.

i) Take some $\tilde{\kappa} > \kappa$ such that "a boundary condition from $(\mathcal{U}^+)^{\Lambda^c}$ (analogously, $(\mathcal{U}^-)^{\Lambda^c}$) is only "slightly felt" on $\{t, \text{dist}(t, \Lambda^c) \geq \tilde{\kappa}\}$ in the gaussian approximation.
ii) Replace $B_{\delta}^{\kappa}(x)$ by $B_{\delta}^{\tilde{\kappa}}(x)$. This is still not the best choice because by fixing x on $B_{\delta}^{\tilde{\kappa}}(x)$, the configuration x may "tempt to go out" of $\mathcal{U}^+$ resp. $\mathcal{U}^-$ outside of $B_{\delta}^{\tilde{\kappa}}(x)$. (This is not the case in our Example but it may happen with an additional interaction $|x_{t+s}+x_{t-s}-2x_t|^2$, $|s| = 1$). See Figure 3.
Take, therefore, a suitable (large) $\tilde{\delta} < \delta$, denote by $\tilde{\mathcal{U}}^{\pm}$ the $\tilde{\delta}$ neighborhood of $\mathfrak{z}^{\pm}$ and define $\tilde{B}(x)$ as the smallest set containing $B_{\delta}^{\tilde{\kappa}}(x)$ such that $x_t \in \tilde{\mathcal{U}}^{\pm}$ for all $t \in \tilde{B}(x)$, dist $(t, (\tilde{B}(x))^c) \leq \kappa$.

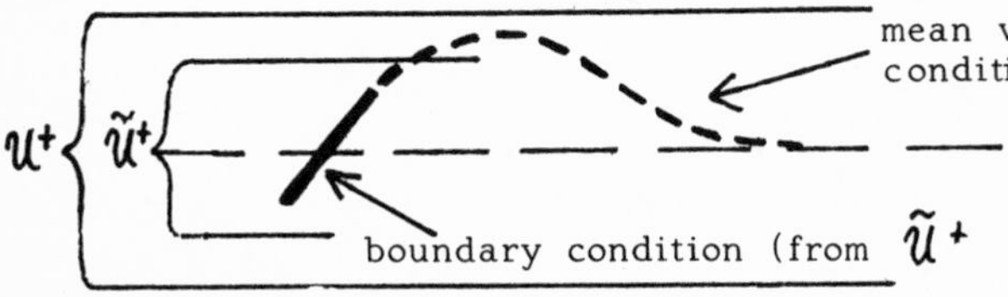

Figure 3

The contours are then defined as restrictions of x to the connected components of $\tilde{B}(x)$.

<u>Note.</u> Most of the argumentation below will use only the following two properties of contours:

a) the support of a contour is a connected set

b) outside of a contour, any almost ground configuration can live (only its type, + or −, being prescribed by the contour).

<u>An Abstract Approach.</u>

It is sometimes useful to consider an "abstract Pirogov Sinai situation" where the notions of almost ground configuration and of contours are given as primitive notions at the very definition of the model.

A configuration is then a collection of three objects:

i) a family of contours $\{\Gamma_i\}$; a compatible one i.e. such that for any component of $\mathbb{Z}^\nu \setminus \bigcup_i \text{supp}\,\Gamma_i$ there is a nonconflicting prescription of its sign (i.e. the prescription whether to live in X^+ or X^-). We denote by $\Lambda^\pm = \Lambda^\pm\{\Gamma_i\}$ the union of all these $\pm$ components. It remains to specify

ii) some $y^+ \in X^+(\Lambda^+)$

where, usually, $X^+(\Lambda^+) = (\mathcal{U}^+)^{\Lambda^+}$ for some $\mathcal{U}^+$, and

iii) some $y^- \in X^-(\Lambda^-) (= (\mathcal{U}^-)^{\Lambda^-})$.

Such a general approach will be really necessary when studying interfaces. Given such a triple $(\{\Gamma_i\}, y^+, y^-)$ we assume that its hamiltonian has the form

$$\mathcal{H}(\{\Gamma_i\}, y^+, y^-) = H(y^+) + H(y^-) + \sum_i \Phi(\Gamma_i) \tag{8}$$

where H is a "ground energy" and Φ is a "contour energy".

In the discrete spin case, an especially transparent formula can be obtained:

$$H(x) = e^+|\Lambda^+| + e^-|\Lambda^-| + \sum_i \Phi(\Gamma_i) \tag{8'}$$

where $e^{+(-)}$ is the mean energy of the ground state $x^{+(-)}$.

The basic assumptions on H and Φ are:

1) Quick (usually, an exponential one) decay of correlations in both the "restricted fields" X^+, X^- endowed with the hamiltonian H. (In Main Theorem, this is a consequence of the condition (1)).

2) A Peierls type condition

$$\Phi(\Gamma) \geq \tau\,|\text{supp}\,\Gamma|\,, \quad \tau \text{ large} \tag{9}$$

(which follows from (2) - after (6) - in that case).

3) We will assume that our hamiltonian (8) is such that a phase transition will occur. We emphasise again that it is not our task there to describe the situation which appears outside of the point of the maximal possible number of coexisting phases. (If some "nonstable" phases are to be expected, one has to modify the method).

§2. Contour models

This is the essential construction of the theory. We present a more general approach (than that of [4]') there. (See [1] for more details). The construction is very natural in the symmetric case (if there is a symmetry between the restricted fields, as well as between the contours separating them). In fact, the notion of a contour model was first introduced in [9] (which is the thorough study of the Ising model).
Nevertheless the nonsymmetric case is the more important and characteristic for the applications of the theory. Our emphasis will be on it, too.

Realizations of the contour ensemble

In addition to the notion of a "compatible family of contours" we introduce also the notion of an "admissible family of contours" i.e. such that dist (supp Γ_i , supp $\Gamma_{i'}$) $\geqslant$ 2 for any $i \neq i'$. We say that a contour Γ is of the type +(or-) if the external component of (supp Γ)^c is signed + (or-). Admissible famillies of contours of one type only will be considered.

In the symmetric case, an admissible family of contours corresponds (by flipping some contours) to some compatible family (uniquely up to +/- symmetry). This has no analogy in the nonsymmetric case but, still, in any case the notion of an "admissible family of external contours of the type +" coincides with the notion of a "compatible family of external contours of some configuration of the type +". (A configuration is of the type + if $x_t \in \mathcal{U}^+$ "outside" of the region "surrounded by" $\tilde{B}(x)$).

The central idea of the Pirogov Sinai theory is then to describe the behaviour of the system of external contours of configurations of the "real" model by studying it as a behaviour of the system of external contours of realizations of a suitable contour ensemble, defined as follows:
A realization of a + contour ensemble is a pair of two objects (compare the notion of a configuration in (8)):
i) an almost ground configuration (on the whole space)

$$y^+ \in X^+ = (\mathcal{U}^+)^{\mathbb{Z}^\nu} ,$$

ii) a family of admissible contours $\{\Gamma_i\}$ of the type +.
Analogously is defined the - contour ensemble.

Graphically, a behaviour of a real configuration (see Figure 4a; there is one contour Γ of the type + and one contour $\tilde{\Gamma}$ of the type - inside Γ)

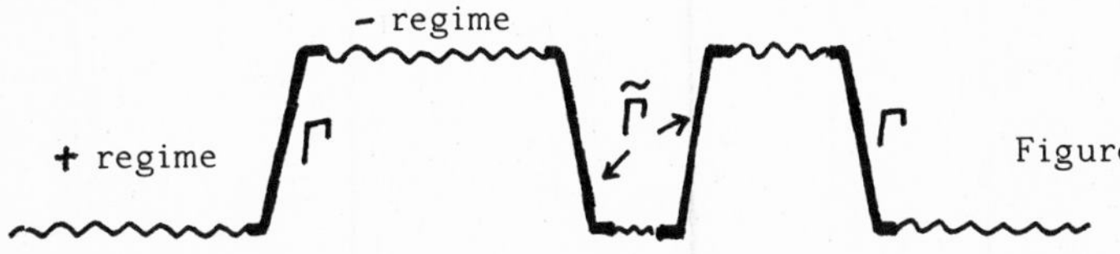

Figure 4 (a)

is modelled by a behaviour of a realization containing Γ and having also the same values on the "exterior" of Γ :

Figure 4 (b)

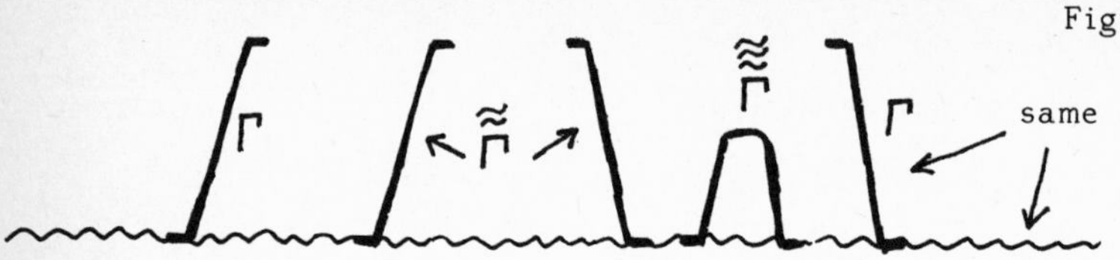

Note. The appearance of another + contours $\tilde{\tilde{\Gamma}}$, $\tilde{\tilde{\tilde{\Gamma}}}$ inside Γ has nothing to do with the "real" behaviour inside Γ – there is no "flipping" in the nonsymmetric case.

Assuming now that the "ground hamiltonian" H remains the same as before, it is a natural idea to try to assign a proper contour weight to any contour of an admissible system $\{\Gamma_i\}$ such that a model "externally equivalent" (in the sense above) to the real one would be obtained.
This is developed in more detail in the next section, where the basic framework of the theory is presented.

Definition of various partition functions. The Contour Weight. Equivalence of ensembles.

We need some geometrical notions describing a structure of contours:

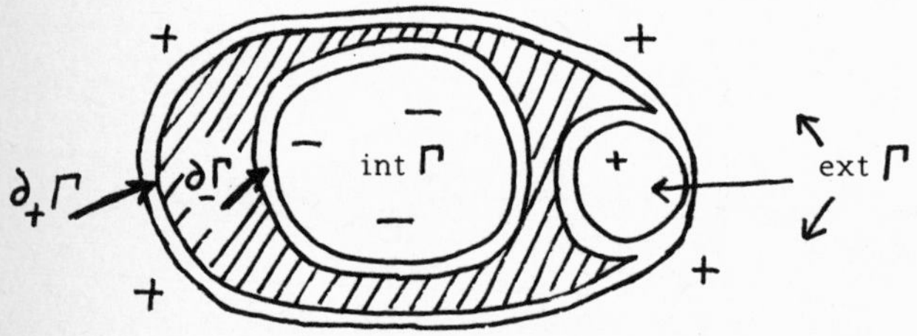

Figure 5

Write $\partial\Lambda = \{t \in \Lambda : \text{dist}(t, \Lambda^c) \leq r\}$.
Given a contour Γ denote by $\mathbb{\Gamma} = \text{supp}\,\Gamma$. Write $\mathbb{\Gamma}^c = \text{ext}\,\Gamma \cup \text{int}\,\Gamma$ and put $V(\Gamma) = \mathbb{\Gamma} \cup \text{int}\,\Gamma$ where (for a contour of the type +) int $\Gamma \equiv \text{int}_-\,\Gamma$ (= the union of all connected components of $\mathbb{\Gamma}^c$ which are signed −). (This is a somewhat unusual, but convenient notion of an externality). Writing $\partial_-\mathbb{\Gamma} = \partial(\text{int}\,\mathbb{\Gamma})^c$, $\partial_+\mathbb{\Gamma} = \partial V(\Gamma)$ we have the decomposition $\partial\mathbb{\Gamma} = \partial_+\mathbb{\Gamma} \cup \partial_-\mathbb{\Gamma}$. Finally we write $\partial_\pm\Gamma = (\Gamma)_{\partial_\pm\mathbb{\Gamma}}$.

Consider the + contour ensemble, for the definiteness. Given a finite Λ and some boundary condition $x_{\partial\Lambda^c} \in (\mathcal{U}^+)^{\partial\Lambda^c}$ denote by

$$Z(\Lambda, x_{\partial\Lambda^c}) = \int \exp\left(-H(x_\Lambda / x_{\partial\Lambda^c})\right) dx_\Lambda \qquad (10)$$

the "diluted" partition function; the integral being taken over all x_Λ such that for any contour Γ of x_Λ ,

$$\text{dist}\,(V(\Gamma), \Lambda^c) \geq 2 \quad . \qquad (10')$$

(Contours of x_Λ are defined by extending x_Λ "naturally" by some $x_{\Lambda^c} \in (\mathcal{U}^+)^{\Lambda^c}$).

We will also need a more restricted notion of

$$Z(\Lambda, M, x_{\partial\Lambda^c}) = \int \exp(-H(x_\Lambda | x_{\Lambda^c}))\, d\, x_\Lambda \qquad (11)$$

where the "forbidden zone" M imposes another restriction

$$\text{dist}\,(V(\Gamma), M) \geq 2 \qquad (11')$$

to (10').

Given a +contour Γ write $V^* = V^*(\Gamma) = V(\Gamma) \setminus \partial_+ \Gamma$ and denote by

$$Z(\Gamma) = \int \exp(-H(x_{V^*} | \partial_+ \Gamma)\, d\, x_{\text{int}\,\Gamma} \qquad (12)$$

(the "crystallic" partition function)

the integral being taken over all possible x_{int} such that Γ is a contour of $\Gamma \cup x_{\text{int}\,\Gamma}$, extended "naturally to the whole $\mathbb{Z}^\nu$".

The following notion is a crucial one:

Define a contour weight $F(\Gamma)$ by the formula

$$F(\Gamma) = \ln Z(V^*, \Gamma, \partial_+ \Gamma) - \ln Z(\Gamma) \quad . (13)$$

Note. This can be interpreted as the "work needed to install the contour". In the symmetric case, this is almost the same quantity as $\Phi(\Gamma)$. (This is no longer true in the nonsymmetric situation).

Finally define a hamiltonian assigned to a realization $\mathcal{N}_\Lambda = (y_\Lambda^+, \{\Gamma_i\})$ by

$$\mathcal{H}(\mathcal{N}_\Lambda | x_{\partial\Lambda^c}) = H(y_\Lambda^+ | x_{\partial\Lambda^c}) + \sum_i F(\Gamma_i) \quad . (14)$$

An analogy of (10) in the contour ensemble is

$$Z(\Lambda, F, x_{\partial\Lambda^c}) = \int \exp(-\mathcal{H}(\mathcal{N}_\Lambda | x_{\partial\Lambda^c})\, d\, \mathcal{N}_\Lambda \qquad (15)$$

(diluted partition function of the contour ensemble) where the integral is taken over all realizations $\mathcal{N}_\Lambda = (x_\Lambda, \{\Gamma_i\})$ satisfying (10') for any $\Gamma = \Gamma_i$.

Theorem. Under (13) we have, for each finite Λ, the relation

$$Z(\Lambda, F, x_{\partial\Lambda^c}) = Z(\Lambda, x_{\partial\Lambda^c}) \qquad (16)$$

both for $x_{\partial\Lambda^c} \in (\mathcal{U}^+)^{\partial\Lambda^c}$ and (correspondingly, in the − contour ensemble) for $x_{\partial\Lambda^c} \in (\mathcal{U}^-)^{\partial\Lambda^c}$.

Note. Then the distribution of $x_{\Lambda \setminus \bigcup_i V^*(\Gamma_i)}$ and of the external contours (of x_Λ, resp. of the system $\{\Gamma_i\}$) is really the same both in the real and the contour ensemble. This is the equivalence of ensembles (used in [1]).

The proof of the theorem is rather straighforward, using some induction over the "level" of Λ (which is the maximal possible cardinality of a "concentric", admissible system of contours in Λ and using the very definition (13) at each stage of the induction. See [1], theorem 4.2.6.

A few words about what picture is to be expected, looking on the external contours of the system. Take some finite Λ and some boundary condition $x_{\partial\Lambda^c} \in (\mathcal{U}^+)^{\partial\Lambda^c}$. There are two characteristic kinds of a behaviour of a typical configuration in Λ (see Fig. 6): either all the contours are "small" and "rare" or the systems jumps - as quickly as possible - to the "stable" regime, thus forming a large contour near the boundary of a given set Λ . The latter behaviour is characteristic for a nonstable phase (of the type +), for large volumes Λ .

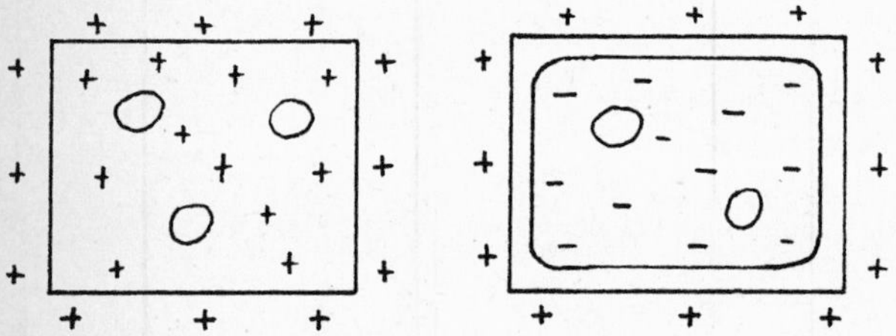

Figure 6

We are interested in the first behaviour and we will study it in the language of the contour model. We will see later that the stability (of both phases) is in fact equivalent to the validity of the following condition (for both types of a contour):

Peierls type condition for the contour weight.

For any contour Γ ,

$$F(\Gamma) \geq \tau |\Gamma| , \quad \tau \text{ large} \qquad . \tag{17}$$

§ 3. Investigation of the contour models.

Using the notations $\Phi(x_\Lambda) = H(x_\Lambda | \emptyset)$ we can write (13) as

$$F(\Gamma) = (\Phi(\Gamma) - \Phi(\partial_+\Gamma)) + (\ln Z(V^*, \Gamma, \partial_+\Gamma) - \ln Z(\text{int}\,\Gamma, \partial_-\Gamma)) \qquad .(18)$$

Clearly, what is needed is a precise estimate of the diluted partitions functions $Z(\Lambda, M, x_{\partial\Lambda^c}) (= Z(\Lambda, M, F, x_{\partial\Lambda^c})$.

Approximating model.

In the situation of Main Theorem, this is the corresponding gaussian model arising if the second order Taylor expansion of the potentials Φ_A (around x_A^+ resp. x_A^-) is taken instead of Φ_A .

(In the discrete spin case, the approximating model is a trivial one, the "constant model" $\{x^+\}$ resp. $\{x^-\}$. Another types of approximating models may be needed in other situations - see the contributions of J. Bricmont and J.Slawny)

Define the approximating partition function

$$Z^{\text{approx}}(\Lambda, x_{\partial\Lambda^c}) = \int_{\text{all } x_\Lambda} \exp\left(- H^{\text{approx}}(x_\Lambda)\right) d x_\Lambda \tag{19}$$

(with H^{approx} just described). Put

$$Z^{\text{rel}}(\Lambda, M, F, x_{\partial\Lambda^c}) = Z(\Lambda, M, F, x_{\partial\Lambda^c})(Z^{\text{approx}})^{-1} \qquad .(20).$$

Estimates of Z^{rel}.

The basic technical problem of the theory is an estimate of the boundary terms of the partition functions (20). Usually, such an estimate is obtained by some expansion techniques (see [1] , § 2 where a selfcontained exposition of the techniques can be found). We summarize the main result:

Note. Actually, this problem is not so central in the symmetric case, where a mere fact of a phase transition can be proven without referring to it - simply by using the usual "symmetric" Peierls argument. In fact, (17) is the basis of a "nonsymmetric Peierls argument" - which is formulated in the context of contour models. (To obtain a more precise information about the structure of phases, the forthcoming estimates are useful even in the symmetric case!).

Theorem. Assume that (17) holds. Take some finite Λ and some $x_{\partial\Lambda^c} \in (\tilde{\mathcal{U}}+)^{\partial\Lambda^c}$ (see Figure 3!). Then

$$\ln Z^{\text{rel}}(\Lambda, M, F, x_{\partial\Lambda^c}) = h^{\text{rel}} |\Lambda| + \Delta(\Lambda, M, F, x_{\partial\Lambda^c}) \qquad (21)$$

where

$$|\Delta| \leq \varepsilon |\partial\Lambda^c \cup M| , \quad \varepsilon \text{ small} \qquad (22)$$

and where h^{rel} is the "relative free energy".

Proof. (See [1]). In the discrete case, this is a consequence of some standard "polymer model" techniques of cluster expansion. In the continuous spin case, we obtain, rather, some "polymer perturbation of a gaussian", more precisely an expression of Z^{rel} of the following type:

$$Z^{\text{rel}} = \int \sum_{\{A_i\}} \prod_i f_{A_i}(x_{A_i})\, d\mu^+(x_\Lambda) \qquad (23)$$

where $\mu^{+(-)}$ is the (positive mass) approximating gaussian field and where the functions f_A satisfy a condition of the type

$$\int |f_A|^2 d\mu^+ \leq \varepsilon^{|A|} q^{d(A)} \qquad \varepsilon \text{ small}, \quad q < 1 \qquad (24)$$

where $d(A)$ is the lenght of a shortest tree on A.

Partition functions of the type (23) are studied in Theorem 2.3.8 of [1] .

Some keysteps and keywords of the proof:

1) expansion of f_A into products of Wick polynomials
2) writing the expectation of $\prod_i f_{A_i}$ as a sum of products of semiinvariants (= truncated expectations) of subfamillies of $\{f_{A_i}\}$
3) estimate of semiinvariants of Wick polynomials, using some diagrammatic expansions; establishing the "tree decay" (in the spirit of [8]).

4) expression of Z^{rel} as a polymer partition function of the type

$$Z^{rel} \doteq \sum_{\{T_i\}} \prod_i k_{T_i} \tag{25}$$

where k_T, defined and estimated by means of 1)-3) satisfy a condition of the type

$$|k_T| \leq \varepsilon\, q^{|T|\, d(T)}, \quad \varepsilon \text{ small}, \quad q < 1 . \tag{26}$$

5) applying the methods of cluster expansions on (25).

In fact, a more precise result is obtained (which is not formulated but essentially proven in [1] , § 2.3.8):

Theorem.

$$\ln Z^{rel} = \sum_{T\cap\Lambda\neq\emptyset} \tilde{k}_T = h^{rel}|\Lambda| + \sum_{T\cap\Lambda\neq\emptyset\ \&\ T\cap\Lambda^c\neq\emptyset} \tilde{\tilde{k}}_T \tag{27}$$

where $\tilde{k}_T = \tilde{k}_T(x_{T\cap\Lambda^c}, F)$ (analogously, $\tilde{\tilde{k}}_T$) are translation invariant and satisfy a condition of the type (26). (This is not needed there but it is substantial in the study of interfaces).

Corollary. Under the condition (17), the relation (18) can be written (if we substitute (16) in its right hand side and use (21)) as

$$F(\Gamma) = \Phi(\Gamma) + \tilde{\Phi}(\Gamma) + \Delta(\Gamma, F) + (h^+(F) - h^-(F))|\operatorname{int}\Gamma| \tag{28}$$

where $h^{\pm}(F)$ is the free energy of the $\pm$ contour model, $\Delta(\Gamma, F)$ (a quantity of the type (21)) satisfies the estimate

$$|\Delta(\Gamma, F)| \leq \varepsilon|\Gamma| \quad , \quad \varepsilon \text{ small}$$

and $\tilde{\Phi}$ is some "counterpart" to Φ, depending only on the hamiltonian H^{approx}. (This term disappears in the discrete case).

Note. This relation seems to be more complicated than the "explicit" relation (18). Moreover $h^+(F)$ must be equal to $h^-(F)$ if the equivalence of ensembles holds. In fact, it is better to solve, instead of (28), the equation

$$F(\Gamma) = \Phi(\Gamma) + \tilde{\Phi}(\Gamma) + \Delta(\Gamma, F) \tag{29}$$

without assuming that F defines an equivalent contour ensemble and then ask whether

$$h^+(F) = h^-(F) \quad . \tag{30}$$

If this is true, the phase transition occurs and both phases are described by the corresponding contour model. If not, then the solution of (29) is formal(!) and we have to modify our strategy.

Solution of (29).

The method of fixed point theorem can be used ([4], [1]). Only the solutions satisfying (17) are obtained (but they are the only reasonable ones).

It is clear that, first, one must check (17) for the contour weight $\Phi(\Gamma) + \tilde{\Phi}(\Gamma)$. This is true in Main Theorem but the proof is again omitted there. It requires some investigation of the structure of contours (preferably, the "wrapped" ones!). See [1], § 4.3 .

In the discrete spin case, $\tilde{\Phi}$ disappears and (17) follows, for $\Phi(\Gamma)$, easily from the Peierls condition of the type (1).

Finally, the degeneracy removing condition of the type (4) (for ℓ only) assures that (30) can be satisfied for a "suitably balanced" original hamiltonian H . Thus, the point of a phase transition is found.

Concluding note (to § 2 and § 3). The method presented in this contribution is, in some sense, an "intersection" of the methods of [4] , [5] but adapted to the continuous spin case (≡ [1]). In the investigation of nonstable phases, it seems to be easier to adapt the method of [5] . Most of the argumentation of the theory (in particular the technical estimates (21), (27)) remain unchanged except that some constructions and estimates characteristic to [5] must be added.

II. INTERFACES

This part of the exposition is a very informal one.

Essentially, one has to combine the discrete version of the method [3] with the techniques outlined in the first part of this exposition.

Note. There will be some new geometrical notions like walls etc. We mentioned yet earlier the need for replacing the "naive" definition of a contour by a more intricate one ("wrapped contours"). This is (technically) even more important there. Nevertheless we will not do it in this introduction because, without going into details of the proof of the technical results of § 2 below no need for the introduction of "wrapped walls" would be seen on this introductory level. (There would be no change on the "conceptional level").

Notations. We again take $X = (\mathbb{R}^k)^{\mathbb{Z}^\nu}$ $(\nu \geq 3)$. Let x^+ , x^- , $x^\pm$, $\tilde{x}^\pm$ be as in the Main Theorem. We take the rescaling (6) everywhere.

Let $x \in X$ be such that e.g. $x = x^\pm$ almost everywhere. The infinite component of $B(x)$ will be called the support of an interface and denoted by the symbol $\mathcal{N}$. The interface Ω is then the restriction of x on $\mathcal{N}$.

Contours (the "naive" ones) are defined as before. We will say that a configuration x (of the type above) is a "canonical" one if it has no contours.

(It is convenient to study a given interface as an interface of a canonical configuration x .)

§ 1. Geometrical structure of an interface.

The geometrical notions of this § are still some adaptations of the notions of [6].

Take a canonical configuration x.

A "perpendicular" $p = \{(t_1, \ldots, t_{\nu-1}, (\cdot))\}$ is called a correct one if the restriction of x on the cylinder $C_p = \{(s_1, \ldots, s_\nu) \; ; \sum_1^{\nu-1} |s_i - t_i| \le \kappa\}$ has the following property: after a suitable vertical shift of x, either

$$|x_t - x^{\pm}_t| \le \delta \quad \text{for all} \quad t \in C_p$$

or (30)

$$|x_t - \tilde{x}^{\pm}_t| \le \delta \quad \text{for all} \quad t \in C_p \quad .$$

(Accordingly, we say that a perpendicular is of the type $x^{\pm}$ or $\tilde{x}^{\pm}$).

Note. We use the same δ as in part I. We recall that δ is large but not "too large" such that $x^{\pm}$ and $\tilde{x}^{\pm}$ would be "clearly distinguishable in (30), and a possible vertical shift would be uniquely determined.

We say that a correct perpendicular p is a standard one if no shift of p is needed in the preceding definition.

Definition of a wall. Remove from Ω all the points belonging to some correct perpendicular. Split the remainder into connected components. A restriction of Ω on any such component $\mathbb{W}$ is called a wall, denoted by W. $\mathbb{W}$ is called the support of W.

The union of all correct perpendiculars can be also splitted into connected components, which are called ceiling cylinders. A ceiling is an intersection of the ceiling cylinder with Ω. The ceilings are not necessarily absolutely "flat" (as in the discrete spin case), but, nevertheless, a unique "height" can be clearly assigned to them, as a height of some perpendicular intersecting them. We do not specify any configuration on a ceiling.

External walls. Taking the projection ("shadow") of $\mathbb{W}$ into the horizontal plane $\mathbb{Z}^{\nu-1} \subset \mathbb{Z}^\nu$ one can define the notion of an external wall analogously as we did it for contours in part I.

The only difference is that while all the + components of (supp Γ)^c were included into ext Γ (for a contour Γ of the type +), we include into the "exterior of a wall W "only those ceilings cylinders "touching $\mathbb{W}$ " which have the same type ($x^{\pm}$ or $\tilde{x}^{\pm}$) and also the same height as the "truly external" one.

This prescription for ext W and int W will be used when specifying whether one wall is "inside" another one. In particular this will be used when defining the notion of an external wall.

Standard walls. We define a height of a wall as the height of the surrounding external ceiling cylinder. Standard walls are those ones having the height 0. The Figure 8 shows two standard walls W, $\tilde{W}$. The wall W elevates the ceiling but does not change its type, while $\tilde{W}$ changes the type of the ceiling.

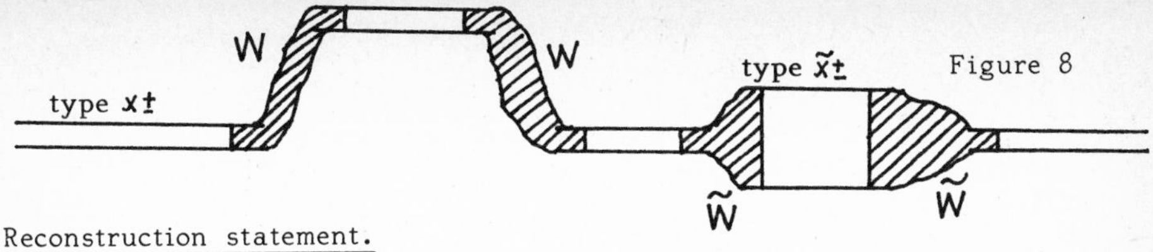

Figure 8

Reconstruction statement.

Call the union of all walls of Ω as the "truncated interface" Ω^*. Ω^* is a "compatible" collection of walls where compatibility means a nonconflicting prescription on

i) the types of the surrounding ceilings ("horizontal" compatibility)

ii) the heights of them .

We claim that there is a one to one correspondence between truncated interfaces and horizontally compatible collections of standard walls.

§. 2. Outline of the further strategy. Wall and aggregate models.

Fix some finite Λ and some truncated interface $\Omega^* = \Omega^*(x_\Lambda)$. Assume that dist $(W, \Lambda^c) \geq 2$ for all walls W of Ω^* and, analogously, for all contours of x_Λ. Assume e.g. that x_Λ is "externally of the type $x\pm$ " and take the partition function

$$Z(\Lambda, \Omega^*) = \int \exp\left(-H(x_\Lambda \mid (x\pm)_{\partial\Lambda^c}\right) d\, x_\Lambda \tag{31}$$

the integral being taken over all x_Λ described above, with a fixed Ω^*. (Notice that we integrate "through" the ceilings, too).

Very roughly this partition function can be expressed as

$$\ln Z(\Lambda, \Omega^*) = \text{const} - \left(f\,|C| + \tilde{f}\,|\tilde{C}| + \sum_i \Phi(W_i)\right) \tag{32}$$

where const depends only on Λ, H ; f resp. $\tilde{f}$ are the "mean interface energies of $x\pm$ resp. $\tilde{x}\pm$ and where $|C|$ (analogously, $|\tilde{C}|$) is the cardinality of the projection, to $\mathbb{Z}^{\nu-1}$, of all the ceiling points of the type $x\pm$. The quantities $\Phi(W)$ are some "wall energies" assigned to each wall W of Ω^* . (They are defined similarly as the contour energies $(\Phi + \tilde{\Phi})(\Gamma)$ in part I, § 2).

We know (from the reconstruction statement above) that Ω^* (and $\{W_i\}$) will be uniquely determined if we prescribe the "standard vertical shifts" of W_i only. Thus, the behaviour of Ω^* is roughly-modelled by some "standard walls model" with a hamiltonian (assigned to any collection of standard walls $\{W_i\}$) of the following type:

$$\mathcal{H} = \sum_i \Phi(W_i) + f\,|C| + \tilde{f}\,|\tilde{C}| \tag{33}$$

This is apparently a Pirogov Sinai type model (7) (the "abstract notion of a contour" being obviously needed there - even if one "factorizes" the notion of a wall by taking its shadow to $\mathbb{Z}^{\nu-1}$). The model is a $\nu-1$ dimensional one and we can, therefore, expect phase transition in it for $\nu \geq 3$.(This will be the phase transition in the type ($x\pm$ or $\tilde{x}\pm$) of the external ceiling).

The basic question now is about an exact substitute for (32).

Note. We will formulate the result only. The methods of its proof are essentially those of [1] , §2.3.8 (as in the proof of (27), with some generalizations needed). Perhaps the only really new problem which arises there is that the notion of an approximating gaussian measure cannot be defined so simply as in part I. Namely to avoid technical problems in the expansions like (27) it is desirable to have an approximating measure which is translation invariant "above" and "below" the interface (at least far from Ω^*). Also desirable is the horizontal translation invariancy around the ceilings where the approximating measure changes "vertically from μ^+ to μ^- ". Clearly, such a measure cannot be realized by a finite range quadratic hamiltonian. It can be, however, realized by an infinite range hamiltonian of the type (we write $M = \Lambda \setminus \Omega$)

$$H^{\text{approx}}(x_M \mid x_{M^c}) = \sum_{N,t:\, N\cap M=\{t\}} (a_N, x_t) + \sum_{N,t,s:\, N\cap M=\{s,t\} \text{ or } \{t\}} (b_N x_t, x_s) \qquad (34)$$

where a_N (analogously, b_N) decay as

$$|a_N| \leq q^{d(N)}, \qquad q < 1 \qquad (35)$$

d (N) being the lenght of a shortest tree on N . (This will be explained in detail in [2] , developing further some simple estimates of §2.2,[1]). Using now some adaptation of the techniques of § 2.3.8 [1] (as in the proof of (27); this will be explained in detail in [2]), we obtain the following result.

Theorem.

$$-\ln Z(\Lambda, \Omega^*) = \text{const}(\Lambda) + f|C| + \tilde{f}|\tilde{C}|$$

$$+ \sum_i \Phi(W_i) + \sum_{\tilde{W}, \tilde{\Lambda}} \tilde{\Phi}(\tilde{\Lambda}, \tilde{W}) + \sum_{T\cap(\Lambda^c\cup\Omega^*)\neq\emptyset} k_T((\Omega^*)_{T\cap\Omega^*}) \qquad (36)$$

where $\Phi(W)$ are the wall energies, $\tilde{\Phi}(\tilde{\Lambda}, \tilde{W})$ are some "counterparts" to Φ which are linear quadratic functions of $\tilde{W}, (x\pm)_{\tilde{\Lambda}}$ where $\tilde{W} = (\bigcup_i W_i)_{\mathbb{W}}$, $\mathbb{W} \subset \bigcup_i W_i$, $\tilde{\Lambda} \subset \Lambda^c$ and k_T are some quantities defined by means of the + and - contour models. They satisfy the following estimates:

i) $$|\tilde{\Phi}(\tilde{\Lambda}, \tilde{W})| \leq q^{d(\tilde{\mathbb{W}}\cup\tilde{\Lambda})} \qquad (37)$$

ii) $$|k_T| \leq \varepsilon^{|T\setminus\Omega^*|} q^{d(T)} \quad (q<1) \qquad (38)$$

The quantities $\Phi(W)$, $\tilde{\Phi}(\tilde{\Lambda}, \tilde{W})$ are translation invariant and the same is true for k_T (if also $T\cap\Omega^*$ and $\Omega^*_{T\cap\Omega^*}$ are shifted).

Note. The quantities $\tilde{\Phi}$ disappear in the discrete spin case.

Another expression of $Z(\Lambda, \Omega^*)$.

Fix Ω^* and $\mathbb{A} = \Omega \cap \Lambda$ and consider the limit $\Lambda \uparrow (\mathbb{A} \cup \Omega^c)$ (see Figure 9). Writing

$$Z^{(\Lambda)}(\Omega^*) = Z(\Lambda, \Omega^*)\, Z(\Lambda, \emptyset)^{-1}, \quad Z(\Omega^*) = \lim_{\Lambda} Z^{(\Lambda)}(\Omega^*)$$

we obtain, in the limit, the following consequence of (36):

$$-\ln Z(\Omega^*) = f|C| + \tilde{f}|\tilde{C}| + \sum_i \Phi(W_i) + \sum_{\tilde{W}, \tilde{\Lambda}} \tilde{\Phi}(\tilde{\Lambda}, \tilde{W}) + \sum_{T \cap \Omega^* \neq \emptyset} k_T \qquad (39)$$

where only those ceiling points are counted which are "above $\mathbb{A}$"; $\tilde{\Lambda} \subset \mathbb{A}$ and where some modified k_T must be taken for those T which intersect $\Omega \setminus \mathbb{A}$.

This can be written, with another k_T, as follows:

$$Z(\Omega^*) = \exp\Big(-\sum_i \Phi(W_i) - \sum_{\tilde{\Lambda}, \tilde{W}} \tilde{\Phi}(\tilde{\Lambda}, \tilde{W})\Big) \sum_{\substack{\{T_i\} :\, T_i \cap \Omega^* \neq \emptyset \,\&\\ \mathrm{dist}(T_i, T_{i'}) \geq 2}} \prod_i k_{T_i} \exp(-f|C| - \tilde{f}|\tilde{C}|) \qquad (40)$$

(see Figure 9)

Figure 9

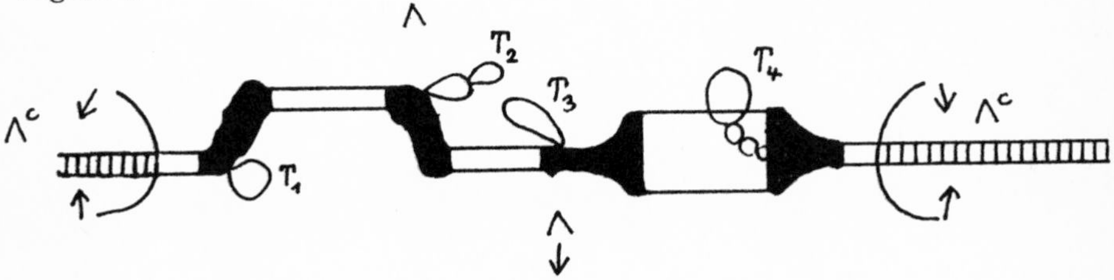

Glueing of W and T. Aggregates.

We will ignore the quantities $\tilde{\Phi}$, for the simplicity of notations. (Writing $\tilde{k} = \exp(-\tilde{\Phi}) - 1$, the quantities $\tilde{k}$ can be handled much the same way as k).

Having the idea of a "wall model" in mind, the relation (40) can be rewritten as follows:

Write $T \sim W$ if the "shadows" of T and W on $\mathbb{Z}^\nu$ have a distance ≤ 1. Split the graph $\sim$ (on the set of all $\{W_i\}$ and $\{T_i\}$ into connected components. The components are called aggregates and we can write (40), using the notion of an aggregate weight

$$\exp(-\Phi(\mathcal{A})) = \prod_i k_{T_i} \exp\Big(-\sum \Phi(W_i)\Big), \quad \mathcal{A} = \{W_i\} \,\&\, \{T_i\} \qquad (41)$$

as

$$Z(\Omega^*) = \sum_{\{\mathcal{A}_k\}} \exp\Big(-\sum_k \Phi(\mathcal{A}_k) - f|C| - \tilde{f}|\tilde{C}|\Big) \qquad (42)$$

the sum being taken over all aggregates whose union contains exactly those walls which are contained in Ω^*.

Aggregate model.

Notice that all the external walls of the system of all walls of some aggregate have the same height. This is called the height of the aggregate $\mathcal{A}$. The exterior of an aggregate is the union of all correct perpendiculars not intersecting the aggregate, having the same type and a same height as the "true exterior".

It is easily seen that the aggregate model with a hamiltonian

$$\mathcal{H} = f\,|C| + \tilde{f}\,|\tilde{C}| + \sum_i \Phi(\mathcal{A}_i) \tag{43}$$

is a model of the type outlined in § 1, part I . This is the exact substitute for the idea of a wall model. Namely any "compatible" collection of aggregates can be uniquely represented as a "horizontally compatible" collection of standard aggregates.

We will not go into the details of this model, and conclude our exposition by few remarks on this model only:

1) If we show that an aggregate model behaves like an "infinite ceiling of the type $X\pm$ (resp. $\tilde{X}\pm$) with small and rare (but uniformly distributed) "islands" surrounded by external aggregates (a characteristic picture obtained by the P.S. theory) then a similar picture is obtained when looking on the behaviour of the system of all walls. (External walls can be reconstructed from external aggregates).

2) We have to check a Peierls type condition for $\Phi(\mathcal{A})$. Such a condition follows from (38) and the following inequality:

Peierls type condition for $\Phi(W)$.

$$\Phi(W) \geq \tau\,|W|\,, \quad \tau \text{ large} \quad . \tag{44}$$

(This is an inequality of the same type as (9). It follows from (2)). The condition for $\Phi(\mathcal{A})$ is then as follows:

$$|\exp(-\Phi(\mathcal{A}))| \leq \varepsilon^{|\hat{\mathcal{A}}|}\, q^{d(\hat{\mathcal{A}})}\,, \quad \varepsilon \text{ small}\,, \quad q<1\,, \tag{45}$$

where $\hat{\mathcal{A}} = (\bigcup_i W_i) \cup (\bigcup_i T_i)$, $\mathcal{A} = \{W_i\}\ \&\ \{T_i\}$

Notice that while walls are connected this is no longer true for aggregates. It turns out. however, that formulating the "abstract P. S. setting" one can replace the assumption "contours are connected and the Peierls condition $\Phi(\Gamma) \geq \tau\,|\Gamma|$, τ large, holds for them" by a more general assumption "contours are some objects, whose energy satisfies a condition

$$|\exp(-\Phi(\Gamma))| \leq \varepsilon^{|\Gamma|}\, q^{d(\Gamma)}, \quad \varepsilon \text{ small}\,, \quad q<1 \quad ," .$$

(The notion of an external contour must be generalized accordingly).

3) The quantity $\exp(-\Phi(\mathcal{A}))$ may be negative, because k_T may be negative. This is unusual but cause no difficulties. (One can also take a variant of (36), (39) with new, positive k_T and some new Φ , $\tilde{\Phi}$).

4) The degeneracy removing condition (4) assures that for suitable values of λ, λ' the coexistence of both the aggregate models really takes place. (The proof of an existence of such λ, λ' is some elaboration of the simple observation that a coexistence of both types of an interface is mainly controlled by a change in the mean ceiling energies f, $\tilde{f}$.

Concluding remark.

We tried to explain that it would be really desirable to have some reasonably general version of the Pirogov Sinai theory, as this approach proved to be useful in many applications. Besides the applications sketched there, there are also other interesting fields which should be investigated from this point of view:

1) Infinite range models (with rapidly decaying interactions)
2) Wetting phenomena
3) Impure models (e.g. with an external random field).

While 1) seems to be accesible without much change the problem 3) is apparently very involved as some of its partial solutions (Fröhlich, Fisher, Spencer and Imbrie - see [10] show.

See the contributions of J. Bricmont and J. Slawny for some further information about the various aspects of the theory (in particular, about 2), and also about some discrete spin models with infinitely many ground states). We want to mention that many authors (Lebowitz, Bricmont, Pfister, Kuroda, and others) contributed to the study of interfaces. See [11] for the references along this "Peierls type" line of study. Finally we note that there are many other advances of the theory e.g. [12]) which are not reflected in our text.

References.

1 Dobrushin R.L., Zahradník M., Phase diagrams for continuous spin systems; to appear in "Math. Problems of Stat. Phys. and Dyn., ed. by R.L. Dobrushin, D. Reidel (1985)

2 Holický P., Kotecký R., Zahradník M., Rigid interfaces for continuous spin systems; to appear

3 Zahradník M., Rigid interfaces; Seminar in Crit. Phenomena, Random Systems, Gauge Theories, Les Houches Summer School 84, ed. by K. Osterwalder and R. Stora, North Holland, to appear (1986)

4 Sinai Ya.G., Theory of Phase Transitions; New York, Pergamen Press 1982

5 Zahradník M., An Alternate Version of P.S. Theory; Comm. Math. Phys. 93,559-581 (1984)

6 Dobrushin R.L., Gibbsian state which describes coexistence of phases for a three dimensional Ising model; Theor. Prob. and Appl. 17,612-639 (1972) in Russian

7 Imbrie J.Z., Phase Diagrams and Cluster expansions for Low Temperature $\mathcal{P}(\varphi)_2$ Models; Comm. Math. Phys. 82,261-304 and 305-343 (1981)

8 Malyshev V.A., Cluster expansions in lattice models of statistical physics and the quantum theory of fields; Uspechi Mat, Nauk 35,212 (1980) (in Russian. English translation in Russ.Math.Surv. 35, 1-62 (1980)

9 Minlos R.A., Sinai Ya. G. "Phase separation" phenomenon at low temperatures, in some lattice models of a gas, I, Matem. Sbornik, 72, No 3, 1967, 1967, 375-448; II Trudy Mosk. Matem. Obsch. 19,1968, 113-178 in Russian

10 Fröhlich J., Disordered Systems; Lectures in Crit. Phenomena, Random Systems, Gauge Theories, Les Houches Summer School 84, ed. by K. Osterwalder and R. Stora, North Holland, to appear (1986)

11 Pfister C.E., in Scaling and Selfsimilarity in Physics, ed. by J. Fröhlich, Birkhauser 1983

12 Dinaburg E.I. Sinai Ya. G., An analysis of ANNI model by Peierls contour method; Comm. Math. Phys. 98,119-144 (1985)

See [1], [4], [11] for a more complete list of references

SPIN GLASSES, EFFECTIVE DECREASE OF LONG-RANGE INTERACTIONS

Aernout C.D. van Enter
SFB 123
Im Neuenheimer Feld 294
6900 Heidelberg, FRG

Abstract: Long-range spin-glass models with random Hamiltonians

$$H = -\sum_{ij} J(i,j)\,|i-j|^{-\alpha} S_i S_j$$

where the J(i,j) are independent, identically distributed random variables with mean $\mathbb{E}\,J(i,j) = 0$, satisfying some moment conditions, in some respects behave like nonrandom models with Hamiltonians

$$H = -\sum |i-j|^{-2\alpha} S_i S_j .$$

This result is obtained through a study of the thermodynamic limit in any dimension, and the absence of long range order and the asymptotic correlation decay in dimension one and two. Hence the effective decrease of the interaction tends to be twice as fast as the original one.

§1 Introduction

Spin glasses are magnetic materials (like iron) which are brought as impurities into a non-magnetic environment (like gold). The magnetic properties of the iron atoms in their metallic host are supposedly described by the dilute RKKY-Hamiltonian

$$H = - \sum_{i,j} \varepsilon_i \varepsilon_j \frac{\cos 2K_F |i-j|}{|i-j|^3} S_i S_j \qquad (1)$$

where the ε_i are (site) random variables with values 0 and 1, describing the dilution. The system is characterized both by randomness (due to the site dilution) and by an oscillating long-range interaction. Experimentally there occurs a transition in the susceptibility.

According to Edwards and Anderson (EA), one can describe spin glasses by putting random bond interactions on a regular lattice [1,2]. The Hamiltonian then becomes

$$H = - \sum_{i,j} J(i,j) S_i S_j \qquad (2)$$

with the J(i,j) as random variables which are independent, with a distribution which only depends on the distance $|i-j|$ and which has zero average. This gives a random bond model for a spin glass.

These are the most generally studied spin-glass models. Among physicists there is still a lot of controversy how they behave and the results which one can prove rigorously are modest. I plan to present some of these results, but first I will mention some problems which are still open.

The first problem is to show that there exists a spin-glass transition for some (non-mean-field) random bond model on a regular lattice in a sufficiently high dimension or with a sufficiently long-range interaction. The conjectured lowest dimension at which there is a transition for the nearest neighbour model has gone down last year, from four to three [3,4].

A spin-glass transition is characterized by a positive Edwards-Anderson order parameter [1,5]

$$q_{EA} = \limsup_{\Lambda\to\infty} \sup_{\alpha} \frac{\sum_{i\in\Lambda} \left\{\omega'_{\alpha}(S_i)\right\}^2}{|\Lambda|} > 0 \qquad (3)$$

Here the α denote the set of Gibbs states (or boundary conditions). Nonanalytic behaviour, due to Griffiths singularities is expected to occur in a much larger domain [6-9].

A second problem is the following. The Sherrington-Kirkpatrick (equivalent neighbour) model has presumably a transition. What happens at low temperatures? Is it possible to give mathematical meaning to the Parisi hypothesis [10-12] and the corresponding complicated low T structure? In some respects the answer is yes [13], though in a rather artificial model. Has this anything to do with the behaviour in the physical dimension d = 3? The latest guess [14,39] is: Probably not that much.

What one can prove at present is that some of these questions are well-defined (for example, there is a well-defined thermodynamic limit [15,16] and a well-defined EA order parameter [1,5]) and that there is no spin-glass phase for some models where the corresponding ferromagnets do have a transition [7-9,17-24]. In the next section I will present some of these results.

As a side remark, I want to mention that there are some spin-glass models which one can treat in a mathematically more detailed way. They are obtained by weakening the EA assumption. Either one can consider random site models [25-29] or random bond models on non-regular (hierarchical) lattices [30,31].

§2 Long-range interactions, effective decrease: results

The models we will treat have the following Hamiltonian

$$H = - \sum_{i,j} J(i,j)\,|i-j|^{-\alpha d} S_i S_j \tag{4}$$

where the J(i,j) are independent, identically distributed random variables with mean $\mathbb{E}\, J(i,j) = 0$ and a cumulant expansion with finite radius of convergence. The S_i denote either classical or quantum spins.

The results which we have obtained are compared in the next table, both with the well-known non-random case and with previous results which have been obtained with the help of Bernstein's inequality for large deviations.

	SG [16c,20-23,38]	Non-random [32-35]	SG (Bernstein) [7, 16a, 17-19]
A) Thermodynamics exist	$\alpha > \frac{1}{2}$	$\alpha > 1$	$\alpha > \frac{1}{2}$
B) Unique Gibbs state d = 1	$\alpha > 1$	$\alpha > 2$	$\alpha > \frac{3}{2}$ (+C^∞ property)
C) No broken rotation symmetry in n-vector models d = 2	$\alpha \geqslant 1$	$\alpha \geqslant 2$	$\alpha \geqslant \frac{3}{2}$
D) McBryan-Spencer bounds in n-vector models	$\alpha > 1$	$\alpha > 2$	$\alpha > \frac{3}{2}$
$\lvert\langle s_0 s_n\rangle\rvert \leqslant$ (d = 1)	Cst $n^{-(\alpha-\frac{1}{2})}$	Cst $n^{-(\alpha-1)}$	Cst n^{-cst}
(d = 2)	"$\exp\left\{-(\ln n)^{1-\varepsilon}\right\}$"	n^{-T}	$\exp\left\{-(\ln n)^{1-\varepsilon}\right\}$

Except for the result in quotation marks which holds in L^2-sense, all the spin-glass results are true almost surely with respect to the random variables $\{J(i,j)\}$. These results are supposedly optimal [24,36].

§3 Proofs

A) Existence of thermodynamics

To prove the existence of the thermodynamic limit of the free energy density one uses [16c] (for free boundary conditions) the subadditive ergodic theorem of Akcoglu and Krengel [37].

For notational simplicity we consider Ising spins in d = 1 with a Gaussian distribution of the J(i,j) with unit variance.

We write

$$\tilde{J}(i,j) = J(i,j)\,|i-j|^{-\alpha}. \tag{5}$$

To apply the theorem, we have to check that the free energy is subadditive and that the averaged free energy density satisfies a uniform lower bound (stability).

Subadditivity of the free energy follows from an argument of Griffiths [40] which applies to all even spin models.

A stability bound for the averaged free energy is derived as follows

$$-\mathbb{E}F_\Lambda = \mathbb{E}\ln Z(H_\Lambda) = \mathbb{E}\ln \operatorname{tr}\exp(-H_\Lambda) < \text{(by Jensen's inequality)}$$

$$< \ln \mathbb{E}\operatorname{tr}\exp(-H_\Lambda) = \ln \operatorname{tr}\prod_{i,j\in\Lambda}\mathbb{E}\exp\tilde{J}(i,j)S_iS_j = \sum_{i,j\in\Lambda}|i-j|^{-2\alpha}$$

$$\text{(which is of order } O(|\Lambda|)) \tag{6}$$

In the last equality we used the fact that the J(i,j) have a

Gaussian distribution and that the S_i are Ising spins. In general there occur some extra terms of higher order, but they do not alter the argument [16c]. This proves the result for free boundary conditions. Now we add an arbitrary boundary condition. Take for Λ the interval $[-|\Lambda|+1,0]$ and take $\sigma = \{\sigma_i\}_{i\in\mathbb{N}}$ with $\sigma_i = \pm 1$. Then the surface energy between Λ and the right half-line is

$$W_\Lambda^\sigma = \sum_{\substack{i\in\Lambda \\ j\in\mathbb{N}}} \tilde{J}(i,j) S_i \sigma_j = \sum_{i\in\Lambda} W_i^\sigma . \tag{7}$$

We estimate its norm by

$$\mathbb{E}\,\|W_\Lambda^\sigma\| = \sum_{i\in\Lambda} \mathbb{E}|W_i^\sigma| = \sqrt{\frac{2}{\pi}} \sum_{i\in\Lambda} \left\{\mathbb{E}(W_i^\sigma)^2\right\}^{\frac{1}{2}} =$$

$$\sqrt{\frac{2}{\pi}} \sum_{i\in\Lambda} \left\{\mathbb{E}\Big(\sum_{j\in\mathbb{N}} \tilde{J}(i,j)\sigma_j\Big)^2\right\}^{\frac{1}{2}} = \sqrt{\frac{2}{\pi}} \sum_{i\in\Lambda} \left\{\sum_{j\in\mathbb{N}} |i-j|^{-2\alpha}\right\}^{\frac{1}{2}} =$$

$$= O\Big(|\Lambda|^{\frac{3}{2}-\alpha}\Big). \tag{8}$$

By the law of large numbers for non-identical random variables it follows that

$$\lim_{\Lambda\to\infty} \frac{W_\Lambda^\sigma}{|\Lambda|} = 0 \qquad \text{J-almost surely} \tag{9}$$

whenever $\alpha > \frac{1}{2}$ and hence

$$f = \lim_{\Lambda\to\infty} \frac{F_\Lambda}{|\Lambda|} \tag{10}$$

exists and is independent of boundary conditions. This finishes the proof of A).

Remark: If $\{\tilde{J}_1(i,j)\}$ and $\{\tilde{J}_2(i,j)\}$ are two independent sets of interactions, an estimate of the form (6), with tr replaced by ω_{J_1} the Gibbs state for $H_1 = -\Sigma\, \tilde{J}_1(i,j)S_iS_j$, shows that the energy density is continuous as a function of the interactions

$$|f(H_1+H_2) - f(H_1)| \leqslant \sum_j \mathbb{E}\, \tilde{J}_2(i,j)^2. \tag{11}$$

B) Uniqueness of the Gibbs state in one dimension

When we prove the absence of transitions, it turns out that thermal fluctuations suppress energetically "bad" configurations.

If we consider the (formal) quantity

$$W = \sum_{\substack{i\leqslant 0\\ j>0}} \tilde{J}(i,j)S_iS_j \tag{12}$$

formula (8) shows that $\|W\| = \infty$ for $\alpha \leqslant \frac{3}{2}$ by choosing the spin configuration appropriate. It turns out that $\|W\| < \infty$ for $\alpha > \frac{3}{2}$. This looks plausible from (8) and was proven by Khanin [17]. This fact is basic for the derivation of the results in the last (Bernstein) column of §2 and restricts its domain of validity. However, if we consider the corresponding free energy expression, which is (6) with H_Λ replaced by W, we see that this is finite whenever $\alpha > 1$. The appropriate free energy-like quantity to consider is a relative entropy [33b,34b]. A weak version of "no phase transition" is the absence of spin-flip symmetry breaking. One proves this by showing that the free energy cost of flipping an infinite half-line (which is a relative entropy) is finite, J-almost surely [20]. Let ω be an extremal Gibbs state and ω' the state which one obtains by flipping all the spins on a half-line. Then, for the relative entropy it holds

$$0 \leqslant S(\omega \mid \omega') \equiv \int d\omega \, \ln {}^{d\omega}/d\omega' = -2\omega(W) \tag{13}$$

We write $\omega = \omega_{H_1 + H_2 + W}$, with H_1 and H_2 living on the left and right half-line respectively. Then the relative entropy can be rewritten:

$$S(\omega|\omega') = -\omega_{H_1 + H_2}(2W \exp W) \quad \frac{Z(H_1 + H_2)}{Z(H_1 + H_2 + W)} \quad . \tag{14}$$

Both factors in this product can be shown to be finite J-almost surely by an estimate of the form (6). For the first term we use the fact that $|\kappa e^{\kappa}| \leqslant \exp(2\kappa) + \exp(-2\kappa)$ and for the second term we use the fact that (variational principle)

$$\frac{Z(H_1 + H_2)}{Z(H_1 + H_2 + W)} \leqslant \exp -\omega_{H_1 + H_2}(W) . \tag{15}$$

The important point to note is that one may interchange the average over the J's which occur in W with the modified thermal expectation $\omega_{H_1 + H_2}$ which does not contain these J's anymore.
By translating the origin one shows that ω and the spinflipped ω have a <u>finite</u> relative entropy with respect to each other, hence are absolutely continuous with respect to each other and thus identical because they are extremal Gibbs. For quantum spins the notion of absolutely continuous measures gets replaced by quasi-equivalent states.

<u>Remark</u>: To prove uniqueness of the Gibbs state, one applies a similar estimate on a large (finite) region surrounding Λ, and shows that the interaction energy between Λ and the far outside is uniformly small [38]. The fact that in an operational sense the Gibbs state is unique, because all J-independent boundary conditions give rise to mutually absolutely continuous Gibbs states was commented upon in [21].

C) Unbroken rotation symmetry in two dimensions

To prove a result of Mermin-Wagner type, one applies to a state ω a constant rotation in an inner region Λ_1, keeps all the spins outside a large region $\Lambda_2 \supset \Lambda_1$ fixed and rotates the spins in $\Lambda_2 - \Lambda_1$, over angles θ_i which linearly interpolate. We write $\omega_{\Lambda_1,\Lambda_2}$ for this transformed state (this idea has been applied by Fröhlich and Pfister [34b] in the nonrandom case). The relative entropy $S(\omega|\omega_{\Lambda_1,\Lambda_2})$ is now bounded by a sum of terms of the form

$$\{1-\cos(\theta_i-\theta_j)\}\omega(S_iS_j\,\tilde{J}(i,j)) =$$

$$\{1-\cos(\theta_i-\theta_j)\}\frac{\omega_{i,j}(S_iS_j\,\tilde{J}(i,j)\exp S_iS_j\,\tilde{J}(i,j))}{\omega_{i,j}(\exp S_iS_j\,\tilde{J}(i,j))} =$$

$$\{1-\cos(\theta_i-\theta_j)\}\frac{\omega_{i,j}(S_iS_j\,\tilde{J}(i,j)\exp S_iS_j\,\tilde{J}(i,j))}{1+\omega_{i,j}\{\exp(S_iS_j\,\tilde{J}(i,j))-1\}} \quad . \tag{16}$$

Short range contributions can be treated separately, along the lines of [34b], hence without loss of generality we can assume that the $\tilde{J}(i,j)$ are small. In fact, here we need a distribution for the $J(i,j)$ with bounded support. However, more general distributions like Gaussians can also be treated. Thus we can make a convergent Taylor expansion of (16) in $\tilde{J}(i,j)$. The modified thermal average $\omega_{i,j}$, in which the term $S_iS_j\,\tilde{J}(i,j)$ is subtracted from the original Hamiltonian does not contain the random variable $J(i,j)$ and hence can be interchanged with the expectation over $J(i,j)$. The result is

$$\mathbb{E}(16) = \{1-\cos(\theta_i-\theta_j)\}O(\,\mathbb{E}\,\tilde{J}(i,j)^2). \tag{17}$$

In this way we derive (by reducing the estimates to those of Fröhlich and Pfister [34b])

$$\mathbb{E}\, S(\omega|\omega_{\Lambda_1,\Lambda_2}) < \infty \tag{18}$$

uniformly in Λ_1, if Λ_2 (dependent on Λ_1) is chosen appropriately. This proves the rotation invariance of ω [21].

D) McBryan-Spencer bounds

To prove McBryan-Spencer upper bounds on correlation functions for n-vector models, one uses the following inequality [35a,b]

$$|\langle S_0 S_n\rangle| \leq \frac{Z(H')}{Z(H)} \exp -(a_n - a_0) \tag{19}$$

where

$$H' = -\Sigma\, \tilde{J}'(i,j) S_i S_j \tag{20a}$$

and

$$\tilde{J}'(i,j) = \cosh(a_i - a_j)\tilde{J}(i,j) \tag{20b}$$

By the variational principle

$$\frac{Z(H')}{Z(H)} \leq \exp -\omega_{H'}(H' - H). \tag{21}$$

The expression $\omega_{H'}(H' - H)$ is estimated in a similar way as the relative entropy in case C) [22].

Following Picco [19] and Messager, Miracle-Sole and Ruiz [35c] we take the following choices for the a_j:

$$a_j = a_{|j|} \quad \text{and } a_j - a_{j-1} = \frac{k}{j}\ , \ j = 1, 2, \ldots \tag{22}$$

for d = 1 and

$$a_{|j|} - a_{|j|-1} = \frac{\bar{k}}{j(\ln^+ j)^{1-\gamma}} \quad , \; |j| = 1,\ldots N \qquad (23)$$

for d = 2.

Again, the rewriting of (21) in terms of modified thermal expectations allows us to reduce the estimates to the non-random case of [35c].

Acknowledgements: Most of this work was done together with J.L. van Hemmen and J. Fröhlich, whom I thank for pleasant collaborations. I want to thank the organizers for their invitation to present it in a pleasant conference.

References

[1] S.F. Edwards and P.W. Anderson: J. Phys. F 5, 965 (1975).

[2] Reviews on different aspects of spin glasses can be found in the proceedings of the Heidelberg Colloquium on Spin Glasses. Ed. J.L. van Hemmen and J. Morgenstern, Springer Lecture Notes in Physics 192 (1983).
See also D. Chowdhury and A. Mookerjee: Phys. Rep. 114, 1 (1984).

[3] J. Morgenstern and A. Ogielski: Phys. Rev. Lett. 54, 928 (1985) (transition d ⩾ 3).

[4] a) R. Fisch and A.B. Harris: Phys. Rev. Lett. 38, 785 (1977).
b) J. Morgenstern and K. Binder: Phys. Rev. Lett. 43, 1615 (1979). (transition d ⩾ 4).

[5] A.C.D. van Enter and R.B. Griffiths: Comm. Math. Phys. 90, 319 (1983).

[6] R.B. Griffiths: Phys. Rev. Lett. 23, 17 (1969).

[7] M. Cassandro, E. Olivieri and B. Tirozzi: Comm. Math. Phys. 87, 229 (1982).

[8] A. Berretti: J. Stat. Phys. 38, 483 (1985).

[9] J. Fröhlich and J. Imbrie: Comm. Math. Phys. 96, 148 (1985).

[10] D. Sherrington and S. Kirkpatrick: Phys. Rev. Lett. 35, 1792 (1975).

[11] G. Parisi: J. Phys. A 13, 1101, 1887 (1980).

[12] G. Parisi: Phys. Rev. Lett. 50, 1946 (1983).

[13] M. Cassandro, E. Olivieri and P. Picco: Rome preprint.

[14] D. Fisher and H. Sompolinski: Phys. Rev. Lett. 54, 1063 (1985).

[15] The thermodynamic limit for short-range random interactions has been treated in

a) R.B. Griffths and L.J. Lebowitz: J. Math. Phys. 9, 1284 (1968).

b) F. Ledrappier: Comm. Math. Phys. 56, 297 (1977).

c) P. Vuillermot: J. Phys. A 10, 1319 (1977).

d) J.L. van Hemmen and R.G. Palmer: J. Phys. A 15, 3881 (1982).

[16] The thermodynamic limit for long-range random interactions has been treated in

a) K.M. Khanin and Ya.G. Sinai: J. Stat. Phys. 20, 573 (1979).

b) S. Goulart Rosa: J. Stat. Phys. A 15, L 51 (1982).

c) A.C.D. van Enter and J.L. van Hemmen: J. Stat. Phys. 32, 141 (1983).

[17] K.M. Khanin: Theor. Math. Phys. 43, 445 (1980).

[18] P. Picco: J. Stat. Phys. 32, 627 (1983).

[19] P. Picco: J. Stat. Phys. 36, 489 (1984).

[20] A.C.D. van Enter and J.L. van Hemmen: J. Stat. Phys. 39, 1 (1985).

[21] A.C.D. van Enter and J. Fröhlich: Comm. Math. Phys. 98, 425 (1985).

[22] A.C.D. van Enter: J. Stat. Phys. 41, 315 (1985).

[23] L. Slegers, A. Vansevenant and A. Verbeure: Phys. Lett. 108 A, 267 (1985).

[24] The existence of transitions in the corresponding ferromagnetic

models is proven in

a) F.J. Dyson: Comm. Math. Phys. 12, 212 (1969).

b) J. Fröhlich and T. Spencer: Comm. Math. Phys. 84, 87 (1982).

c) H. Kunz and C.E. Pfister: Comm. Math. Phys. 46, 245 (1976).

[25] D.C. Mattis: Phys. Lett. 56 A, 421 (1976).

[26] J. Luttinger: Phys. Rev. Lett. 37, 778 (1976).

[27] a) J.L. van Hemmen: Phys. Rev. Lett. 49, 409 (1982).

b) J.L. van Hemmen, A.C.D. van Enter and J. Canisius: Z. Phys. B 50, 311 (1983).

c) J.L. van Hemmen, contribution to 2a).

[28] a) J.P. Provost and G. Vallée: Phys. Rev. Lett. 50, 598 (1983).

b) F. Benamira, J.P. Provost and G. Vallée: J. de Phys. 46, 1269 (1985).

[29] D. Amit, H. Gutfreund and H. Sompolinski: Jerusalem preprint.

[30] P. Collet and J.P. Eckmann: Comm. Math. Phys. 93, 379 (1984).

[31] P. Collet, J.P. Eckmann, V. Glaser and A. Martin: J. Stat. Phys. 36, 89 (1984).

[32] A treatment of the thermodynamic limit can for example be found in

a) D. Ruelle: Statistical Mechanics, Benjamin New York (1969).

b) R.B. Israel: Convexity in the theory of lattice gases, Princeton University Press, Princeton N.J. (1979).

[33] Absence of phase transitions for one-dimensional models is proven (among others) in

a) D. Ruelle: Comm. Math. Phys. 9, 367 (1968).

b) H. Araki: Comm. Math. Phys. 44, 1 (1975).

c) J. Bricmont, J.L. Lebowitz and C.E. Pfister: J. Stat. Phys. 21, 573 (1979).

d) M. Cassandro and E. Olivieri: Comm. Math. Phys. 80, 255 (1981).

[34] Two-dimensional models where the rotation symmetry is not broken

are treated in

a) N.D. Mermin and H. Wagner: Phys. Rev. Lett. 17, 1133 (1966).

b) J. Fröhlich and C.E. Pfister: Comm. Math. Phys. 81, 277 (1981).

and references mentioned there.

[35] McBryan-Spencer estimates are given in

a) O. McBryan and T. Spencer: Comm. Math. Phys. 53, 299 (1977).

b) J. Glimm and A. Jaffe: Quantum Physics §16.3, Springer-Verlag New York, Heidelberg, Berlin (1981).

c) A. Messager, S. Miracle-Sole and J. Ruiz: Ann. Inst. Henri Poincaré 40, 85 (1984).

d) K.R. Ito: J. Stat. Phys. 29, 747 (1982).

e) S.B. Shlosman: Theor. Math. Phys. 37, 1118 (1978).

[36] G. Kotliar, P.W. Anderson and D.L. Stein: Phys. Rev. B 27, 602 (1983).

[37] a) M.A. Akcoglu and U. Krengel: J. Reine Angew. Math. 323, 53 (1981).

b) U. Krengel: Ergodic theorems §6.2, de Gruyter, Berlin, New York (1985).

[38] M. Campanino, A.C.D. van Enter and E. Olivieri: in preparation.

[39] J. Fröhlich: private communication.

[40] R.B. Griffiths: Phys. Rev. 176, 655 (1968).

ANALYTICITY IN SOME MODELS OF QUANTUM STATISTICAL MECHANICS

Huzihiro ARAKI
Research Institute for Mathematical Sciences
Kyoto University, Kyoto 606, JAPAN

Abstract Application of C*-algebra approach to proof of analyticity of correlation functions for some models of quantum statistical mechanics is reviewed. Analyticity of correlation functions of ground states with respect to the asymmetry parameter γ and the external magnetic field strength λ for the XY-model on one dimensional lattice and that of equilibrium states with respect to the inverse temperature β and the coupling constants J_1 and J_2 for the Ising model on two dimensional lattice are obtained for all real values of parameters except the singularity at the critical lines.

The same method also gives an exact decay rate for the exponential clustering of the correlation functions of the latter model as a function of parameters.

§1. Introduction

The C*-algebra approach is by now a well-established mathematical framework for quantum statistical mechanics of spin lattice systems. Recent study shows that somewhat abstract theory of C*-algebras can actually be used in finding physically interesting quantitative properties of some (exactly soluble) models of spin lattice systems such as the one-dimensional XY-model and the two-dimensional Ising model. In the present article, we review the results in [9] and [5] about analytic properties of the correlation functions.

For the one-dimensional lattice with its sites lebelled by integers $j \in \mathbb{Z}$, we consider the C*-algebra $\mathfrak{A}$ generated by the algebras $\mathfrak{A}_j$ of 2×2 matrices, $j \in \mathbb{Z}$ (mutually commuting for different j's), where elements of $\mathfrak{A}_j$ are linear combinations of the identity matrix and Pauli spin matrices $\sigma_\alpha^{(j)}$, $\alpha=x,y,z$. The X-Y model is the dynamical system $(\mathfrak{A}, \alpha_t)$ defined by

$$\alpha_t(A)=\lim_{N\to\infty} e^{itH(N)}Ae^{-itH(N)}, \quad A\in\mathfrak{A} \tag{1.1}$$

$$H=-J\{(1+\gamma)\sum\sigma_x^{(j)}\sigma_x^{(j+1)}+(1-\gamma)\sum\sigma_y^{(j)}\sigma_y^{(j+1)}+2\lambda\sum\sigma_z^{(j)}\} \tag{1.2}$$

where $H(N)$ denotes H with summation restricted to $\mathfrak{A}_j$, $-N\leq j\leq N$.

All ground states of the dynamical system $(\mathfrak{A},\alpha_t)$ are obtained in [8]. For the (finite-dimensional) subalgebra $\mathfrak{A}(N)$ generated by $\mathfrak{A}_j$, $-N\leq j\leq N$, there is a unique eigenstate φ_N of $H(N)\in\mathfrak{A}(N)$ with the lowest eigenvalue. The limit

$$\varphi_{\lambda,\gamma}(A)=\lim_{N\to\infty}\varphi_N(A), \quad A\in\bigcup_N\mathfrak{A}(N) \tag{1.3}$$

exists and determines a state $\varphi_{\lambda,\gamma}$ of $\mathfrak{A}$. If either $|\lambda|\geq 1$ or $\gamma=0$, $\varphi_{\lambda,\gamma}$ is the unique ground state of $(\mathfrak{A},\alpha_t)$. If $|\lambda|<1$ and $\gamma\neq 0$, $\varphi_{\lambda,\gamma}$ is an average of two pure ground states $\varphi_{\lambda,\gamma,+}$ and $\varphi_{\lambda,\gamma,-}$ and their mixtures exhaust all ground states of $(\mathfrak{A},\alpha_t)$ except for soliton states in the case of $(\lambda,\gamma)=(0,\pm 1)$.

For a state φ of a dynamical system $(\mathfrak{A},\alpha_t)$, correlation functions are

$$\varphi(\alpha_{t_1}(A_1)\cdots\alpha_{t_n}(A_n)), \quad A_\ell\in\mathfrak{A}, \quad t_\ell\in\mathbb{R}. \tag{1.4}$$

<u>Theorem 1.1</u>. <u>The correlation functions (1.4) are real analytic functions of the parameters</u> (λ,γ) <u>for</u> $\varphi=\varphi_{\lambda,\gamma}$ <u>except for the critical lines</u> $\lambda=\pm 1$ <u>and</u> $\gamma=0$, <u>and for</u> $\varphi=\varphi_{\lambda,\gamma,\pm}$ <u>in the region</u> $|\lambda|<1$, $\gamma\neq 0$ <u>of the broken symmetry.</u>

In the case of two-dimensional Ising model with the Hamiltonian

$$H=-\sum_{i,j}[J_1\xi_{i,j}\xi_{i+1,j}+J_2\xi_{i,j}\xi_{i,j+1}], \tag{1.5}$$

$\xi_{i,j}=\pm 1$ being the spin variable at the lattice site (i,j) and J's being nonzero real parameters, it is known [1,10] that there is a unique equilibrium state ψ_β for $|\beta|<\beta_c$ while there are exactly two extremal (and ergodic) equilibrium states $\psi_{\beta,+}$ and $\psi_{\beta,-}$ for $|\beta|>\beta_c$, where β_c is the unique solution of

$$(K_1^*\equiv)(1/2)\log\{\coth|\beta J_1|\}=\beta|J_2|(\equiv K_2). \tag{1.6}$$

The correlation functions are

$$\psi(\prod_{(i,j)\in I} \xi_{i,j}) \tag{1.7}$$

for a finite subset I of lattice sites.

<u>Theorem 1.2.</u> <u>The correlation functions (1.7) are real analytic functions of</u> β, J_1 <u>and</u> J_2 <u>for</u> $\psi=\psi_\beta$, $|\beta|<\beta_c$ <u>and for</u> $\psi=\psi_{\beta,\pm}$, $|\beta|>\beta_c$.

§2. <u>The CAR algebra and the spin algebra</u>

The Jordan-Wigner transformation

$$c_j=T_j(\sigma_x^{(j)}-i\sigma_y^{(j)})/2,\quad c_j^*=T_j(\sigma_x^{(j)}+i\sigma_y^{(j)})/2,\quad (j\in\mathbb{Z}), \tag{2.1}$$

bridges the spin algebra $\mathfrak{A}$ and the CAR algebra $\mathfrak{A}^{CAR}$ generated by c's and c*'s satisfying the canonical anticommutation relations (CAR's), where T_j is supposed to serve the role of the product of $\sigma_z^{(j)}$, j ranging from the left end of the one-dimensional lattice up to the site $j-1$. The difference with the case of a finite lattice is that T's are no longer an element of $\mathfrak{A}$. Following the method introduced in [4], we extend the algebra $\mathfrak{A}$ to

$$\hat{\mathfrak{A}} = \mathfrak{A}+T\mathfrak{A} \tag{2.2}$$

by adding $T=T_1$ satisfying $T=T^*=T^{-1}$ and $TAT=\Theta_-(A)$ for all $A\in\mathfrak{A}$. Here Θ_- is the automorphisms of $\mathfrak{A}$ satisfying

$$\Theta_-(\sigma_\alpha^{(j)}) = \begin{cases} \sigma_\alpha^{(j)} & \text{for } j>0 \\ \Theta(\sigma_\alpha^{(j)}) & \text{for } j\leqq 0 \end{cases} \tag{2.3}$$

for $\alpha=x,y,z$ and Θ is another automorphism of $\mathfrak{A}$ satisfying

$$\Theta(\sigma_z^{(j)})=\sigma_z^{(j)},\quad \Theta(\sigma_x^{(j)})=-\sigma_x^{(j)},\quad \Theta(\sigma_y^{(j)})=-\sigma_y^{(j)} \tag{2.4}$$

for all $j\in\mathbb{Z}$. The C*-algebra $\hat{\mathfrak{A}}$ is the crossed product of $\mathfrak{A}$ by Θ_--action of the group Z_2 of two elements. The operator T_j in (2.1) is T multiplied by $\sigma_z^{(j)}\cdots\sigma_z^{(j-1)}$ for $j>1$ and by $\sigma_z^{(0)}\cdots\sigma_z^{(j)}$ for $j<1$.

The automorphism Θ can be extended to $\hat{\mathfrak{A}}$ such that $\Theta(T)=T$. It maps $\mathfrak{A}^{CAR}$ onto itself and counts the even-oddness of c's and c*'s: $\Theta(c_j)=-c_j$, $\Theta(c_j^*)=-c_j^*$. The algebra $\mathfrak{A}$ and $\mathfrak{A}^{CAR}$ are split into a sum of Θ-even and Θ-odd parts (denoted by suffices $\pm$) with the following relations.

$$\mathfrak{A}_+=\mathfrak{A}_+^{CAR}\ ,\quad \mathfrak{A}_-=T\mathfrak{A}_-^{CAR}. \tag{2.5}$$

In particular, this gives a specific isomorphism of the Θ-even part $\mathfrak{A}_+$ of the spin algebra $\mathfrak{A}$ onto the even part $\mathfrak{A}_+^{CAR}$ of the CAR algebra $\mathfrak{A}^{CAR}$.

Since $H(N)$ is Θ-even, $H(N)\in\mathfrak{A}_+=\mathfrak{A}_+^{CAR}$. Therefore the limit (1.1) for $A\in\mathfrak{A}^{CAR}$ makes sense within $\mathfrak{A}^{CAR}$. In fact the limit exists and can be described as follows. Let $f=\{f_j;\ j\in\mathbb{Z}\}$ and $g=\{g_j;j\in\mathbb{Z}\}$ be in $\ell_2(\mathbb{Z})$, and $h=\binom{f}{g}$ be in $\mathcal{K}=\ell_2(Z)\oplus\ell_2(Z)=\ell_2(Z)\otimes\mathbb{C}^2$. For each h, we define

$$B(h)=\sum_{j\in Z}(f_jc_j^*+g_jc_j). \tag{2.6}$$

It satisfies $B(h_1)^*B(h_2)+B(h_2)B(h_1)^*=(h_1,h_2)$ and $B(h)^*=B(\Gamma h)$, where $\Gamma\binom{f}{g}=\begin{pmatrix}\bar{g}\\ \bar{f}\end{pmatrix}$, and generates $\mathfrak{A}^{CAR}$. Any unitary U on $\mathcal{K}$ commuting with Γ induces an automorphism $\alpha(U)$ of $\mathfrak{A}^{CAR}$ satisfying

$$\alpha(U)B(h)=B(Uh),\qquad (h\in\mathcal{K}). \tag{2.7}$$

The dynamics α_t of $\mathfrak{A}^{CAR}$ is given by

$$\alpha_t=\alpha(e^{iKt}) \tag{2.8}$$

for some $K=K(\lambda,\gamma)$ satisfying

$$K^*=K,\qquad K\Gamma=-\Gamma K. \tag{2.9}$$

Combining α_t for $\mathfrak{A}$ and $\mathfrak{A}^{CAR}$, we see that the limit (1.1) exists for $A=T=(c_1+c_1^*)\sigma_x^{(1)}$ and hence for any $A\in\hat{\mathfrak{A}}$ defining a dynamics α_t on $\hat{\mathfrak{A}}$.

§3. Ground states of the XY model

A method of obtaining all ground states of $(\mathfrak{A}, \alpha_t)$ from the knowledge of ground states of $(\mathfrak{A}^{CAR}, \alpha_t)$ via ground states of $(\mathfrak{A}_+, \alpha_t)$ has been developed in some generality [8,7].

In the present model, the operator K above does not have an eigenvalue 0. This implies that $(\mathfrak{A}^{CAR}, \alpha_t)$ has the unique (Fock) ground state φ_E characterized by

$$\varphi_E(B(h_1)^*B(h_2))=(h_1, Eh_2). \tag{3.1}$$

where $E=E(\lambda,\gamma)$ is the spectral projection of $K=K(\lambda,\gamma)$ for $(0,+\infty)$. If $(\lambda,\gamma)\neq(0,\pm 1)$, K has a continuous spectrum. This implies that the restriction of φ_E to $\mathfrak{A}_+^{CAR} = \mathfrak{A}_+$ is the unique ground state of $(\mathfrak{A}_+, \alpha_t)$ and hence

$$\varphi_{\lambda,\gamma}(A)=\varphi_E((A+\Theta(A))/2) \tag{3.2}$$

is the unique Θ-invariant ground state of $(\mathfrak{A}, \alpha_t)$, necessarily coinciding with (1.3). The equality of (1.3) with (3.2) persists for $(\lambda,\gamma)=(0,\pm 1)$.

By the equivalence criterion for cyclic representations of $\mathfrak{A}_+$ associated with Fock states (in terms of the Hilbert-Schmidt norm and the Z_2-index), it was shown in [8] that $\varphi_{\lambda,\gamma}$ is the unique ground state of $(\mathfrak{A}, \alpha_t)$ (and hence is pure) for $|\lambda|\geqq 1$ and for $\gamma=0$, while there exist mutually disjoint pure ground states $\varphi_{\lambda,\gamma,\pm}$ satisfying

$$\varphi_{\lambda,\gamma}(A)=(\varphi_{\lambda,\gamma,+}(A)+\varphi_{\lambda,\gamma,-}(A))/2 \tag{3.3}$$

for $|\lambda|<1$, $\gamma\neq 0$.

§4. Strategy for the proof of analyticity

For any two projections E_j satisfying $E_j^*=E_j=E_j^2$ and $E_j+\Gamma E_j\Gamma=1$ $(j=1,2)$, there exists a unitary U_{12} commuting with Γ and satisfying $U_{12}E_2U_{12}^*=E_1$ and hence satisfying

$$\varphi_{E_2}(A)=\varphi_{E_1}(\alpha(U_{12})A) \tag{4.1}$$

for all $A\in\mathfrak{A}^{CAR}$. Since $\Theta=\alpha(-1)$ always commutes with $\alpha(U)$, $\alpha(U_{12})$

is an automorphism of $\mathfrak{A}_+^{CAR} = \mathfrak{A}_+$. We use the following criteion for the extendability of $\alpha(U)$ from $\mathfrak{A}_+$ to an automorphism $\hat{\alpha}$ of $\mathfrak{A}$.

Theorem 4.1. (Evans and Lewis) *The automorphism (U) extends from $\mathfrak{A}_+$ to an automorphism $\hat{\alpha}$ of $\mathfrak{A}$ if and only if there exists a unitary u in $\mathfrak{A}_+$ satisfying*

$$[\alpha(U)\Theta_-\alpha(U^*)\Theta_-](A)=uAu^* \tag{4.2}$$

for all $A \in \mathfrak{A}^{CAR}$. Then u can be chosen to satisfy $u\Theta_-(u)=1$ and $\hat{\alpha}$ is given by

$$\hat{\alpha}(A_+ + A_- T)=\alpha(U)(A_+)+\alpha(U)(A_-)uT \tag{4.3}$$

for $A_\pm \in \mathfrak{A}_\pm^{CAR}$ (and hence for any $A=A_+ + A_- T \in \mathfrak{A}$).

The strategy for the proof of analyticity is as follows. For given (λ,γ) not on the critical lines $|\lambda|=1$ and $\gamma=0$, and for (λ',γ') in a sufficiently small neighbourhood of (λ,γ), we use Theorem 3.1 to show that, for $E_1=E(\lambda,\gamma)$ and $E_2=E(\lambda',\gamma')$, $\alpha(U_{12})$ has an extension from $\mathfrak{A}_+$ to an automorphism $\hat{\alpha}$ of $\mathfrak{A}$. The formulas (3.2) and (4.1) imply

$$\varphi_{\lambda',\gamma'}(A)=\varphi_{\lambda,\gamma}(\hat{\alpha}(A)). \tag{4.4}$$

We show that $\hat{\alpha}(A)$ is real analytic in (λ',γ') for a certain class of A including $A=\alpha_{t_1}(A_1)\cdots\alpha_{t_n}(A_n)$ with strictly localized $A_1,\cdots,A_n$. Then we obtain the real analyticity of (1.4) for all (λ,γ) not on critical lines.

The real analyticity of $\hat{\alpha}(A)$ is proved in 3 steps:

Step 1. $E_{\lambda,\gamma}$ is explicitly given (see Section 4) and is real analytic in (λ,γ) (except on critical lines). For $\|E_1-E_2\|<1$, there is a canonical construction of U_{12} given in [9] (based on the angle operator between 2 projections, see Appendix of [3]), such that U_{12} is real analytic in (λ,γ) and (λ',γ') for the present case.

Step 2. Let $(\theta_- f)_j=\epsilon_j f_j$ with $\epsilon_j=1$ for $j>0$ and $\epsilon_j=-1$ for $j\leq 0$. Let $(\theta_- h)=\begin{pmatrix}\theta_- & f\\ \theta_- & g\end{pmatrix}$ for $h=\binom{f}{g}$. Then $\Theta_-=\alpha(\theta_-)$ and the condition (4.2) amounts to the condition that $\alpha(V)$ for $V=U_{12}\theta_- U_{12}^*\theta_-$ is inner and its implementer is in $\mathfrak{A}_+^{CAR}$. The criterion for this given in Section 8 of [8] is that $V-1$ is in the trace class and $\det V=1$.

The trace class condition will be discussed in the next section. The determinant condition is automatically satisfied if $\|V-1\| < 2$ due to the property $\Gamma V\Gamma=V$ which implies $\det V=(-1)^{\nu}$ with ν being the multiplicity of the eigenvalue -1 for V. $\|V-1\| < 2$ is satisfied if $\|E_1-E_2\|$ is sufficiently small $(<\sqrt{3}/2)$ so that $\|U_{12}-1\| < 1$.

When $\|V-1\| < 2$, $V-1=e^{iH}$ for $H=H^*$, $\|H\|<\pi$ and if $V-1$ is in the trace class, H is real analytic in (λ,γ) with values in the trace class. Then u in Theorem 4.1 is given by $\exp i(B,HB)/2$ with the bilinear Hamiltonian (B,HB) ([3]) and is real analytic in (λ,γ).

Step 3. If U is real analytic in (λ,γ), then $\alpha(U)(B(h))= B(Uh)$ is real analytic in (λ,γ) and so is $\alpha(U)(A)$ for any polynomial A of B's, which includes all strictly localized A. Hence $\hat{\alpha}(A)$ given by (4.3) is real analytic in (λ,γ) for strictly local A. Due to the exponential increase of the dimension of local algebra with the size of localization region, this real analyticity extends to all A with sufficiently fast decreasing exponential tail, i.e. $A=\sum B_n$ with B_n supported in the interval $[-n,n]$ and $\|B_n\| \leqq C_p \exp -pn$ for sufficiently large positive p. (See Lemma 3.3 of [5].) In [2], $\alpha_t(A)$ is shown to be such an operator with any choice of p. (cf. Theorem 4.2 of [2].)

§5. Trace class estimate

We introduce the Fourier series expansion for $f \in \ell_2(\mathbb{Z})$ and $h \in \ell_2(\mathbb{Z})\oplus\ell_2(\mathbb{Z})$

$$\tilde{f}(\theta)= \sum_{n\in\mathbb{Z}} e^{in\theta} f_n, \quad \tilde{h}(\theta)=\begin{pmatrix}\tilde{f}(\theta)\\ \tilde{g}(\theta)\end{pmatrix} . \tag{5.1}$$

The operator K in (2.8) is then the multiplication of

$$\tilde{K}(\theta)=2\begin{pmatrix}\cos\theta-\lambda & -i\gamma\sin\theta\\ i\gamma\sin\theta & -\cos\theta+\lambda\end{pmatrix} \tag{5.2}$$

and $E=E(\lambda,\gamma)$ is the multiplication of

$$\tilde{E}(\theta)=(2\mu(\theta))^{-1}(K(\theta)+\mu(\theta)) \tag{5.3}$$

$$\mu(\theta)=2[(\cos\theta-\lambda)^2+\gamma^2\sin^2\theta]^{1/2} \tag{5.4}$$

where $\mu(\theta)\neq 0$ for all real θ if (λ,γ) is not on critical lines. The crucial property is that $\tilde{E}(\theta)$ is holomorphic in λ, γ, and $z=e^{i\theta}$ in the neighbourhood of real (λ,γ) not on critical lines and $|z|=1$. This property persists for U_{12} which is again a multiplication of $\tilde{U}_{12}(\theta)=\hat{U}_{12}(e^{i\theta})$.

By a shift of integration contour $\gamma_r=\{z;|z|=r\}$ for

$$(U_{12})_{j,k}=(2\pi i)^{-1}\oint_{\gamma_1}\hat{U}_{12}(z)z^{j-k-1}dz, \tag{5.5}$$

we obtain the exponential decrease of $(U_{12})_{j,k}$ for large $|j-k|$, where $(U_{12}h)_j=\sum_k (U_{12})_{j,k}h_k$. Since $q_\pm\equiv(1\pm\theta_-)/2$ are projections for $j>0$ and for $j\leqq 0$, the identity

$$U_{12}\theta_-U_{12}^*\theta_- - 1=U_{12}(q_+-q_-)U_{12}^*(q_+-q_-)-U_{12}(q_++q_-)U_{12}^*(q_++q_-)$$

$$=-2(U_{12}q_+U_{12}^*q_-+U_{12}q_-U_{12}^*q_+) \tag{5.6}$$

and the exponential decrease of $(U_{12})_{j,k}$ (and hence that of $(U_{12}^*)_{j,k}$) for large $|j-k|$ imply that $q_+U_{12}^*q_-$, $q_-U_{12}^*q_+$ and hence (5.6) are in the trace class due to the following Lemma.

Lemma 5.1.

$$\|X\|_{tr}=\inf\sum_{\ell,m}|(e_\ell,Xe_m')|\leqq\sum_{\ell,m}|(e_\ell,Xe_m)| \tag{5.7}$$

where the infimum is over all orthonormal bases $\{e_\ell\}$ and $\{e_m'\}$.

The inequality follows from the following estimate for the polar decomposition $X=u|X|$:

$$\|X\|_{tr}=\mathrm{tr}|X|=\sum_m|(e_m', u^*Xe_m')|=\sum_m|\sum_\ell(e_m',u^*e_\ell)(e_\ell,Xe_m')|\leqq\sum_{\ell,m}|(e_\ell,Xe_m')|$$

due to $|(e_m',u^*e_\ell)|\leqq 1$.

The trace class estimate above also much simplifies the Hilbert-Schmidt class estimate of

$$E(\lambda,\gamma)-\theta_-E(\lambda,\gamma)\theta_-=2(q_+E(\lambda,\gamma)q_-+q_-E(\lambda,\gamma)q_+)$$

in [5] necessary for judging equivalence of φ_E and $\varphi_{\theta_-E\theta_-}$.

§6. The Ising model

By the transfer matrix method, the correlation function (1.7) can be expressed as Schwinger functions of a ground state of a C*-dynamical system $(\mathfrak{A}, \alpha_t^\beta)$ with the same spin algebra $\mathfrak{A}$ as before and the dynamics α_t^β given in [6]:

$$\psi(\xi(k_1,I_1)\cdots\xi(k_n,I_n)) = \varphi(\alpha_{ik_1}^\beta(\sigma_x(I_1))\cdots\alpha_{ik_n}^\beta(\sigma_x(I_n))) \tag{6.1}$$

where $\xi(k,I)=\prod_{j\in I}\xi_{k,j}$, $\sigma_x(I)=\prod_{j\in I}\sigma_x^{(j)}$ and the formula holds for $\psi=\psi_\beta$ and $\psi_{\beta,\pm}$ with $\varphi=\varphi_\beta$ and $\varphi_{\beta,\pm}$. Here φ_β is the unique Θ-invariant ground state of $(\mathfrak{A},\alpha_t^\beta)$, which is pure for $|\beta|>\beta_c$ and is an average of mutually disjoint pure ground states $\varphi_{\beta,\pm}$ for $|\beta|>\beta_c$ ([6]).

The identification (6.1) has been attained for ψ_β and $_\beta$ in [6] and for $\psi_{\beta,\pm}$ and $\varphi_{\beta,\pm}$ by using the preceding result along with the cluster property in [5]. The operators $\alpha_{ik}^\beta(\sigma_x(I))$ and their products are of the type discussed in Step 3 of Section 4 (with any p). Thus the machinery explained for the ground states of the XY model works also for the present case and Theorem 1.2 follows. Details are refered to [5].

By using the exact information on the spectrum of the time translation generator (in the representation associated with 's), an exact decay rate for the exponential clustering of the correlation functions can be obtained. A typical result is as follows (Proposition 6.5 (2) and (3) of [5]):

Theorem 6.1. (1) For any polynomial F_1 and F_2 of ξ's,

$$\lim_{\ell\to\infty} e^{|\ell|\delta}|\psi(F_1\tau_{(\ell,0)}(F_2))-\psi(F_1)\psi(F_2)|=0 \tag{6.2}$$

where $\tau_{(\ell,0)}$ is a lattice translation, $\delta=4(|K_2|-K_1^*)$ and $\psi=\psi_{\beta,\pm}$ if $|\beta|>\beta_c$ and $\delta=2(K_1^*-|K_2|)$ and $\psi=\psi_\beta$ if $|\beta|<\beta_c$.

(2) There exists polynomials F_1 and F_2 of ξ's for any given $\varepsilon>0$ such that

$$\lim_{\ell\to+\infty} e^{\ell(\delta+\varepsilon)}|\psi(F_1,\tau_{(\ell,0)}(F_2))-\psi(F_1)\psi(F_2)|=\infty \tag{6.3}$$

where δ and ψ are the same as (1).

References

[1] Aizenmann, M., Commun. Math. Phys. 73, 83(1980).
[2] Araki, H., Commun. Math. Phys. 14, 120(1969).
[3] Araki, H., Publ. RIMS, Kyoto Univ. 6, 384(1970).
[4] Araki, H., Publ. RIMS, Kyoto Univ. 20, 277(1984).
[5] Araki, H., Analyticity of correlation fuctions of the Ising model in two dimensions, Prepreint RIMS-525, submitted to Commun. Math. Phys.
[6] Araki, H., and Evans, D.E., Commun. Math. Phys. 91, 489(1983).
[7] Araki, H., and Matsui, T., Progr. in Phys. 10, 17(1985). (Statistical Physics and Dynamical Systems, eds. Fritz, J., Jaffe, A., and Szász, D., Birkhäuser).
[8] Araki, H., and Matsui, T., Commun. Math. Phys. 101, 213(1985).
[9] Araki, H., and Matsui, T., Analyticity of ground states of the XY-model, Preprint RIMS-522. Lett. Math. Phys. in press.
[10] Higuchi, Y., p.517 in Random Fields I, ed. Fritz, J., Lebowitz, J.L., and Szász, D., Colloquia Societatis Janos Bolyai, vol.27 (North-Holland, Amsterdam-Oxford-N.Y., 1981).
[11] Matsui, T., Explicit formulas for correlation functions of ground states of the 1 dimensional XY model, Ann. Inst. H. Poincaré Sect. A, in press.

K-THEORY OF C*-ALGEBRAS IN SOLID STATE PHYSICS

Jean BELLISSARD
Université de Provence
et Centre de Physique Théorique
MARSEILLE (FRANCE)

1)- C*algebras : an appealing but incomplete tool ?

The subject I would like to present here today pretends to use C*-algebras in concrete problems of modern physics, a mathematical tool which have been liable to controversies among the mathematical physicists for the last twenty years. When D. Testard, R. Lima and I started our program about solving some mathematical problems in the study of disordered systems, five years ago, we naively believed that using C*-algebras was just like using any other tool of mathematical physics like functional analysis or probability theory. However when we began to explain to the experts what we were able to do, not very much I confess, we realized that before trying to convince them, it was necessary to get really hard results. Since we knew perfectly well the limitations of these technics, we understood quite soon that what could be solved with it, was not as spectacular as was implicitly demanded by both the believers and the sceptics in the community. We spontaneously adopted what can be qualified as a schizophrenic attitude : on the one hand we presented an external work as reliable as possible, and on the other hand we kept unpublished for a very long time our results always present in the back of our mind as a guide in performing our program.

What I will explain today have been presented almost in essence at an informal meeting in Marseille during the spring of 1981, apart from the chapter on the quantum Hall effect. Actually some traces of this work can be found in the recent literature. Several results were partially quoted in seminars, conferences [14], reviews [94], or even in some papers [17,56]. We finally published the first part very recently [19]. Reading this last paper will explain the way it has been received by the scientific community. At first sight, the scheme seems very simple and so efficient that it looks miraculous. And when miracles appear the world is immediately divided into two irreducible components : those who are true believers and those who just reject the claims. Fortunately, or perhaps unfortunately depending upon the side one prefers to choose, there is no miracle, at least in physics. In our problem, even though the scheme looks very simple, when one starts the proofs, one realizes immediately how heavy they are : all the machinery of C*-algebras and algebraic geometry is needed, and one looses quite rapidly the original physical intuition ! Moreover many technical details remain unsolved or require very complicated and winding proofs. Clearly C*-algebras are very appealing mathematical tools but they must be used together with other technics to become efficient.

One can say that C*-algebras started their career in mathematical physics in the early sixties with the Haag-Kastler axioms [48] of what is known today as the Structural Field Theory. The hope was that the miraculous aspect of these objects would solve the mysterious difficulties encountered in Relativistic Quantum Field Theory. Twenty years later it is quite clear that the flu went the other way : more have been learned from the physics to understand the mathematical tool than the converse. Moreover, very few came from the original motivation namely the understanding of the particle physics. On the contrary, like for many advances in physics during the last thirty years, the concrete understanding came from fields connected with Condensed Matter Physics. For example, the KMS condition, one of the major step in the Tomita-Takesaki theory [M29] and in the classification of type III factors by A. Connes [25], was introduced in the field by R. Haag, H. Hugenholtz & M. Winnink [47]. This improvement came after several mathematical physicists who started there work with the axiomatic Quantum Field Theory, had changed their mind and decided to study the Statistical Mechanics [M24]. The attempt of J. Glimm and A., Jaffe [45] in using C*-algebras in constructive quantum field theory was soon abandoned and replaced by the probabilistic approach[M10,M27].

However eight years ago a breakthrough was performed by A. Connes [26,27,30] who associated to any foliated manifold a canonical C*-algebra, playing the role of the basic object needed to apply the methods of algebraic geometry and to get topological invariants as well as a cohomology theory. He contributed in giving a concrete content to the K-theory, and created the cyclic cohomology, to take into account this new situation [28-34]. He also emphasized that non commutativity of the algebras was closely related to the ergodicity of the foliation, or equivalently to the fact that the set of leaves was not endowed with a non trivial topology.

At about the same time, algebraic geometry entered in quantum field theory with the contribution of mathematicians to the classification of instantons in gauge field theories [M15]. Later on, they also interpreted the anomalies (as Adler's one) in term of topological obstructions. More recently the supersymmetries offered the mathematicians a formal tool for calculating simply the formulæ appearing in the Index theory. Nevertheless the previous works are all performed on the ground of a semi classical quantization by mean of hypothetical Feynman path integrals: the algebraic geometry concerns usual fibre bundles, on usual manifolds. Nothing in this approach is essentially non commutative. The reason is that nobody knows how to built a theory for describing gauge fields in four dimensions, and one has the hope that something of the topological obstructions will remain in the final theory if any. However literally speaking there is an apparent contradiction between the fact that topological obstructions require smooth configurations of gauge fields, whereas it is well known in constructive field theory that almost surely no configuration is smooth.

Topological invariants actually reflect the properties of the observable algebra, more than those of the configurations. The Quantum Hall Effect (QHE) may be a good laboratory to test such an idea. What will be presented here concerns only the ordinary QHE, namely a one particle theory. The quantum mechanics of such a system is well understood. Moreover its has been demonstrated that, when the Fermi energy belongs to a gap, and at zero temperature, the Hall conductivity can be interpreted (in e^2/h units) as the Chern character of some fibre bundle above the Brillouin zone, at least when the magnetic flux through a unit cell, correctly normalized, is rational [99,9]. When it is irrational, the C*-algebra of observables becomes non commutative in a non trivial way : it is type II as was remarked a long time ago by A. Grossmann [46]. It is no longer possible to speak about Brillouin zone, and the previous interpretation of the Hall conductivity becomes meaningless. Nevertheless, thanks to the A. Connes proposal, the C*-algebra of observable can be interpreted as the algebra of continuous functions on an hypothetical Brillouin zone which does not exist as a geometrical object but has sufficiently many continuous "non-commutative" functions to be studied by the technics of the algebraic geometry. In particular, the Chern character still exists and varies continuously with the magnetic field, in such a way that the semiclassical approach gives the correct answer.

It is widely admitted that the fractional QHE comes from many bodies interactions. A correct mathematical theory of it should require a Fock space, and a second quantized algebra of observable. It is therefore a (non relativistic) quantum field theory. Nevertheless there is a structural question here related to the calculation of the Hall conductivity : somewhere there must be some topological invariant. Is this related to the differential algebra introduced by Arveson [4] and used in the definition of the cyclic cohomology [32,33]? Is the vacuum a closed current in the sense of Connes ? Is the Hall conductivity given by a Chern character in the many body problem ?

Besides the internal interest of the physical problem, which goes far beyond the analysis I will give here, it seems that investigating easier situations which are better understood in the physical set up than the intricate questions in particle physics, is an efficient way of improving our understanding of the mathematical tool, and gives the hope that in the next future, paradoxes will be solved. For this reason I feel that mathematical physicists should take more seriously the mathematical study of simple , realistic models of condensed matter physics where topological invariants exist in order to understand much better what to do in more ambitious questions like the structure of gauge field theories.

Acknowledgments : This work benefited from many discussions both encouraging or discouraging and from several helps to visit many Institutes, giving me the opportunity to meet many experts. The list of contributors to these exchanges would be too long to be reproduced here, and they have been quoted in previous works. Let me thank them collectively here. Let me however express my gratitude to A. Connes whose constant enthusiasm, efficiency and patience in explaining the properties and the importance of C* algebras, convinced me to start this program and to go beyond the formalism it offered at the beginning.

2)- Why are C*-algebras needed to describe disordered systems ?

A real sample used in experimental solid state physics is made of a finite assembly of atoms bound together by electrostatic interactions. However, even though finite and relatively small (few millimeters in size, sometime smaller in crystallography), any sample contains so many atoms that it is better described via the use of the infinite volume limit. On the other hand, it appears homogeneous at large scale while at atomic scale the disorder breaks any translation invariance.

When studying electronic properties of such a crystal, it is commonly admitted that a one electron approximation is usually quite good. Collective properties of the electrons gas are described through the model of a perfect Fermi gas. The disorder appears simply as external forces coming from the random positions of the atoms or impurities. In this set up, the Schrödinger operator for an electron, acting on the space $L^2(\mathbb{R}^D)$ (D being the dimension of the crystal), is given by an effective hamiltonian of the form :

$$H = -(\hbar^2/2m)\Delta + \Sigma_i \; v_i(x-x_i) \qquad (1)$$

where v_i is the potential created by the ith atom or impurity and x_i denotes its position.

The disorder may have several sources. One is given by the randomness of the position of the atoms. This randomness may come from the occurrence of many defects due to the way the sample has been prepared. It may as well come from structural reasons, namely the thermodynamical equilibrium favors the occurrence of some disorder : this is the case for amorphous materials (glasses) or quasi-crystals. Another source of disorder is also given by the impurities which modify the atomic potential at random positions.

In any case, to describe such a system in full generality, one usually introduces a probability space (Ω, Σ, μ) , the points of which labelling the hamiltonian and describing in an implicit way the configuration of the material. In other words H becomes a function of $\omega \in \Omega$. For obvious mathematical reasons, one demands that H be a measurable function of ω, at least in the strong resolvent sense (we recall that Borel sets are the same for the norm and the strong topology in the algebra of bounded operator in a separable Hilbert space). To describe the macroscopic homogeneity of the material we just remark that translating the electron in the sample is equivalent to translating the atoms backward and since the sample looks almost the same at any place, this is just changing the configuration ω in Ω. Therefore, there must be an action $\omega \rightarrow T_x\omega$ of the translation group $\mathbb{R}^D$ on Ω such that, if U(x) is the unitary operator representing the translation by x in the Hilbert space $L^2(\mathbb{R}^D)$ then :

$$(2) \qquad U(x) \,.\, H_\omega \,.\, U(x)^* = H_{T_x\omega} \qquad \text{(covariance)}$$

This action will be at least measurable and will satisfy the group property $T_xT_y = T_{(x+y)}$. As we shall see in the end of this section,we may assume Ω to be actually a compact space and the $\mathbb{R}^D$ action to be given by a group of homeomorphisms. The probability measure will serve later on in dealing with "self averaging" quantities.

The framework can be applied to the study of the phonon spectrum as well. For indeed, phonons are elementary mechanical excitations of the atoms around their equilibrium position. In the approximation where there is no phonon-phonon interaction, they are described quantum mechanically as a free Bose field, and only the one phonon hamiltonian must be considered : we must solve the corresponding classical problem. The classical eigenmodes are just eigenvalues of a discrete Schrödinger-like operator with random coefficients. Again the randomness comes from the disorder inside the sample. In any case this operator will obey a covariance condition given by (2).

Even though in practice a sample is given from the beginning and will not change at low temperature during the experiment, the physical energy operator is not the operator H_ω alone corresponding to the given configuration of the disorder. Otherwise it would imply the choice of an origin of coordinates inside the sample, preventing the use of its homogeneity : there is no physical reason "a priori" to prefer one origin instead of one another ! Therefore the observable algebra must contain not only H_ω but also the whole family of its translated. Since in general two elements of this family do not commute **the observable algebra will be non commutative in a fundamental way**.

When performing calculations in quantum mechanics, we need to compute several operators starting from the basic observables. In particular, we will need operators defined through series expansion and the question of convergence will be addressed. Therefore we need to define on the set of observables an algebraic structure which will allow us to make computations, and also a topology which will be essential in proving convergence of infinite series. Clearly the algebra obtained will be a *-algebra since we need to distinguish real numbers from the complex ones. However there is a wide choice of possible topologies depending upon the technical point of view we will choose. Nevertheless there is a canonical choice namely **a topology which is of purely algebraic origin**. There are two kinds of such topological *-algebras : the C*-algebras and the W*-algebras. For in the former case the norm of an element A is nothing but the spectral radius (a purely algebraic object) of A*A. On the other hand a W*-algebra is a C*-algebra with a predual (which is unique) [M25]. Equivalently it is a weakly closed *-subalgebra of $B(\mathcal{H})$. The von Neumann theorem [M6], shows that it coincides with its bicommutant (again an algebraic property even though it may depend upon the representation). As we shall see the W*-algebra built out of the translated of the original hamiltonian is too large to contain any relevant informations. For this reason we shall prefer the C*-algebra generated by the hamiltonian and its translated. In a sense this algebra is the smallest object which contains all the relevant physical informations.

The next step in describing the formal framework consists in answering the question whether it is possible to recover the nature of the probability space Ω from the physical datas. Since the translated of the hamiltonian must belong to the algebra of observables, it is natural to built Ω out of the family of self adjoint operators we obtain in this way. For this reason we introduce the following mathematical criterion to describe what we call "homogeneity".

<u>Definition</u> : A bounded operator A on $L^2(\mathbb{R}^D)$ is called homogeneous if the family of its translated $\{U(x)AU(x)^* ; x \in \mathbb{R}^D\}$ has a compact strong closure.

◇

Strong compactness means that for any finite family of vectors $\varphi_1,\dots,\varphi_N$, and any $\varepsilon > 0$, there is a finite set $x_1,\dots,x_M$ in $\mathbb{R}^D$, such that each vector $U(x)\,A\,U(x)^*\varphi_i$ $(x \in \mathbb{R}^D)$ stays within the distance ε of one of the vectors $U(x_j)\,A\,U(x_j)^*\varphi_i$ $(1 \leq j \leq M)$ in $L^2(\mathbb{R}^D)$. In other words, after translation, the operator A reproduces itself everywhere up to ε. This is why we called it homogeneous.

Actually this notion of homogeneity is quite weak. Indeed the operator of multiplication by a continuous function vanishing at infinity is homogeneous according to our definition (it corresponds to the potential created by finitely many impurities). Another example is given by the following :

<u>Proposition 1</u> : if $V \in L^\infty(\mathbb{R}^D)$ then the resolvent of the Schrödinger operator $H = -\Delta + V$ is homogeneous. ◇

To prove this result, we use a Neuman series expansion and the expression of the Green function of the free laplacian to show that the mapping $V \to (z-H)^{-1}$ is strongly continuous if we endow $L^\infty(\mathbf{R}^D)$ with the weak topology. Since any ball in $L^\infty(\mathbf{R}^D)$ is weakly compact and since translating the resolvent is equivalent to translating the potential alone, the set of translated of the resolvent lies in a strongly compact set.

The main interest of this definition lies in the following construction. The hull $\Omega = \mathrm{Hull}(A)$ of a homogeneous operator A is the strong closure of the set of its translated. By hypothesis it is a compact space. Moreover it is translation invariant. Thus $\mathbf{R}^D$ acts on Ω through a group of homeomorphisms which will be denoted by $\{T_x\,;\, x\in\mathbf{R}^D\}$: any $\omega \in \Omega$ is a bounded operator as the strong limit of some sequence $A_k = U(x_k)AU(x_k)^*$ and we have $T_x\omega = U(x)\omega U(x)^*$. Therefore, the map $(x,\omega) \to T_x\omega$ is continuous. In this way, associated to any homogeneous operator, we get a dynamical system which will also be called the hull of A.

From the proof of proposition 1 it follows that if V belongs to $L^\infty(\mathbf{R}^D)$ the hull of the resolvent of $H = -\Delta + V$ is homeomorphic to the weak closure of the set of translated of V in $L^\infty(\mathbf{R}^D)$. For instance if V is continuous and vanishes at infinity, the hull is homeomorphic to the one point compactification of $\mathbf{R}^D$, namely a D-sphere, and $\{\infty\}$ is the set of non wandering points of the flow generated by the translations.

More interesting is the case for which V is almost periodic in the sense of Bohr [M4]. Generalizing this case leads to the following definition :

<u>Definition</u> : A bounded operator A on $L^2(\mathbf{R}^D)$ is called almost periodic if the family of its translated $\{U(x)AU(x)^*\,;\, x\in\mathbf{R}^D\}$ has a compact norm closure.

◇

By mimicking the construction of the hull of an almost periodic function on $\mathbf{R}$, one can show that the hull of an almost periodic operator is an abelian compact group, and $\mathbf{R}^D$ acts as $\omega \to \omega + y(x)$ where y is a continuous homomorphism from $\mathbf{R}^D$ into Ω with a dense image. Following Mackey's notion of virtual group [71,72], such a dynamical system will be called an "abelian virtual group".

Let us note that the previous construction is universal because any kind of dynamical system can be obtained in this way. For indeed if Ω is a compact space endowed with an $\mathbf{R}^D$ action through a continuous group T of homeomorphisms, and if v is any continuous function on Ω, then for each ω_0 in Ω, the function $x \to v(T_{-x}\omega_0) = V(x)$ is uniformly continuous and the hull of the Schrödinger operator it defines is homeomorphic the dynamical system (Ω,T).

In our purpose, for most interesting examples, taking the W*algebra generated by the family of translated of a homogeneous operator A in $L^2(\mathbf{R}^D)$ will give rise to the full set of bounded operators. Therefore no specific property of the physical system we want to investigate will be seen out of it. For this reason we shall prefer to take the smallest possible *algebra which contains A and all its translated, and which is closed in the norm topology : let C*(A) be this C*-algebra.

Using a "3ε argument", one can show, that if A is homogeneous, so does any element of C*(A), for the product is strongly continuous on bounded sets. Moreover, if $\omega \in \Omega$, and if $(x_k)_{k>0}$ is a sequence in $\mathbf{R}^D$ converging to ω in Ω (we consider $\mathbf{R}^D$ as a subset of ω here), then for any $B \in C^*(A)$ the sequence $U(x_k)\, B\, U(x_k)^*$ converges strongly to some operator B_ω which depends only upon ω and B. The map $\pi_\omega : B \in C^*(A) \to B_\omega$ is a *representation of C*(A) into the space of bounded operators on $L^2(\mathbf{R}^D)$. Moreover, the map $\pi : \omega \in \Omega \to \pi_\omega$ is pointwise strongly continuous and it satisfies the following covariance condition :

$$(2) \qquad U(x)\, \pi_\omega(B)\, U(x)^* = \pi_{T_x \omega}(B) \qquad \omega \in \Omega,\ x \in \mathbf{R}^D,\ \ B \in C^*(A)$$

As a matter of fact, it cannot be more than strongly continuous in general for we have :

<u>Proposition 2</u> : the representation map π is pointwise norm continuous if and only if A is almost periodic, or equivalently if its Hull (Ω,T) is an abelian virtual group.

◇

The proof is just an elementary application of the definition. We now address the question whether one can compute explicitly the algebra C*(A). To each dynamical system (Ω,T) is associated a natural C*algebra, namely the cross-product of $C(\Omega)$, the algebra of continuous functions on Ω, by $\mathbf{R}^D$ through T [M23]. What is the connection between C*(A) and $C^*(\Omega,T)$?

Let us recall the construction of $C^*(\Omega,T)$. The vector space $C_c(\Omega \times \mathbf{R}^D)$ of continuous functions on $\Omega \times \mathbf{R}^D$ with compact support is endowed with a structure of *-algebra as follows :

$$(3) \qquad a.b(\omega,x) = \int d^D y\ \ a(\omega,y)\, b(T_{-y}\omega, x-y) \qquad a^*(\omega,x) = a(T_{-x}\omega,-x)^*$$

We consider now on $L^2(\mathbf{R}^D)$ the family of *representation π_ω defined as :

$$(4) \qquad \pi_\omega(a)\, \psi(x) = \int d^D y\ \ a(T_{-x}\omega, y-x)\, \psi(y)$$

$\pi_\omega(a)$ is a bounded operator. A C*norm is then given by :

(5) $$\| a \| = \sup_{\omega\in\Omega} \| \pi_\omega(a) \|$$

Then $C^*(\Omega,T)$ is nothing but the completion of $C_c(\Omega\times\mathbf{R}^D)$ under this norm. If $a \in C^*(\Omega,T)$, $\pi_\omega(a)$ is a strongly continuous function of ω and satisfies the covariance condition (2). Therefore, since Ω is compact, it is a homogeneous operator. We will say that A is affiliated to $C^*(\Omega,T)$ if there is ω_0 in Ω, and $a \in C^*(\Omega,T)$ such that $A = \pi_{\omega_0}(a)$.

Not every homogeneous operator A is affiliated to the C*algebra of its hull. Two conditions are required for this: first of all A must be "diagonal dominant" namely it must be the norm limit of a sequence A_n of operator for which there is $R_n > 0$, such that $\langle\psi| A_n \varphi\rangle = 0$ whenever the distance between the supports of ψ and φ is greater than R_n. Secondly A must be regular, namely it is the norm limit of operators with a continuous kernel. For instance the operator of multiplication by V is never affiliated with C*(Hull(V)) even if it is smooth. However we get :

<u>Proposition 3</u>: If $V \in L^\infty(\mathbf{R}^D)$ then $H = -\Delta + V$ is affiliated to $C^*(\Omega,T)$ where (Ω,T) is the Hull of the resolvent of H.

◇

To prove this result it is sufficient to compute the kernel associated to the operator $A(f)=f(-\Delta)Vf(-\Delta)$ where f is some smooth function with an L^1 positive Fourier transform. This operator is strongly continuous with respect to V if $L^\infty(\mathbf{R}^D)$ is endowed with the weak topology. Since the weak closure of the set of translated of V is homeomorphic to the Hull of the resolvent of H this shows that A is actually affiliated to C*(Hull(H)). Now, it is clear also that if we now take for f the Fourier transform of $(1+p^2)^{-1/2}$, A is a norm limit of operators of the form A(g) with g rapidly decreasing. Then it is still in the C*algebra. Taking V=1 we get the resolvent of $-\Delta$ which is therefore affiliated to C*(Hull(H)), and using a Neuman expansion of the resolvent of H we get the result.

3)- The tight binding representation :

In this section we want to take into account that the samples studied previously look like discrete lattices, even if there is some disorder in them. As we shall see, there is a mathematical counterpart, namely the Poincaré section [M26], which transforms a continuous flow into a discrete one. Correspondingly there will be a C*algebra for the discrete flow. The converse operation is called the suspension of a discrete flow (or of the C*algebra).

Let us consider the model given in §2-eq(1). Let us assume for simplicity that there is only one species of atoms. This is not actually a restriction if all species have an homogeneous spatial distribution. We call R the smallest distance between any pair of atoms in the sample. We can think of H as depending on R by setting :

$$(1) \qquad V(x) = \Sigma_i \, v(x-Rx_i))$$

where v is the atomic potential. If the disorder is not too big, we can label the atom by $i \in \mathbf{Z}^D$: it is not too faraway from the ideal position it would occupy in a perfect crystal. Now, v is attractive, and therefore the atomic hamiltonian $H_a = -\Delta + v$ must have bound states in order to bind the electrons. Let E be the energy of such a bound state and let ψ be the corresponding wave function. We will also call δE the energy distance between E and the energy level closest to E. If we choose the energy to be zero at infinity, where the atomic potential vanishes, the wave function ψ decreases exponentially fast at infinity, and the rate of decreasing is $|E|$ [M1]. Therefore if R is big enough, the functions $\psi_i(x) = \psi(x-Rx_i)$ are approximate eigenfunctions of $H = -\Delta + V$, and E is the corresponding eigenvalue. Actually this is not quite correct unless we take the limit $R\to\infty$. To take into account the corrections let us consider the matrix L on $\mathbf{Z}^D$ defined by :

$$(2) \qquad L_{ij} = \langle\psi_i \mid \psi_j\rangle = \delta_{ij} + O(e^{-\mathbf{E}\mathbf{R}}) \qquad \text{as } R\to\infty$$

This matrix is non negative, and there is R_0 such that if $R > R_0$, $\|L-\mathbf{1}\| < 1$ and therefore $L^{-1/2}$ is well defined. Now we set :

$$(3) \qquad \varphi_i = \Sigma_j \, (L^{-1/2})_{ji} \, \psi_j \qquad \Rightarrow \qquad \langle\varphi_i|\varphi_j\rangle = \delta_{ij}$$

Thus we get an orthonormal basis of the subspace $\mathcal{H}$ of $L^2(\mathbf{R}^D)$ spanned by the ψ_i's. Moreover, $\|\varphi_i - \psi_i\| = O(e^{-\mathbf{E}\mathbf{R}})$ as $R\to\infty$.

Let now P be the projection onto $\mathcal{H}$. Then for R large enough the spectrum of PHP is contained in the open interval J of width $\delta E/2$ centered at E, which does not meet the spectrum of (1-P)H(1-P). To study the spectrum of H which is contained in this interval, there is a well known method, the first step of which goes back to Schur [101]. It was developed by Feschback and is nowadays called the projection method [101]. Let us define the following energy dependent operator (where Q=1-P):

$$(4) \qquad H(z) = PHP + PHQ\,(z-QHQ)^{-1}QHP$$

As a function of z it is holomorphic on the complement of the spectrum of QHQ, and therefore it is holomorphic in a neighbourhood of the interval J. The main result about it is the following :

Proposition 4 : (i) An energy ε belongs to the part of the spectrum of H contained in J if and only if ε belongs to the spectrum of $H(\varepsilon)$.
(ii) ε is an eigenvalue of H with eigenvector φ if and only if ε is an eigenvalue of $H(\varepsilon)$ with eigenvector $P\varphi$. Then φ can be recovered from the formula :

$$(5) \qquad Q\varphi = (\varepsilon-QHQ)^{-1}QHP\varphi$$

◊

From the definition of ψ_i it is simple to show that for any Φ in $L^2(\mathbf{R}^D)$:

$$(6) \qquad \langle\Phi|H\psi_i\rangle = E\,\langle\Phi|\psi_i\rangle + \Sigma_{j\neq i}\,\langle\Phi|v(.-Rx_j)\psi_i\rangle = E\,\langle\Phi|\psi_i\rangle + O(\|\Phi\|\,e^{-ER})$$

In particular it follows from (2),(3) & (6) that QHP is a bounded operator and that the correction $PHQ\,(z-QHQ)^{-1}QHP$ is bounded by $O(e^{-2ER})$ as $R\to\infty$ uniformly for $z\in J$. Now, the matrix of H(z) in the basis $(\varphi_i)_i$ is diagonal dominant namely :

$$(7) \qquad \langle\varphi_i\,|\,H(z)\varphi_j\rangle = O(e^{-RE|x_i-x_j|}) \qquad \text{as } R\to\infty$$

Let us remark that L_{ij} is actually a function of x_i-x_j, and therefore, there is a function φ in $L^2(\mathbf{R}^D)$ such that $\varphi_i = U(x_i)\varphi$.

What we have done here was to replace the unbounded operator H on the continuum $\mathbf{R}^D$ by an infinite matrix acting on the sequences indexed by the lattice of atoms. As far as we are concerned with the local properties of the spectrum, the two problems are equivalent.

The next step is to introduce a knew C*algebra which will contain the previous matrix. Following A. Connes [27], we need the notion of transversal. Indeed let us consider in Ω the set :

Ξ = closure of $\{T_{x_i}\omega_0 \,;\, i \in \mathbf{Z}^D\}$ where $\omega_0 \in \Omega$ represents V.

Ξ is a "transversal" in the following sense (which differs slightly from the Connes definition, to take into account the topology of Ω):

<u>Definition</u> : A closed subset Ξ of Ω is called a transversal if for any continuous function f on $\mathbf{R}^D$ with compact support, the map

(8) $$(\omega,f) \to \nu^{\omega}(f) = \sum_{x:T_{-x}\omega\in\Xi} f(x)$$

is continuous, for the topology of uniform convergence on compact sets.

◊

It is not hard to check that if Ξ is a transversal, for any $\omega \in \Omega$ the set $L(\omega) = \{x \in \mathbf{R}^D \,;\, T_{-x}\omega \in \Xi\}$ is actually discrete. In addition, given a bounded set Λ in $\mathbf{R}^D$, if $N(\omega,\Lambda)$ denotes the number of points (or atoms) of $L(\omega)$ in Λ, then by subadditivity, the limit :

(9) $$\rho = \lim_{\Lambda \uparrow \mathbf{R}^D} |\Lambda|^{-1} \sup_{\omega} N(\omega,\Lambda)$$

exists. It is a measure of the density of the atoms in the sample. Now let g be a bounded continuous function decreasing at infinity in such a way that $\|g\| < \infty$ where :

(10) $$\| g \| = \sum_{q\in\mathbf{N}} q^{D-1} \sup_{|x|\geq q} |g(x)|$$

Then one can show that (9) implies $\sup_{\omega} |\nu^{\omega}(g)| \leq \text{const.}\|g\|$, and therefore, since such a g can be approximated in this norm by continuous functions with compact support, $\nu^{\omega}(g)$ is also continuous with respect to ω. Let B be the Banach space of such g's. Now let us choose v in $L^{\infty}(\mathbf{R}^D)$ and such that $\| v \| < \infty$ (where the supremum is replaced by the essential one in (10)). For any continuous function f with compact support, the convolution $f * v$ belongs to B, which shows that $\langle V_{\omega} | f \rangle = \langle \nu^{\omega}(v) | f \rangle = \nu^{\omega}(f * v)$ is continuous in ω. Again since $\| v \|$ is finite, V_{ω} is essentially bounded, which shows that it is weakly continuous with respect to ω.

As we already quoted, the flow on Ω can be restricted to Ξ by mean of the "first return map" introduced by H. Poincaré [M21] in the 19th century to study the existence and the stability of periodic orbits of the planets. In our many dimensional framework, the first return map does not exist, but is replaced by the notion of groupoid of the transversal Ξ [27]. Let $\Gamma(\Xi)$ be the (closed) subset of $\Omega \times \mathbf{R}^D$ made of the elements $\gamma = (\omega,x)$ such that $\omega \in \Xi$ and

$T_{-x}\omega \in \Xi$. It is a locally compact groupoid in the sense that $\gamma=(\omega,x)$ and $\gamma' = (\omega',x')$ can be multiplied provided $\omega' = T_{-x}\omega$; in this case we get $\gamma\circ\gamma' = (\omega,x+x')$. A *algebra can be constructed through continuous functions on $\Gamma(\Xi)$ with compact support with the following rules :

$$(11) \quad a.b(\omega,x) = \Sigma_{y\in L(\omega)}\, a(\omega,y)b(T_{-y}\omega,x-y) \qquad a^*(\omega,x) = a(T_{-x}\omega,-x)^*$$

As in the previous section, let $C^*(\Xi)$ be the completion of this algebra with respect to the C^* norm $\|a\| = \sup_{\omega\in\Xi} \|\eta_\omega(a)\|$ where η_ω is the *representation on the space $l^2(L(\omega))$ of square summable sequences indexed by $L(\omega)$ and defined by :

$$(12) \qquad \eta_\omega(a)\,\psi\,(x) = \Sigma_{y\in L(\omega)}\, a(T_{-x}\omega,y-x)\,\psi(y)$$

These representations satisfy also a covariance condition of a different kind. Namely, for $\gamma=(\omega,x)$ in $\Gamma(\Xi)$, let $U(\gamma)$ be the unitary operator from $l^2(L(T_{-x}\omega))$ into $l^2(L(\omega))$ given by $U(\gamma)\,\psi\,(y) = \psi(y-x)$.Then we get :

$$(13) \qquad U(\gamma)\,\eta_{T_{-x}\omega}(a) = \eta_\omega(a)U(\gamma)$$

The main application of this formalism is given by :

<u>Proposition 5</u> : Let $h(z)$ be the function on $\Gamma(\Xi)$ defined from (7) by :

$$(14) \quad h(z;\omega,x) = \langle \varphi \mid H_\omega(z)\, U(x)\varphi \rangle$$

Then $h(z)$ belongs to $C^*(\Xi)$ for any z for which $H(z)$ is defined. Moreover the matrix of $H_\omega(z)$ in the basis $\{\varphi_i\}$ and of $\eta_\omega(h(z))$ in the basis $\{x_i\}$ coincide.

◇

We will not give the proof of this fact, for (14) is just another way of writing $H_\omega(z)$.

To finish this section let us give one example of such reduction, which is sometime called a "lattice approximation" or also in Solid State Physics a "tight binding representation" because when (4) is valid, the electrons are essentially localized near the bottoms of the potential and are tightly bound to the atoms. Our first example will be a quasi periodic Schrödinger operator in one dimension [see in particular 16]; let v be a smooth function on **R** decreasing rapidly to zero at infinity. The potential is given by :

(15) $$V_\xi(x) = \Sigma_{n\in\mathbf{Z}}\, v(x-x_n) \qquad x_n = n + g(\xi)-g(\xi-n\alpha)$$

where g is a continuous function of one variable ξ periodic of period one and α is an irrational number. Then g can be seen as a continuous function on the one dimensional torus $\mathbf{T} = \mathbf{R}/\mathbf{Z}$. To built the hull we define on $\mathbf{T}\times\mathbf{R}$ the map $F(\xi,t) = (\xi+\alpha, t+f(\xi))$, where $f(\xi) = 1+g(\xi+\alpha)-g(\xi)$. Then the function $\mathbf{V}(\xi,x) = V_\xi(x)$ given by (15) is F invariant and continuous on $\mathbf{T}\times\mathbf{R}$. Therefore it defines a continuous function on the quotient space $\Omega = \mathbf{T}\times\mathbf{R}/F$. A flow on Ω is defined by taking the quotient of the translation $T_y:(\xi,x)\to(\xi,x+y)$ (which commutes to F), and we get $V_\xi(x-y) = \mathbf{V}(T_{-y}(\xi,x))$. This shows that Ω is actually the hull. The set $\mathbf{T} = \{(\xi,x)\in\Omega\ ;\ x=0\}$ is a transversal, and the first return map is nothing but the rotation R_α by α. Then it is easy to check that the C* algebra of this transversal $C^*(\mathbf{T}, R_\alpha)$ is generated by two unitaries U and V such that $UV = e^{i2\pi\alpha}VU$ as was firstly proposed by M. Rieffel [86]. The reduced hamiltonian H(z) has a matrix indexed by $\mathbf{Z}$ which can be seen as an element of $C^*(\mathbf{T}, R_\alpha)$. A special case is provided if we can approximate H(z) by a tridiagonal matrix, and if in addition only the first Fourier coefficients are big, one gets :

(16) $$h(z) = a(U + U^*) + b(V + V^*) \qquad a,b \in \mathbf{R}$$

This operator is called the Almost Mathieu operator and has many interesting properties [M16,M28,16,18,23,41,53,94], as we shall se later on.

Our second example is given by the recently discovered quasicrystals [91]. The potential seen by an electron in a quasicrystal can be described in much the same way. However the precise structure of the hull is usually quite complicated : the quasicrystal of AlMn originally discovered had a symmetry of order five. In the original work of D. Schechtman, I. Blech, D. Gratias, J.V. Cahn [91], it was recognized that 6 integers were necessary to label the positions of the peaks of intensity in the X-ray picture. A model has been proposed recently [39,40,65,70] which describes the diffraction pattern observed experimentally. Even though the method leading to its construction was already known before [64], this model have the advantage of being very simple and efficient. Since this crystal exhibits a diffraction pattern having the icosahedral symmetry, and since 6 integers are needed to label the positions of what one believes to be the atoms (there is still a problem in interpreting the nature of the white spots on the pictures), we will look at a 6 dimensional representation of the icosahedral group. This group leaves the 6 dimensional hypercubic lattice $\mathbf{Z}^6$ invariant. Moreover it has two 3-dimensional invariant subspaces. Let Π and Π' be the corresponding eigenprojections. Of course since there is no translation invariant 3-dimensional lattice having a symmetry of order 5, these eigensubspaces are not rationally oriented with respect to the lattice $\mathbf{Z}^6$. Let Δ be the 6-dimensional closed hypercube of unit length, and $\Omega = \Pi'\Delta$. It has been shown that Ω is a polyhedron called a triacontahedron [40] which has axis of order 2, 3 and 5. The model proposed by Duneau & Katz consists in selecting those points $m \in \mathbf{Z}^6$ such that $\Pi' m \in \Omega$, and to project them on the orthogonal subspace, namely $x_m = \Pi m$. Since $\mathrm{Im}(\Pi)$ is 3-dimensional, we will identify it with $\mathbf{R}^3$ and the x_m's with the positions of atoms. By construction the icosahedral group commutes with Π and Π', and therefore the set of atoms is a (non translation invariant) lattice invariant by this group. If we want to take into account a generic translation in $\mathbf{R}^6$ we get the following electronic potential for the quasi crystal :

$$(17) \qquad V_\xi(x) = \sum_{m \in \mathbf{Z}^6} v(x - \Pi(\xi + m))\, \chi_\Omega(\Pi'(\xi + m))$$

where χ_Ω is the characteristic function of Ω. Using the previous method, the Hull is the 6-dimensional torus $\mathbf{T}^6 = \mathbf{R}^6/\mathbf{Z}^6$. Then $V_\xi(x)$ is given by a discontinuous function on this torus (because of the characteristic function of Ω) as a function of ξ, and by an action of $\mathbf{R}^3$ in the Π direction. The dynamical system that one obtains is an abelian virtual group. The tight binding representation of our hamiltonian will give rise to the C*algebra of the transversal given by the image of the set $\Omega \times \{0\}$ in this torus.

In principle the construction is so general that any kind of flow on a compact manifold could be described. The question is to know whether other examples are to be found in experimental physics. The proposal of Sadoc and Mosseri [76], recently studied in detail by D. Nelson [see 78], seems to indicate that indeed other kinds of flows appears in the description of the short range order in liquids and glassy materials.

4)- <u>The density of states and Shubin's formula</u> :

The result of the previous analysis is that a disordered system is described by a self adjoint operator which is affiliated to the C*algebra of a dynamical system given by some compact space Ω, the hull, on which there is an action of $\mathbf{R}^D$. As we showed in the previous section, this continuous dynamical system can be replaced by a discrete one, which, in many cases reduces to the action of $\mathbf{Z}^D$ on some compact space Ξ. Before proceeding, let us note that even in physical situations, the groups $\mathbf{R}^D$ or $\mathbf{Z}^D$ can be replaced by a more general locally compact group. For instance in describing the local structure of glassy systems, Kleman & Donnadieu [61] found that a description through hyperbolic geometry is relevant. In this case we have a SL(2,$\mathbb{R}$) action or even the action of a Fuchsian group [M8]. On the other hand, theoretical physicists like to use a special lattice called the "Bethe lattice" [67], to describe the thermodynamics of lattice models in the limit where the dimension goes to infinity. It is not difficult to convince oneself that the Bethe lattice can be seen as a sublattice of the Poincaré disc, and is generated from one point by a free group with a finite number of parameters. Therefore it appears also as a piece of hyperbolic geometry. Another example is given by disordered systems in which spin effects are taken into account, or systems having more symmetries or quasi symmetries than just the translation one as it is the case for quasicrystals.

For this reason let us consider a compact space Ω on which acts a locally compact group G. In much the same way it is easy to construct a C* algebra for this system. We will avoid the construction (see [M23]) for it is formally the same as before. Let us simply remark that it has an identity operator if and only if G is discrete. There is however one property of G which is essential in what follows : it must be **amenable**. Amenability means that in G, there is an increasing sequence $(\Lambda_n)_{n>0}$ of open sets of finite Haar measure (like hypercubes in $\mathbb{R}^D$) the union of which covering the group, the frontier of which has a negligible measure compare to their volume in the limit $n \to \infty$. In other words **surface effects are negligible compare to volume effects**. More precisely, if $x \in G$, Δ denotes the symmetric difference, and $|\Lambda|$ the volume of Λ with respect with a left Haar measure one demands :

$$(1) \qquad \lim_{n\to\infty} |\Lambda_n \Delta (x.\Lambda_n)|/|\Lambda_n| = 0$$

Such a family of open sets of finite volume is called a "Følner sequence" [M12]. As a matter of fact this condition precisely eliminates the hyperbolic geometry, for it is known that in the Bethe lattice, surface effects are not negligible compare to volume ones !

Given one self adjoint operator on $L^2(G)$ affiliated to the C*algebra of (Ω,G), one defines its integrated density of states (if it exists) as follows :

$$(2) \qquad \mathbf{N}_\omega(E) = \lim_{n\to\infty} |\Lambda_n|^{-1} \#\{\text{eigenvalues of } H_\omega|_{\Lambda_n} \leq E\}$$

Let us mention that the first derivative $d\mathbf{N}_\omega(E)/dE$ of this quantity is accessible to the experiment. This is done either around the Fermi level by measuring the current-voltage characteristics of a diode in which the sample has been put, or by mean of X-ray scattering, which permits to investigate the band structure at smaller energies.

For this definition to be meaningful, it is obviously necessary that the restriction of H to bounded regions exists and have a discrete spectrum. Actually if H is affiliated to $C^*(\Omega,G)$, this is always true. Indeed if $a \in C^*(\Omega,G)$, and if χ is the operator of multiplication by the characteristic function of some bounded open set Λ on $L^2(G)$, then for any $\omega \in \Omega$, $\chi\pi_\omega(a)\chi$ is compact. Thus, the restriction of the resolvent of H on χ is compact. This is enough to show that the resolvent of the restriction is also compact. However the restriction of the resolvent is not the resolvent of the restriction. The main difficulty comes from the definition of the restriction of H to Λ. If H is a Schrödinger operator, this requires to specify the boundary conditions. In general, if G is not discrete, then H must be unbounded and bounded from below, at least locally, in order that its restriction have only finitely many eigenvalues near $-\infty$. On the other hand, the domain of H must at least contain "sufficiently many" functions supported by Λ : the set $D(\Lambda)$ of functions in the domain of H supported by Λ must be dense in $L^2(\Lambda)$. Moreover it is also necessary that $\bigcup_n D(\Lambda_n)$ be a core for H in order to recover H as the limit of the family of its restrictions in the strong resolvent sense. If this is true, since H is bounded from below locally, its restriction on $D(\Lambda)$ has self adjoint extensions (the Friedrichs one for instance). A boundary condition is just the assignment of one self adjoint extension. The independence with respect to the boundary conditions is related to the fact that boundary effects are negligible compare to volume effects, namely to the amenability of G. Yet, we do not know what are the necessary and sufficient conditions on H to insure the existence of the density of states in general. However, there are many concrete cases where this has been proved (see conditions (A) below).

When the integrated density of states of H exists, it is a G-invariant Borel function of ω. In particular, it is almost surely constant with respect to an ergodic G-invariant probability on Ω. On the other hand, it is obviously increasing with respect to E. It has been proved by F. Delyon & B. Souillard [36] that for Schrödinger operators on a lattice $\mathbf{Z}^D$, this function is continuous. Their argument is again related to the amenability of $\mathbf{Z}^D$, and we suspect that for reasonable hamiltonians, the integrated density of states is indeed continuous with respect to the energy. Let us note however that there are C* algebras having projections the kernel of which being continuous with compact support (see the appendix) and therefore it is possible to exhibit examples of hamiltonian in these algebra, having a discontinuous density of states. There are also models studied by physicists, like the laplacian on a Serpinsky lattice [1,85], which admit a discontinuous density of states. Thus we ask the following questions :

<u>Question 1</u> : What are the necessary and sufficient conditions on a self adjoint operator H affiliated to $C^*(\Omega,G)$ to get the convergence of the integrated density of states ?

Question 2 : What are the self adjoint operators H affiliated to $C^*(\Omega,G)$, having an integrated density of states continuous with respect to the energy, when G is amenable ?

Before giving a more precise result, let us introduce another tool. A G-invariant probability measure μ on Ω defines a trace on $C^*(\Omega,G)$ in the following way (where "e" is the neutral element in G) :

$$(3) \qquad \tau_\mu(a) = \int d\mu(\omega)\ a(\omega,e) \qquad a \in C_c(\Omega \times G)$$

where $C_c(\Omega x G)$ is the space of continuous functions with compact support on $\Omega x G$. For indeed, it is a linear map, such that $\tau_\mu(a^*a) \geq 0$ and that $\tau_\mu(ab) = \tau_\mu(ba)$. It can be extended to a (weakly dense) ideal J of $C^*(\Omega,G)$, which is often denoted by $L^1(C^*(\Omega,G),\tau_\mu)$. By the Birkhoff ergodic theorem, if μ is G-ergodic, then for any Følner sequence we get :

$$(4) \qquad \tau_\mu(a) = \lim_{n\to\infty} |\Lambda_n|^{-1} \int_{\Lambda_n} dx\ a(x^{-1}.\omega, e) \qquad \text{for } \mu \text{ a.e. } \omega$$

Using the covariance condition , this is also equivalent to :

$$(5) \qquad \tau_\mu(a) = \lim_{n\to\infty} |\Lambda_n|^{-1} \mathrm{Tr}(\chi_{\Lambda_n} \pi_\omega(a)) \qquad \text{for } \mu \text{ a.e. } \omega$$

Hence such a trace is nothing but a **trace per unit volume**, and the limit converges for μ almost all ω. Let us note the following properties of the trace :

Proposition 6 : (i) If G is amenable, there is always at least one G-invariant ergodic probability measure μ on Ω.
(ii) For any G-invariant ergodic probability measure μ on Ω, and any $a \in C_c(\Omega x G)$, the limit (5) exists for μ almost all ω in Ω, and defines a trace on $C^*(\Omega,G)$.
(iii) If the support of μ is dense in Ω, this trace is faithful.
(iv) If G is discrete, this trace is bounded and normalized, i.e., $\tau_\mu(\mathbf{1}) = 1$ if $\mathbf{1}$ is the identity operator.
(v) If the dynamical system (Ω,G) is uniquely ergodic, namely if there is a unique G-invariant ergodic probability measure μ on Ω, then the limit in (5) is uniform with respect to ω.

◇

That (i) is true comes from a characterization of amenable groups by the existence of invariant means [M12]. Namely, on the C*algebra of continuous bounded functions on G, there is a G-invariant state M, if and only if G is amenable. Such a state is called an invariant mean. Therefore, if f is a continuous function on Ω, the function $x \in G \to f(x^{-1}.\omega) = F(x)$ is continuous and bounded on G, and the quantity $\mu(f) = M(F)$ is a positive linear map such that $\mu(1) = 1$. By Riesz's theorem, μ is a probability measure on Ω.

(ii) is just a consequence of Birkhoff's ergodic theorem [M13]. (iii) expresses the fact that any continuous positive function f on Ω such that $\mu(f) = 0$, vanishes μ-almost everywhere, and , by continuity, everywhere. The proof of (v) is a classical result in the study of dynamical systems (see [M5,M11,52]). ◇

The main interest of the trace, is to give a simple mathematical tool which describes what is called by physicists the **self averaging observables** namely, those physical quantities which are independent of the disorder. They are obtained through averaging a local observable over the sample. It has also been used, without notifying the formal definition of a trace, by R. Johnson & J. Moser [55], in dealing with the rotation number of a one dimensional Schrödinger equation (see section 6 below). The first non trivial result using this formalism was provided by L. Coburn, Moyer and I. Singer [24] who proposed to use C*algebras for extending the index theorem to pseudodifferential operators with almost periodic coefficients. This paper plaid a very important role in the original work of A. Connes when he generalized the index theorem for pseudodifferential operators on a foliated manifold, differentiating along the leaves. Following this proposal Shubin [93] in the mid seventies, studied the C*algebra of zeroth order pseudodifferential operators with almost periodic coefficients. He also studied the properties of the trace and realized that there is a connection between the integrated density of states and the trace, a result described below. Before giving this formula, we recall that if τ is a trace on a C*algebra $\mathcal{A}$, then through the GNS construction [M19], there is a representation π on the Hilbert space $L^2(\mathcal{A},\tau)$ (the completion of $\mathcal{A}$ under the Hilbert norm $\|a\|_2 = \{\tau(a^*a)\}^{1/2}$) such that if η is the canonical map from $\mathcal{A}$ into $L^2(\mathcal{A},\tau)$, we have :

$$(6) \qquad \tau(a^*bc) = \langle \eta(a) | \pi(b)\eta(c) \rangle \qquad a,b,c \in \mathcal{A}$$

We denote by $\mathcal{M}_\tau$ the W*algebra generated by $\pi(\mathcal{A})$ i.e. $\pi(\mathcal{A})''$. Then τ extends to $\mathcal{M}_\tau$. Moreover, if h is a self adjoint element of $\mathcal{A}$, then $\pi(h)$ has a spectral decomposition and the spectral projections belong to $\mathcal{M}_\tau$. Then we have :

Shubin's formula : Let H be a self adjoint operator on $L^2(G)$ affiliated to $C^*(\Omega,G)$, and μ be a G-invariant ergodic probability measure on Ω. We denote by $\chi(H \le E)$ the spectral projection of H on energies smaller than or equal to E. Then, under one of the condition (Ai) below, the density of states exists and for all E, and μ almost all ω one gets :

(7) $$\mathbf{N}_\omega(E) = \tau_\mu(\chi(H \leq E))$$

◇

From what we already explained, this formula is intuitively obvious (see eq. 2 & 5). However we need somewhere to control the restriction of H to a finite volume : as we said, if f is a Borel function on the spectrum of H, the restriction of f(H) is not equal to the image through f of the restriction of H This is hopefully true only asymptotically in the infinite volume limit. For this reason there is no general proof of this formula yet. However it is true under one of the conditions below :

(A1) G is discrete and H is bounded (see the appendix).

(A2) G = $\mathbf{R}^D$ and H is a pseudodifferential operator with quasi periodic coefficients (see Shubin [93], an extension of this work to the case where Ω is any smooth manifold should be easy))

(A3) G = $\mathbf{R}^D$ and H is a Schrödinger operator (see Bellissard, Lima, Testard [19])

The first case corresponds to the "lattice approximation", while the others correspond to the real continuum case. Treating the general case is the object of question 1.

5)- Cantor spectra, gap labelling theorem and K_0-group :

Using the covariance property of a homogeneous operator, one gets some elementary informations about its spectrum.

Proposition 7 : Let $(\Omega,\mathbf{R}^D,T)$ (resp. $(\Omega,\mathbf{Z}^D,T)$) be a dynamical system, and μ be a T-invariant ergodic probability measure on Ω. Let $H=\{H_\omega ; \omega\in\Omega\}$ be a covariant μ-measurable family of self adjoint operators on a Hilbert space $L^2(\mathbf{R}^D)$ (resp. $l^2(\mathbf{Z}^D)$). Then :

(i) there is a closed subset Σ of $\mathbf{R}$ such that for μ-almost all ω in Ω, the spectrum of H_ω is Σ.

(ii) If E is an isolated point of Σ it is an eigenvalue of infinite multiplicity of H_ω, μ-almost surely.

(iii) If $\omega \to H_\omega$ is stongly continuous, Σ is the union of spectra of $\{H_\omega ; \omega$ in the support of $\mu\}$.

(iv) If $\omega \to H_\omega$ is stongly continuous, and if Ω admits periodic points for T in the support of μ, then the set Σ cannot be nowhere dense.

◇

The proof of (i) and (ii) is classical [see 66]. For (iii)and (iv) it suffices to remark that if ω belongs to the support of μ, it may be approximated by a sequence of points ω' for which the spectrum of $H_{\omega'}$ is precisely Σ. Since the spectrum does not increase by strong limit [M18], Σ contains the spectrum of H_ω. If ω is a T-periodic point, H_ω admits a band spectrum [M7,M22,100]: it cannot be nowhere dense. ◇

The previous result applies in particular to the Schrödinger operators on $\mathbf{R}^D$ or $\mathbf{Z}^D$ with a random potential to show that the spectrum is actually connected [see 66]. In contrast, it has been shown that several examples of Schrödinger operators with almost periodic potential, have a nowhere dense spectrum. According to the previous result, this is impossible for a flow having a periodic orbit in the support of an invariant ergodic probability measure. However, this property is actually generic in one dimension for limit periodic potentials, as was shown firstly by J. Moser [75], and by J. Avron & B. Simon [11]. This is also true for a generic set of pairs (α,μ) for the almost Mathieu operator (see §3 eq.16) acting on $l^2(\mathbf{Z})$ as [18,see 23,53] :

$$(1) \qquad \psi(n+1) + \psi(n-1) + 2\mu\cos 2\pi(\xi-n\alpha)\psi(n) = H\psi(n) \qquad n\in\mathbf{Z}$$

It is also believed to be generic for Schrödinger operator with an almost periodic potential. Cantor spectra have been found also for different examples of Jacobi matrices in one dimension. For instance if in (1) the cosine is replaced by a characteristic function of length α, numerical works indicate that this happens [63,79,80]. There is also a class of models [12,13,15,77], having the Julia set of a polynomial as spectrum.

Nobody gave any example of non almost periodic Schrödinger operator in one dimension having a Cantor spectrum, even though there is no reason "a priori" that Cantor spectrum is impossible for non almost periodic potentials. One interesting question in this respect is the following :

Question 3 : Let C be a Cantor set contained in a bounded interval of $\mathbb{R}$. Let μ be a probability measure on C the support of which being C itself. Let H be the operator of multiplication by x in $L^2(C,d\mu)$ and let $(p_n)_{n\geq 0}$ be the orthonormal basis generated by the Schmidt procedure from the set of monomials. Then the matrix of H in this basis is tridiagonal [M9]. When is this matrix almost periodic ?

Let us note however that the previous property is only generic. Non generic examples of one dimensional Schrödinger operators on $\mathbb{R}$ with finitely many bands in their spectrum have been constructed, using the inverse scattering method, in connection with the KdV equation, by Dobruvin, Matveev and Novikov [38].

The situation in more than one dimension is still unclear. There is only one known class of examples of Schrödinger operators, namely the discrete Laplace operators on "Serpinsky gaskets", which have a Cantor set as set of limit points of their spectrum. This was shown by R. Rammal [85, see 1]. However, for the periodic case, Skriganov [95] shown that in 2 dimensions, at high energy the spectrum is connected. We believe that this property survives even for a large class of homogeneous potential, because of the "band overlapping". However it may be perfectly possible that at low energy, there is some Cantor spectrum.

Nevertheless, the question remains in these cases to label the gaps, when the spectrum is a Cantor set. The solution to this problem is a question of taste : it depends whether one is a physicist or a mathematician. If one is a physicist, one will observe that the integrated density of states is locally constant outside the spectrum : it is constant on each gap (see fig 1 below). Since it is a strictly increasing function of the energy on the spectrum, two gaps separated by some non empty piece of the spectrum correspond to different values of the density of state. Thus a given gap J can be labelled by the value $\mathcal{N}(E)$ for $E \in J$.

A mathematician will not be very happy with this method, for it seems rather arbitrary. He would like some more canonical way. Thinking a little bit about the problem, one realizes that there is a mathematical object which can be affected to a gap J of the spectrum, namely the eigenprojection $P(J) = \chi(H \leq E)$ for any $E \in J$. Since E does not belong to the spectrum, the function $x \to \chi(x \leq E)$ is continuous on the spectrum and this eigenprojection

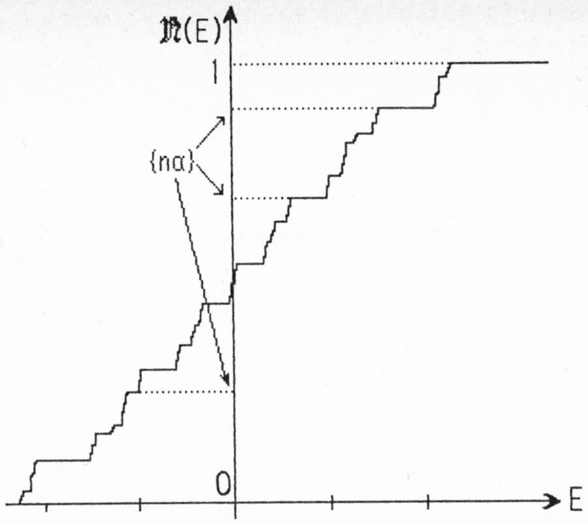

Fig. 1 : The integrated density of states for a one dimensional Schrödinger operator on a lattice or for any self adjoint bounded operator affiliated to the irrational rotation algebra.

still belongs to the C*algebra of the Hull ! This not a trivial property, for in a separable C* algebra projections are rather exceptional : for instance the projections of the C*algebra $\mathcal{K}$ of compact operators on a separable Hilbert space are all the finite dimensional ones. However giving the eigenprojection of the gap is certainly too much an information. For the spectrum does not change under unitary transformations. Thus only the equivalence class of such projections under unitary transformations is needed to label the gap. More precisely [M19,M25] :

Definition : Let $\mathcal{A}$ be a C* algebra with a unit, and P, Q be two projections in $\mathcal{A}$. Then P and Q are equivalent, and we will denote this relation by $P \approx Q$, if there are two elements U and V in $\mathcal{A}$ such that :

$$(2) \qquad P = UV \qquad\qquad Q = VU$$

If P and Q are self adjoint then one can take $V = U^*$. ◇

One could ask the question why we have chosen such an equivalence relation. For indeed as far as the spectrum as a set is concerned, any automorphism of the C*algebra would leave it invariant. Identifying two projections obtained one from the other through such an automorphism would diminish the set of equivalence classes in general. Our choice is motivated by the remark that unitary transformations are those automorphisms which are approximated by local automorphisms, when H is approximated by its restrictions on bounded sets.

The next problem is to compute the set $\mathcal{P}$ of equivalence classes. Actually nobody knows how to do that in general ! However as remarked by

Grothendieck [M3], there is some structure in this set provided one accepts to weaken a little bit our notion of equivalence. For indeed, if [P] denotes the equivalence class of the projection P, and if P,Q are orthogonal to each other (namely PQ = QP = 0) then $P+Q = P\oplus Q$ is also a projection the equivalence class of which depending only upon [P] and [Q]. Therefore we get an addition in $\mathfrak{P}$ defined by :

(3) $[P]+[Q] = [P'\oplus Q']$ for any $P'\approx P$ and $Q'\approx Q$ with $P'Q = Q'P = 0$

Again this addition is not defined for any pair of projections, for it may happen that there is not enough room in $\mathcal{A}$ to rotate this pair in such a way that the two projections become orthogonal. However, if we enlarge $\mathcal{A}$ by adding the union over $n \geq 0$, of the $n\times n$ matrices with entries in $\mathcal{A}$, this will work. Indeed, we will identify P with the matrix

$$(4) \qquad P \approx \begin{vmatrix} P & 0 \\ 0 & 0 \end{vmatrix} \approx \begin{vmatrix} 0 & 0 \\ 0 & P \end{vmatrix} \qquad \text{in } \mathcal{A}\otimes M_2$$

Hence in $\mathcal{A}\otimes M_2$ it is always possible to transform P and Q in $\mathcal{A}$ into a pair of equivalent orthogonal projections. However this will not be true for any pair in $\mathcal{A}\otimes M_2$. Thus we will replace $\mathcal{A}\otimes M_2$ by $\mathcal{A}\otimes M_2\otimes M_2$ and if we keep going we eventually end up with the algebra $\mathcal{A}\otimes\mathcal{K}$. Now this is enough for $\mathcal{K}\otimes\mathcal{K}$ is isomorphic to $\mathcal{K}$ and therefore $\mathcal{A}\otimes\mathcal{K}$ is isomorphic to $\mathcal{A}\otimes\mathcal{K}\otimes\mathcal{K}$. For this reason $\mathcal{A}\otimes\mathcal{K}$ is called the **stabilized algebra**. To conclude, the set $\mathfrak{P}$ of equivalence classes of projections in the stabilized algebra is endowed with an addition which is commutative. Then by "abstract nonsense", Grothendieck told us that there is a canonical group which can be constructed out of $\mathfrak{P}$ exactly like we construct $\mathbf{Z}$ out of $\mathbf{N}$: this group called $K_0(\mathcal{A})$, is the set of classes of pairs ([P],[Q]) where the equivalence is given by :

(5) $\exists\, [R]\in P \quad [P]+[Q']+[R] = [P']+[Q]+[R] \quad \Leftrightarrow \quad ([P],[Q]) \approx ([P'],[Q'])$

Then [P]-[Q] is identified with the class of the pair ([P],[Q]), and the addition extends to a group law. Moreover let us note the following result :

Proposition 8 [M19] : If $\mathcal{A}$ is a separable C*algebra, the group $K_0(\mathcal{A})$ is abelian and countable ◇

Now we remark that any trace τ on $\mathcal{A}$ satisfies the two following properties :

(6) (i) if $P \approx Q$ then $\tau(P) = \tau(Q)$
(ii) if $PQ = QP = 0$ then $\tau(P \oplus Q) = \tau(P) + \tau(Q)$

This shows that τ defines a map τ_* on $K_0(\mathcal{A})$ which is actually a real valued homomorphism through the formula :

$$\tau_*([P]) = \tau(P) \tag{7}$$

Moreover if τ is faithful (namely $a \geq 0$ and $a \neq 0$ implies $\tau(a) > 0$) then τ_* is one-to-one.

Coming back to our original problem we see that if J is a gap the number $\tau(P(J))$ which, by Shubin's formula, is identical to the value of the integrated density of states on the gap, is equal to the number $\tau_*([P])$ and therefore :

Gap labelling theorem 1 : Let H be a self adjoint operator affiliated to $C^*(\Omega,G)$, and such that its density of states exists with respect to a G-invariant ergodic probability measure μ on Ω . Then the values of the density of states on a gap of the spectrum belongs to the positive part of the countable subgroup of $\mathbb{R}$ given by $\tau_{\mu*}(K_0(C^*(\Omega,G)))$. ◇

Corollary 1(homotopy invariance) : Let H be as before, let J be a gap in the spectrum of H, and let $\tau(J)$ be the value of the integrated density of states on J. Then $\tau(J)$ is invariant under small perturbations of H in the norm resolvent convergence. ◇

Even though this construction may appear rather complicate, it is actually very natural. For instance, the **stabilized algebra of a dynamical system is isomorphic to the algebra of its suspension** [29,30,87]! Thus stabilizing is nothing more than the **inverse operation** described in the section 3, namely the **tight binding representation** .

The obvious question now is how to compute the K-group ! Precisely this was one of the breakthrough in the early eighties, to give a calculation of these groups. The first important result was given in 1979 by Pimsner and Voiculescu [81,82] and it concerns the irrational rotation algebra :

Theorem 1 : If α is an irrational number, let $\mathcal{A}_\alpha$ be the algebra generated by two unitaries U and V such that $UV = e^{2i\pi\alpha} VU$, (i.e. $\mathcal{A}_\alpha = C^*(\mathbf{T},\mathbf{Z})$, where $\mathbf{Z}$ acts on $\mathbf{T}$ through the rotation R_α). Then :
(i) on $\mathcal{A}_\alpha$ there is a unique trace τ which is faithful (prop.6).
(ii) the group $K_0(\mathcal{A}_\alpha)$ is isomorphic to $\mathbf{Z}^2$. Its generators are given

by the equivalence classes of the identity $\mathbf{1}$, and of the "Rieffel projection" P_R [86].

(iii) The subgroup $\tau_*(K_0(\mathcal{A}_\alpha))$ is equal to $\mathbf{Z}+\alpha\mathbf{Z}$.

(iv) If h is any self adjoint element of $\mathcal{A}_\alpha$, and J is a gap in its spectrum, there is a unique $n \in \mathbf{Z}$ such that the integrated density of states takes on the value $n\alpha-[n\alpha]$ on J. ◇

The previous argument leading to the gap labelling theorem1, was presented in a meeting at Marseille in April 1981, and announced in the IAMS conference in Berlin in August 1981[14]. On the other hand it can also been found in the review by B. Simon [94] on almost periodic Schrödinger operators. This result was also proven in a direct way by Delyon and Souillard [35] in the very special case of the almost Mathieu operator (with an arbitrary continuous function V instead of the cosine, as a potential in (1)). They actually used a suspension technics together with the work of Johnson and Moser [55] who gave a version of the gap labelling theorem for the one dimensional Schrödinger operator on $\mathbf{R}$ (see below).

Let us also mention that the result of Pimsner and Voiculescu is more general in that they actually computed the K-group for a wider algebra, which is generated by the translation operator U in $l^2(\mathbf{Z})$ together with the operator χ of multiplication by $\chi_{(0,\alpha]}(x-n\alpha)$. Since this new algebra contains the irrational rotation algebra, what they proved was that the K-group of the latter was not bigger than $\mathbf{Z}^2$. To prove that it was equal, they used the first work of M. Rieffel [86] who produced an example of projection P_R, in the form of a tridiagonal matrix, the trace of which was equal to α. Nevertheless this extension of theorem 1 is important to notice for the Schrödinger operator (1) with $\mu\chi$ as a potential became popular recently because it represents the operator generating the **phonon spectrum of a one dimensional quasi crystal**.

However, even though important, the previous theorem was just a first step toward a general theory. The next breakthrough was performed by A. Connes few weeks after he received the work by Pimsner and Voiculescu, and generalized it in giving a geometrical interpretation together with explicit formulæ to compute the image of the K-group by the trace homomorphism. Actually, as soon as 1977, A. Connes had proved a generalization of the index theorem for pseudodifferential operators on a foliated manifold [26,27,31], and this earlier result contained in essence what we are going to explain now.

Definition : Let G be a Lie group of dimension D, acting freely and smoothly on a smooth compact manifold Ω. Let $X_1,\dots,X_D$, be vector fields generating the G-action on Ω. Let also μ be a G-invariant ergodic probability measure on Ω (if G is not amenable μ may or may not exist). The Ruelle-Sullivan current [88] is defined as the linear form on the space of D-differential forms on Ω as follows :

(8) $$\langle \mathfrak{J} \mid \eta \rangle = \int_{\Omega} d\mu \langle \eta \mid X_1 \wedge \ldots \wedge X_D \rangle$$

◇

Then $\mathfrak{J}$ is closed and is positive (i.e. $\langle \mathfrak{J} \mid \eta \rangle \geq 0$ if η is positive on each G-orbit) and this is a characterization of the Ruelle-Sullivan current [30]. To say that $\mathfrak{J}$ is closed means that it is zero on exact forms $\eta = d\varphi$ (φ a (D-1)-differential form) and this is equivalent to say that μ is invariant by G. Therefore $\mathfrak{J}$ defines a linear map denoted by $[\mathfrak{J}]$, from the D-cohomology group $H^D(\Omega,\mathbf{R})$ into $\mathbf{R}$.

Now the main result by A. Connes [30] in our context is the following :

Gap labelling theorem 2 : Let Ω,G as before. Then the countable subgroup of $\mathbf{R}$ given by the image under the trace homomorphism of the K_0-group of $C^*(\Omega,G)$ is equal to the image of the integer D-cohomology group $H^D(\Omega,\mathbf{Z})$ by the homology class $[\mathfrak{J}]$ of the Ruelle-Sullivan current.

◇

In practice this means that we must compute a set of D independent D-cycles in Ω, generating the homology group $H^D(\Omega,\mathbf{R})$. Then, by duality, we must exhibit a set of D linearly independent closed and non cohomologous differential forms of degree D, the integrals of which on the previous cycles being integers. Then we evaluate $\mathfrak{J}$ on this last set of form using (8). The subgroup of $\mathbf{R}$ generated by the numbers that we obtain in this way gives the answer.

Let us consider the special case where Ω is a torus of dimension $\nu > D$, and $\mathbf{R}^D$ acts on it via constant vector fields $\alpha_1,\ldots,\alpha_D$. Let α be the $\nu \times D$ matrix the columns of which being given by the coordinates of the α_i's. To get a free action, this matrix must be "irrational", namely there is no non zero vector m in $\mathbf{R}^D$ with integer coordinates, such that αm be a vector with integer coordinates. Then the dynamical system is an abelian virtual group, and it corresponds to the hull of some Schrödinger operator on $\mathbf{R}^D$ with a quasi periodic potential. We know explicitly a set of generating D-forms on $\mathbf{T}^\nu$, namely $d\omega_{i_1} \wedge \ldots \wedge d\omega_{i_D}$ where the $d\omega_i$'s are the differential of the coordinate functions. There is also a unique invariant measure in this case, namely the Haar measure $d\omega_1 \wedge \ldots \wedge d\omega_\nu$ on the ν-torus. The evaluation of (8) becomes elementary and this gives :

Corollary 1 [19] : In the quasi periodic previous case, the integrated density of states of a self adjoint operator H affiliated to $C^*(\mathbf{T}^\nu, \mathbf{R}^D)$ takes on positive values on the gap of H belonging to

(9) $$L = \Sigma_{(i)} \mathbf{Z}\alpha^{(i)}$$

where the $\alpha^{(i)}$'s are the minors of maximal rank of the matrix $\boldsymbol{\alpha}$.

◇

A special case is given by D=1. Then $\boldsymbol{\alpha}$ is a column matrix, namely a vector in $\mathbf{R}^\nu$, and the minors of maximal rank are just the coordinates of this vector. A typical potential admitting this dynamical system as a Hull, has the form :

(10) $$V(x) = v(\omega-\alpha x) = \Sigma_{m\in\mathbf{Z}^\nu} e^{2i\pi(\omega-x\alpha)m} v(m) \qquad x \in \mathbf{R}$$

where the right hand side represents the Fourier expansion of $v \in C(\mathbf{T}^\nu)$. Thus the coordinates of $\boldsymbol{\alpha}$ represents the independent frequencies of V. The "frequency module" is defined as the group generated by the elementary frequencies, and we get as a corollary [see 55] :

Corollary 2 : (i) Let $H = -d^2/dx^2 + V$ be a Schrödinger operator on $\mathbf{R}$ with a quasi periodic potential V. Then the integrated density of states on a gap takes on values in the frequency module of V.
(ii) If H is any self adjoint pseudodifferential operator on $\mathbf{R}$ with quasi-periodic coefficients, the same result holds, with the frequency module of the coefficients replacing the one of the potential.

◇

The result (i) was first obtained in 1981 by Johnson and Moser [55, see 54], who used all the properties of a second order differential operator to prove it. The other result (ii) is a consequence of Shubin's analysis, and of the gap labelling theorem 2.

The virtue of our approach is that it extends as well to the tight binding approximation :

Proposition 8 : (i) Let (Ω,G) be a dynamical system, where G is connected, and let Ξ be a transversal. Then the stabilized algebras of $C^*(\Omega,G)$ and of $C^*(\Xi)$ are isomorphic (one says that they are Morita equivalent). Consequently their K-group are isomorphic.
(ii) If $G=\mathbf{Z}^D$, there is a compact space Ω' together with a $\mathbf{R}^D$-action such that $(\Omega,\mathbf{Z}^D)$ is isomorphic to the dynamical system of a transversal of Ω'. The system $(\Omega',\mathbf{R}^D)$ is called the suspension of $(\Omega,\mathbf{Z}^D)$. ◇

The proofs of this result are scattered in the literature. The concept of Morita equivalence is quite old [see 87], and the connection with the C*algebra of a transversal is the result of works by M. Rieffel [87], and A. Connes [30]. The suspension construction was well known from the experts in dynamical systems for a long time but we can find an exposition of it in [72(remark),96]. As a consequence we get [see also 42]:

Corollary 3 [19] : (i) Let H be a self adjoint operator on $\mathbf{Z}^D$ affiliated to the C*algebra of $(\mathbf{T}^\nu,\mathbf{Z}^D)$ where $m \in \mathbf{Z}^D$ acts on $\mathbf{T}^\nu$ via the translation by αm where α is a $\nu \times D$ matrix the minors of which being rationally independent. Then the integrated density of states of H on a gap takes on values in the set of linear combinations with integer coefficients of the minors of α of any order (with the convention that 1 is a minor of zeroth order).

(ii) Let H be the discrete laplacian on $\mathbf{Z}^D$ with a quasi periodic potential V, the hull of which being the previous dynamical system. Then (i) apply to it. In particular if D=1, the integrated density of states of H on a gap takes on values in the set $\mathbf{Z} + L$ where L is the frequency module of V

◇

To finish with this section let us give few more results which were noticed in [14,17]. As we noticed in proposition 7, to get a nowhere dense spectrum we need flows without periodic orbits. This is actually not sufficient. It is also necessary that the image of the K-group by the trace be dense in $\mathbf{R}$. Let us mention one example, noticed by A. Connes, of flow for which this does not happen. A. Connes [29] gave a series of sufficient condition for a discrete dynamical system $(\Omega,\mathbf{Z})$ where Ω is a manifold, to insure that $C^*(\Omega,\mathbf{Z})$ is simple without projection. In this case, obviously, the spectrum of any self adjoint element of $C^*(\Omega,\mathbf{Z})$ has no gap.

Example 1 : let Γ be a discrete subgroup of $SL(2,\mathbf{Z})$ with a compact fundamental domain, and let Ω by the compact space $SL(2,\mathbf{R})/\Gamma$. Let φ be a minimal diffeomorphism of Ω (namely such that any orbit is dense). The horocycle flow provides such an example. Then the K_0-group for this discrete flow is equal to $\mathbf{Z}$. In particular since the flow is minimal, its C*algebra is simple [29,43], and any trace τ is faithful. Therefore there is no non trivial projection otherwise there would be a non zero projection P with $\tau(P) \in \mathbf{Z}$. Since $\mathbf{Z}$ is discrete and Ω compact, τ is normalized, and we must have $0 \leq \tau(P) \leq 1$. Thus $\tau(P) = 1$ and $\tau(\mathbf{1}-P) = 0$ which implies $P=\mathbf{1}$. Let now v be a continuous function on Ω, and H be the operator on $l^2(\mathbf{Z})$ given by :

$$H\psi(n) = \psi(n+1) + \psi(n-1) + v(\varphi^{-n}\omega)\,\psi(n) \tag{11}$$

Then H has no gap in its spectrum. ◇

The next example [17] goes just in the opposite direction:

Example 2: let us consider the Schrödinger operator on $l^2(\mathbf{Z})$:

$$H\psi(n) = \psi(n+1) + \psi(n-1) + \lambda\chi_{(0,\beta]}(x-n\alpha)\,\psi(n) \tag{12}$$

where $1,\alpha,\beta$ are rationally independent. It was shown in [17] that the gap labelling theorem was different from what was obtained from the Pimsner Voiculescu theorem. The reason is that in the corresponding C*algebra of the Hull, there is a new projection, namely $\chi_{(0,\beta]}$ the trace of which being β. For this reason the K-group is strictly wider and its image by the trace contains $\mathbf{Z} + \mathbf{Z}\alpha + \mathbf{Z}\beta$. The same phenomena would appear if instead of taking the irrational rotation by α we chose a Denjoy diffeomorphism of the circle [M14,84].

Let us note the last problem :

Question 4 : Thank to the gap labelling theorem, a necessary condition for a self adjoint operator H, affiliated to $C^*(\Omega,G)$ to have a Cantor spectrum is that the image of the K-group by the trace homomorphism is dense in the real line. Another necessary condition is that Ω be free of periodic orbits. Let us suppose that these two conditions hold. Has a generic self adjoint element of $C^*(\Omega,G)$ a Cantor spectrum ?

6)- The case of flows : Johnson's approach.

When investigating the properties of one dimensional Schrödinger operators with random potential, R. Johnson [56] gave another version of the gap labelling theorem, using the notion of rotation number already described by R. Johnson & J. Moser [55], and by M. Herman [52]. He actually found an explicit formula which turned out to be nothing but the Connes formula given in §5 eq.8. However, being more precise, this approach allows us to understand the origin of the "Ruelle-Sullivan" current.

Let us consider a compact metric space Ω, together with a continuous flow $t\in\mathbb{R}\to T_t\omega$, and let μ be an ergodic invariant measure on it. Let now $\omega\to M(\omega)$ be a continuous map on Ω with values in the set of 2x2 matrices with real entries. We now consider the ordinary differential equation :

$$(1) \qquad dX/dt = M(T_{-t}\omega)X(t) \qquad X(t)\in\mathbb{R}^2$$

Without loss of generality we may assume that $M(\omega)$ is traceless for all ω. The solution of (1) can always be written as :

$$(2) \qquad X(t) = \Phi(\omega,t)\,X(0)$$

where $\Phi(\omega,t)$ is a 2x2 real matrix with determinant one satisfying the cocycle properties :

$$(3) \qquad \Phi(\omega,0) = \mathbf{1} \qquad \Phi(\omega,t+s) = \Phi(T_{-s}\omega,t)\,\Phi(\omega,s)$$

We will say that (1) admits an **exponential dichotomy** [89,90] if the trivial fiber bundle $\Omega\times\mathbb{R}^2$ splits into a direct sum of non zero subbundles $E^+\oplus E^-$ such that :

(i) if $(\omega,X)\in E^\pm$, then $(T_{-t}\omega, \Phi(\omega,t)X)\in E^\pm$ for all $t\in\mathbb{R}$ (invariance of $E^\pm$)

(ii) there are positive constant K and r independent of (ω,t) such that if $(\omega,X)\in E^\pm$ then $\|\Phi(\omega,t)X\| \leq K e^{\pm(-rt)}$

An example of such flow is provided by a one dimensional Schrödinger equation (where V is a continuous function on Ω) :

$$(4) \qquad -d^2\psi/dt^2 + V(T_{-t}\omega)\psi(t) = E\,\psi(t)$$

where we set :

$$(5) \qquad X(t) = \begin{bmatrix} X_1(t) = \psi(t) \\ X_2(t) = \psi'(t)/\sqrt{E} \end{bmatrix} \qquad M(\omega) = \begin{vmatrix} 0 & \sqrt{E} \\ -\sqrt{E}+V(\omega)/\sqrt{E} & 0 \end{vmatrix}$$

It admits an exponential dichotomy provided E does not belong to the spectrum of the self adjoint operator described formally by (4) on $L^2(\mathbf{R})$ [56,90]. More precisely, for each $\omega\in\Omega$ there is a solution $u^+(\omega,t)$ (resp. $u^-(\omega,t)$) of (4) unique up to a multiplicative constant, decreasing exponentially at $+\infty$ (resp. $-\infty$) together with its first and its second derivative.

Let $X^\pm(\omega,t)$ be solutions of (1) in $E^\pm$, written in the form :

$$(6) \qquad X^\pm(\omega,t) = r^\pm(\omega,t)\, \mathbf{e}(\theta^\pm(\omega,t))$$

where $\mathbf{e}(\theta)$ is the unit vector of components $\cos\theta$ and $\sin\theta$. We remark that the line passing through $X^\pm$ is uniquely determined by $\theta^\pm$ (modulo π). Thanks to the uniqueness of $X^\pm$, it is easy to see that the angle $\theta^\pm(\omega,t)$ is actually a function of $T_{-t}\omega$ only, which we will denote by $\theta^\pm$ again.

The rotation number of this solution is defined as the amount of angle per unit length namely :

$$(7) \qquad \rho^\pm = \lim_{t\to\pm\infty} \{\theta^\pm(T_{-t}\omega) - \theta^\pm(\omega)\}/t$$

Let us assume that Ω is a manifold and that the flow T on it is defined through the vector field ξ. By using the Birkhoff ergodic theorem, and provided $\theta^\pm$ is differentiable in ω, we get, for μ-almost every ω_0 :

$$(8)\ \rho^\pm = \lim_{t\to\pm\infty} (1/t)\int_0^t ds\ \partial\theta^\pm(T_{-s}\omega_0)/\partial s$$
$$= \int d\mu(\omega)\, \langle d\theta^\pm|\xi\rangle(\omega) = \langle \mathfrak{J}\ |\ d\theta^\pm \rangle$$

where $\mathfrak{J}$ is precisely the Ruelle-Sullivan current for the dynamical system. Let us remark that $d\theta^\pm$ is closed but not exact in general for $\theta^\pm$ is defined modulo π, and in particular, its integral on any loop (or 1-cycle) in Ω is an integer multiple of π. Therefore the cohomology class $[d\theta^\pm/\pi]$ belongs to $H^1(\Omega,\mathbf{Z})$. Thus we get :

<u>Theorem 2</u> (R. Johnson) : (i) The rotation numbers of the flow (1) satisfies $\rho^+=\rho^-$, and $\rho^\pm/\pi$ belongs to the image of the integer cohomology group $H^1(\Omega,\mathbf{Z})$ by the homology class of the Ruelle-Sullivan current. Moreover it is a homotopy invariant.

(ii) If the flow (1) comes from the Schrödinger equation (4), where E belongs to a gap of the spectrum, then the rotation number $\rho^\pm/\pi$ is equal to the integrated density of states and satisfies:

$$(9) \qquad \mathcal{N}(E) = \rho^\pm/\pi = \sqrt{E}/\pi \int d\mu(\omega)\, \{1 + \cos^2(\theta^\pm(\omega))\, V(\omega)/E\}$$

◇

Sketch of the proof : The first part is a consequence of the previous reasoning. The second is an application of the Sturm theorem : the number of zeroes of a solution of (4) in a finite interval [-L,L] with some boundary conditions and the number of the corresponding eigenvalues smaller than or equal to E, differ at most by 2. On the other hand the number of zeroes of $\psi^{\pm}$ in [-L,L] differs from $\{\theta^{\pm}(T_{-L}\omega)-\theta^{\pm}(T_L\omega)\}/\pi$ by at most 1. Dividing by 2L and letting L go to infinity gives the identity between the rotation number divided by π and the integrated density of states. From (5) we get :

$$(10) \qquad d\theta^{\pm}/dt = \sqrt{E}\,\{1 + \cos^2(\theta^{\pm})\ V(T_{-t}\omega)/E\} = \langle d\theta^{\pm}|\xi\rangle(T_{-t}\omega)$$

which gives the last formula. ◇

The previous analysis for flows extends to the case of discrete maps, namely :

$$(11) \qquad X(n+1) = M(T^{-n}\omega)\,X(n) \qquad n\in\mathbf{N}$$

The hypothesis will be now the following: M is a 2x2 real matrix valued continuous function with determinant one which is homotopic to the identity matrix. We then replace the discrete flow (Ω,T) by its suspension $(S\Omega,ST)$ where $S\Omega$ is the quotient of $\Omega\mathrm{x}\mathbf{R}$ by the equivalence relation $(\omega,t)\approx(T\omega,t-1)$, and ST is the quotient of the flow $(\omega,s)\to(\omega,t+s)$. In much the same way, we replace the discrete flow $(\omega,X)\to(T^{-1}\omega,M(\omega)X)$ on the trivial bundle $\Omega\mathrm{x}\mathbf{R}^2$ by its suspension. Since M is homotopic ti the identity, $S(\Omega\mathrm{x}\mathbf{R}^2)$ is homeomorphic to $S\Omega\mathrm{x}\mathbf{R}^2$. If (Ω,T) admits an exponential dichotomy, the suspension admits also an exponential dichotomy. Thus the previous analysis goes through. However we can now define the rotation number directly from (11) as was proposed by M. Herman [52], namely : any matrix in $SL(2,\mathbf{R})$ defines a diffeomorphism on the projective space $P^1(\mathbf{R})$, the set of lines in $\mathbf{R}^2$ which can be identified with the unit circle or with the torus $\mathbf{T}=\mathbf{R}/\mathbf{Z}$. Thus we get a mapping $(\omega,\theta)\in\Omega\mathrm{x}\mathbf{T}\to f(\omega,\theta)\in\mathbf{T}$ which is continuous and such that $f_\omega:\theta\in\mathbf{T}\to f(\omega,\theta)\in\mathbf{T}$ is a diffeomorphism of $\mathbf{T}$. Let us denote by f again the lifting of f on $\Omega\mathrm{x}\mathbf{R}$. It satisfies : $f(\omega,x+1) = f(\omega,x)+1$ $(x\in\mathbf{R})$. The rotation number is then defined as :

$$(12) \qquad \rho = \pi \lim_{n\to\infty} \{f_{T^{-n+1}\omega}\circ\ldots\circ f_\omega(x) - x\}/n$$

That this limit exists form μ-almost every ω and uniformly in x was one of the results of [52]. It is then easy to see that this rotation number is the same as the one defined through the suspension.

Again also, this approach can be used to investigate a one dimensional discrete Schrödinger operator of the form :

(13) $$H\psi(n) = t(T^{-n-1}\omega)\psi(n+1) + t(T^{-n}\omega)\psi(n-1) + v(T^{-n}\omega)\,\psi(n) = E\,\psi(n)$$

where v,t are continuous real functions on Ω and t is positive (actually it is sufficient that t be complex and non vanishing). Then the equation $H\psi = E\psi$ is equivalent to (11) with :

(14) $$M(\omega) = \begin{vmatrix} (E-v(\omega))/t(T^{-1}\omega) & t(\omega)/t(T^{-1}\omega) \\ 1 & 0 \end{vmatrix}$$

In this case also the density of states coincides with ρ/π. The results in examples are therefore identical to those given in the previous chapter.

Let us finish this section by describing an application of this framework to some problem in classical mechanics.

Let F be a monotone twist mapping of the annulus $A=\mathbf{T}\times[0 1]$, namely an homeomorphism which preserves the ends of A, $\mathbf{T}\times\{0\}$ and $\mathbf{T}\times\{1\}$, which is area preserving and such that if $y < y'$ then $F(x,y) < F(x,y')$. The restrictions of F to the ends define two diffeomorphism of the circle the rotation numbers of which will be called $\rho^{(0)}$ and $\rho^{(1)}$. It follows that $\rho^{(0)} < \rho^{(1)}$. One example is given by the "standard map" :

(15) $$F(x,y) = (x+y+k\sin 2\pi x,\ y+k\sin 2\pi x)$$

The Aubry-Mather theorem [7,8,57,58,73] shows that for any irrational ρ in the interval $(\rho^{(0)},\rho^{(1)})$, there is a closed invariant subset $M(\rho)$ such that the restriction of F on it is conjugate through a homeomorphism, to the restriction of a diffeomorphism f of the circle of rotation number ρ to its (unique) minimal invariant set. For indeed [see M14], any diffeomorphism f of $\mathbf{T}$ has a unique minimal invariant set which is the full circle if it is smooth enough (depending on ρ), whereas it is a Cantor subset of $\mathbf{T}$ otherwise, as was firstly shown by Denjoy [37]. Thus $M(\rho)$ is homeomorphic either to a circle or to a Denjoy set. If ρ is rational the situation is more involved [see 73].

Let F denote again the lifting of F on $B=\mathbf{R}\times[0,1]$. Since F is area preserving, using the monotone twist property it can be shown [73] that there is a C^1 periodic function $h(x,x')$ on $\mathbf{R}^2$, called the generating function of F, such that $y'=\partial h/\partial x'(x,x')$ and $y=-\partial h/\partial x(x,x')$. If we set $(x_n,y_n) = F^n(x,y)$ the sequence $\{x_n\}$ satisfies the following non linear equation :

(16) $$\partial h/\partial x'(x_{n-1},x_n) + \partial h/\partial x(x_n,x_{n+1}) = 0$$

which gives for the standard map :

(17) $$x_{n+1} + x_{n-1} - 2x_n + k\sin 2\pi x_n = 0$$

If now $(x,y) \in M(\rho)$ the Aubry-Mather theorem means that there are f,g the lifting of two homeomorphisms of $\mathbf{T}$, and $\xi\in\Sigma$, the minimal set of f, such that $x_n = g(f^n(\xi))$ and that the rotation number of f is ρ.

The linear stability of $M(\rho)$ is described through the linear equation governing the evolution of an infinitesimal change $\delta x_n = \psi(n)$ in x_n, namely, if h is in C^2, through the equation (13) with $\Omega = \Sigma$, $T = f$, $E=0$ and :

$$(18) \qquad t(\xi) = \partial^2 h/\partial x \partial x'(g\circ f^{-1}(\xi), g(\xi))$$

$$v(\xi) = \partial^2 h/\partial x'^2(g\circ f^{-1}(\xi), g(\xi)) + \partial^2 h/\partial x^2(g(\xi), g\circ f(\xi))$$

The rotation number of this discrete system is called here the amount of rotations. It has been studied by J. Mather [74] in the special case where ρ is a rational number and $\{x_n\}$ is periodic, where it is connected to the Morse index. He found a result which can be interpreted in term of K-theory. Let us give the result in the case where now ρ is irrational.

If E=0 does not belong to the spectrum of H, a property which is likely to be generic if one believes that the spectrum of H is a Cantor set, then the amount of rotation is nothing but the integrated density of states (if one normalizes the angle to 1 instead of π). Therefore the gap labelling theorem 1 shows that it belongs to the image by the (unique) trace on $C^*(\Sigma,f)$ of the K_0 group of this algebra. The uniqueness of the trace comes from the uniqueness of an f-invariant measure on $\mathbf{T}$ [M14]. The K-group of this algebra has been computed by I. Putnam, K. Schmidt & C. Skau [84]. If μ is the unique f-invariant probability measure on $\mathbf{T}$, μ is actually supported by Σ. If Σ is a Cantor set then $\mathbf{T}\backslash\Sigma$ is the union of at most countably many intervals which are the "gaps" of Σ. Then Q(f) is the set of values of $g(\xi) = \int_0^\xi d\mu(\eta)$ when ξ varies through the gaps of Σ. Clearly Q(f) is countable, and it is known that if $R(\rho)$ denotes the rotation by ρ, then $g\circ f = R(\rho)\circ g$, showing that Q(f) is also $R(\rho)$ invariant in $\mathbf{T}$ identified with [0,1). Therefore there is a family of real numbers $\{y(i);\ 1 \le i \le n(f)\}$ (where n(f) may be infinite), such that Q(f) equals the union of the $\{y(i)+\mathbf{Z}\rho\}$'s. Actually the y's are uniquely defined if we assume that the differences $y(i)-y(j)$ are rationally independent of ρ. Moreover, Q(f) is defined up to a solid rotation. Therefore we may assume $y(1)=0$. Then the result of the analysis of [84] is the following :

<u>Proposition 9</u> : The C*-algebra $C^*(\Sigma,f)$ is simple and has a unique trace. The trace of any projection in it belongs to the set :

$$(19) \qquad \{\mathbf{Z} + \mathbf{Z}\rho + \mathbf{Z}y(2) + \dots + \mathbf{Z}y(n(f))\} \cap [0,1]$$

If 0 does not belong to the spectrum of the linearisation H(F), the amount of rotation of the map F on the Cantorus $M(\rho)$ belongs also to this set. ◇

7)- The quantum Hall effect :

The Hall effect was discovered and discussed 1879 by E.H. Hall [49], and has been one of the most striking and useful phenomena in solid state physics for a century, since it was one of the first effect giving informations on the microscopical properties of metals. For instance it gives the sign of the charge of the particles carrying the current. This was the way one discovered that the electric current can be carried either by electrons or by holes depending on the material used. It gives also a method to measure the velocity of the carriers [21] and the charge carriers density. On the other hand, it is nowadays commonly used in several kind of devices in electronic, computer, electronic typewriters, automobile engines, etc.

In 1980, a century after the discovery of this effect, a new breakthrough was performed with the discovery [62] of its quantum counterpart which leads to such a precise experiment that the standard for the Ohm, the unit of resistance, is to be defined from it quite soon. Undoubtely, there will be other applications of this effect when the technology will be controlled in order to produce cheap devices.

Let us briefly recall the nature of the classical Hall effect, in order to go step by step toward the understanding of the quantum one. If a thin strip of a metal is submitted to a uniform magnetic field perpendicular to it, one observes a potential difference in the direction perpendicular to the electric current and the magnetic field (see fig. 2). Measuring this voltage leads to the value of the Hall resistance given by the ratio of this Hall voltage and the intensity of the current. To compute it, let us assume that the permanent regime is established. Two forces act on the unit volume of charged particle.

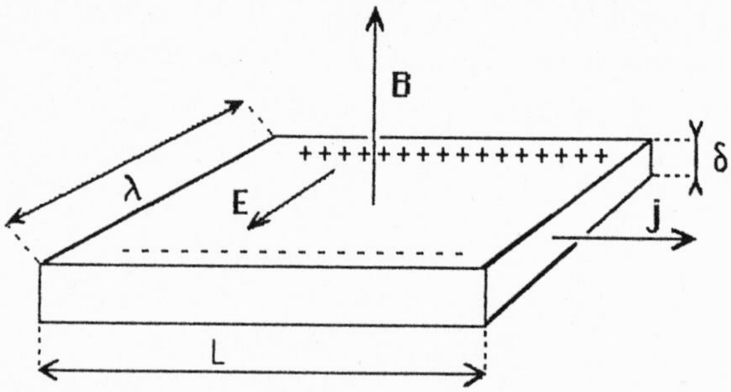

Fig. 2 : The classical Hall effect. The sample must be a thin strip of pure metal. The magnetic field B is perpendicular to the strip. The electric current j is stationary and oriented along the strip in the direction of the longest side. The magnetic forces push the charges as indicated. In the stationary state, they create an electric field E perpendicular to the directions of the magnetic field and the current. One measures the Hall voltage in the E-direction.

The first one is the Lorentz force, perpendicular to the magnetic field and to the current density. Its intensity is given by :

$$(1) \qquad F_{magn.} = jB/c$$

Here c is the speed of light, j is the current density namely the ratio $I/\lambda\delta$, where I is the intensity of the current, δ the thickness and λ the width of the strip. B denotes the magnitude of the magnetic field in the direction perpendicular to the strip. This force is oriented in the direction opposite to the electric field (see fig.2). The other force is created by the charges accumulated on the edges of the strip. They create an electric field E and the force is given by :

$$(2) \qquad F_{elec.} = nqE = nqV_H/\lambda$$

where n is the density of charge carriers, q the electric charge of them, and V_H is the Hall voltage. If we equal the two forces, we get the Hall resistance and the Hall resistivity :

$$(3) \qquad \text{Hall resistance} \quad R_H = V_H/I = B/(n\delta qc)$$

$$\text{Hall resistivity} \quad \rho_H = E/j = B/(nqc)$$

We see that a high resistance is obtained when the strip is very thin, and the carrier density is low enough. Moreover the sign of the Hall voltage is determined by the sign of the charges carrying the current. They are negative for the gold or the copper, but positive for the iron. This was noted by Hall in 1880. Another consequence is that the speed v of the carriers is obtained from the remark that j = nqv which implies v = cE/B (L. Boltzmann 1880 [21]). The only characteristic of the sample entering in this formula is the charge carrier density. This explain why this effect is so universally used.

Let us also remark that in two dimensions, namely in the limit where δ tends to zero, we do not define the resistivity nor the current density in the same way in order to take into account the divergence when $\delta \to 0$. In particular one immediately sees that the two dimensional resistivity is also measured in Ohm whereas the three dimensional one is measured in Ohm x m . Therefore a two dimensional measurement of the Hall resistivity will give a measurement of the Ohm standard without any reference to any unit of length, an advantage only if the accuracy of the Hall measurement is high enough. This was accomplished a century after Hall, with the experiment of Von Klitzing, Pepper and Dorda [62] leading to what is called nowadays the Quantum Hall effect (see below).

The next step is the L.D. Landau theory [68] developed in 1930. In a pure metal, like the copper or the gold, the charge carriers can be considered as free independent particles. They constitute thermodynamically a perfect Fermi gas, and therefore at relatively low temperature, their energy lies below the Fermi level. The motion of each particle is governed by the Schrödinger equation :

(3) $$H\psi(\mathbf{x}) = \{-i\hbar\partial/\partial\mathbf{x} - q\mathbf{A}/c\}^2/2m\ \psi = E\psi(\mathbf{x})$$

where **A** are the three components of the vector potential created by the magnetic field, $q = \pm e$ the carrier charge, and m their effective mass [M20]. In the two dimensional approximation, the motion in the direction perpendicular to the strip is frozen out. On the other hand if the sample is large enough, one can consider it as infinitely extended, and it is therefore identified with $\mathbf{R}^2$. The previous operator can be written as :

(4) $$H = \{K_1^2 + K_2^2\}/2m \qquad K_i = \{-i\hbar\partial/\partial x_i - qA_i/c\} \qquad i=1,2,$$

We remark that the Y's satisfy the following canonical commutation relations :

(5) $$[K_1, K_2] = i\hbar qB/c\ \mathbf{1}$$

Therefore H appears as the hamiltonian for a harmonic oscillator, and the energy spectrum is now :

(6) $$E_n = \hbar\omega_c(n+1/2) \qquad n = 0,1,2,\dots \qquad \omega_c = |q|B/mc$$

ω_c is called the cyclotronic frequency and corresponds to the rotation frequency of the classical motion of a particle in the magnetic field B. When B->0 the spectrum becomes continuous.

The current is now given by the vector valued operator :

(7) $$\mathbf{j} = q\,\mathbf{v} = (iq/\hbar)\,[H, \mathbf{x}] = q\,\mathbf{K}/m$$

Let us assume now that at time zero, an electric field **E** is turned on. Since we consider the infinite volume limit, the two directions in the plane of the sample are equivalent. Let us choose the 1-axis parallel to **E**, and let ε be the modulus of **E**. The evolution of **j** after t=0, is now governed by the modified hamiltonian $H(\varepsilon) = H + q\varepsilon\, x_1$ namely :

(8) $$\mathbf{j}(t) = e^{iH(\varepsilon)t/\hbar}\ \mathbf{j}\ e^{-iH(\varepsilon)t/\hbar}$$

Using the Fermi-Dirac statistics, its thermodynamical average at inverse temperature $\beta=(kT)^{-1}$ is then given by :

(9) $$\langle\mathbf{j}(t)\rangle_\beta = \lim_{\Lambda\to\mathbf{R}^2} |\Lambda|^{-1}\ \mathrm{Tr}(\chi_\Lambda\{\mathbf{1} + e^{\beta(H-E_F)}\}^{-1}\mathbf{j}(t))$$

where χ_Λ is the indicator function of the finite box Λ.

We remark that at zero temperature, each Landau level corresponds to a density of states equal to the quantity B/Φ, where $\Phi = hc/e$ is a quantum of magnetic flux. This can be seen by replacing **j** by **1** and the Fermi distribution by the eigenprojection on the given level in the right hand side of (9). If $(n\delta)$ is the charge carriers density per unit area, we see that the number of level which can be filled is $\nu = (n\delta)\Phi/B$, a quantity called the **band filling**.

It turns out that the right hand side of (9) can be computed exactly: it is made of a time periodic function of t at the cyclotronic frequency. The constant part is parallel to the 2-axis, and its amplitude is :

$$(10) \qquad |\mathbf{j}_{av}| = \lim_{\tau \to \infty} |\tau^{-1} \int_0^\tau dt \, \langle \mathbf{j}(t) \rangle| = (e^2/h)\varepsilon\, \nu(\beta)$$

$$\nu(\beta) = \sum_{n\geq 0} \{1 + e^{\beta(\hbar\omega_c(n+1/2) - E_F)}\}^{-1}$$

The resistivity is now a 2x2 matrix such that $\mathbf{E} = \boldsymbol{\rho}\mathbf{j}$ and $\boldsymbol{\rho}^{-1} = \boldsymbol{\sigma}$ is the conductivity. From the previous calculation, it follows that both matrices are antidiagonal, which means that :

$$(11) \quad \boldsymbol{\rho} = \begin{vmatrix} 0 & \rho \\ \rho & 0 \end{vmatrix} \qquad \boldsymbol{\sigma} = \begin{vmatrix} \sigma_{xx} = 0 & \sigma_{xy} = \sigma \\ \sigma_{xy} = \sigma & \sigma_{yy} = 0 \end{vmatrix} \qquad \sigma = e^2/h . \nu(\beta)$$

In particular, this free system is at the same time a perfect conductor and a perfect insulator in the electric field direction ! In absence of a magnetic field the situation would be just the opposite. The diagonal coefficients of $\boldsymbol{\rho}$ which in this case vanish, are called **magnetoresistivity**.

Actually a real sample is never strictly two dimensional. The motion in the magnetic field direction is governed by the kinetic part $p_3^2/2m$ with a zero boundary condition on the upper and the lower sides of the strip. Since this part commutes with H, it simply adds a discrete set of eigenvalues to each Landau level. They exhibit therefore a slight broadening but do not change the conclusion.

On the formulæ (10) & (11) one sees that the factor $\nu(\beta)$ admits two limiting behaviors. On the one hand at reasonably high temperature (i.e. for β small), the discrete sum in (10) can be replaced by an integral if $\beta\hbar\omega_c$ is small. However βE_F is generally quite big. In the limit where $\beta\hbar\omega_c \to 0$ and $\beta E_F \to \infty$ we get the classical result $\sigma = e^2/h\,\nu = (n\delta)ec/B$, whereas in the limit of low temperature (i.e. $\beta \to \infty$), $\nu(\beta)$ converges to a step function taking on integer values (see fig.3 below).

There are two kind of experimental devices [M2] which are used in the measurement of the Quantum Hall effect. The first one is the metal-oxyde-semiconductor field effect transistor (MOSFET) the other one is a

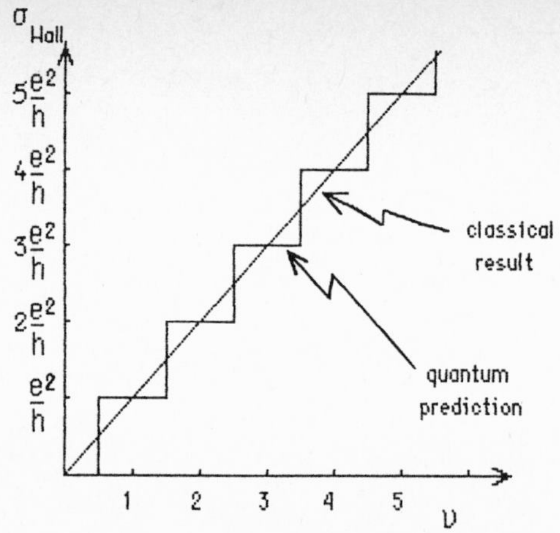

Fig. 3 : The classical Hall conductivity (high temperature) and the quantum one (at zero temperature) for a two dimensional free Fermi gas as a function of the band filling $\nu=(n\delta)hc/eB = (n\delta)\Phi/B$. Experimentally ν can be varied either by changing the magnetic field B or by changing the charge carriers density n.

heterojunction (e.g. Al_xGa_{1-x}-As-AsGa or InP-$In_xGa_{1-x}As$). In these devices one creates a thin layer, called an inversion layer, at the interface between two components, where electrons are confined and have a two dimensional motion. In a heterojunction, the principle is a little bit different [M2,97] but leads also to the existence of a thin layer of electrons with a low density. However they are practically better to use nowadays for the interface is free of impurities, and the mobility of the electrons in the inversion layer is usually much higher. Thus quantum effects are likely to appear at higher temperature.

The previous theory of the free Fermi gas must be improved in practice to take into account the influence of the crystalline structure of the substrate on the electrons. In particular, the effective potential will give rise to a broadening of the Landau levels. There is no reason that the previous result, namely the quantization of the Hall conductivity at low temperature may survive. On the other hand the direct conductivity σ_{xx} does not vanish usually, and exhibits some oscillations when varying the magnetic field or the charge density carrier : this is the de Haas-Shubnikov effect [M20]. However numerical calculations, based upon approximate theories, taking into account the disorder, and performed during the seventies [M2,2] predicted that such a quantization may survive. The question was to know with what accuracy.

In 1975, the japanese group of S. Kawaji [M2] had already observed some deviation from the classical law (11) at low temperature. The samples used at this time improved rapidly afterward, and in 1980 two groups S. Kawaji and K. Wakabayashi [59,60] in Japan and K. von Klitzing, G. Dorda and M. Pepper [62] from the Max-Planck Institut at Grenoble observed that the Hall resistance admits some plateaux when varying the band filling. The von Klitzing group also observed that these plateaux corresponded to integer multiples of e^2/h for the Hall conductivity, with an accuracy better than 10^{-5} (see fig. 5 below)! Since this time the experiment has been performed by several groups, and the accuracy is better than 10^{-7}! Since e^2/h is a physical

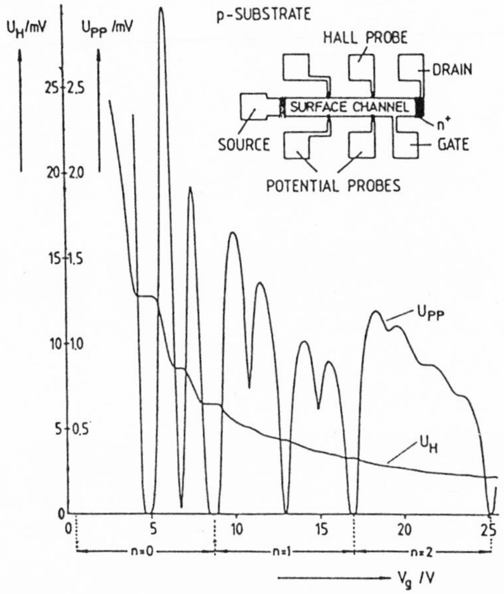

Fig. 4 : Recording of the Hall voltage, U_H, and the voltage drop between the potential probes, U_{pp}, as a function of the gate voltage at T=1.5 K. The magnetic field is 18T. The oscillation up to the Landau level n=2 is shown. The Hall voltage and U_{pp} are proportional to ρ_{xy} and ρ_{xx} respectively ($\rho=\sigma^{-1}$ see eq. 11). The inset shows a top view of the device with a length of L=400μm, a width of W=50μm, and a distance between the potential probes of L_{pp}=130μm. (Taken from ref. [62])

constant related to the fine structure constant, the Hall effect provides an independent way of checking the validity of the quantum electrodynamics. As explained already, it also provides a measurement of a universal unit of resistance ($R_H = e^2/h \approx 6453{,}2$ ohms) and a new Ohm standard.

The main surprise was not the Hall effect itself but the extremely high accuracy of the result. The idea was that it may have a very deep and universal origin. The first explanation was given by Laughlin [69] who related this quantization phenomena to the gauge invariance of the hamiltonian describing the electron. More precisely, he considered a sample having the form of a ring (see fig. 5 below) of perimeter L and width δ. He assumed that a magnetic field of constant modulus was crossing the loop perpendicularly to it. The Hall current is then given by the adiabatic derivative of the total electronic energy U of the system with respect to the magnetic flux through

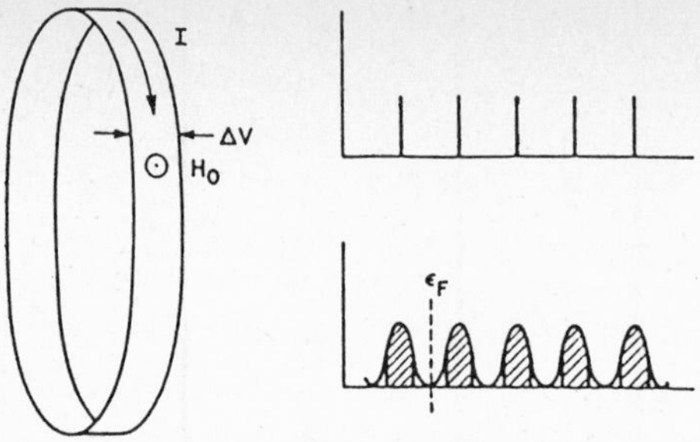

Fig.5 : Left : diagram of metallic loop. Right : density of states without (top) and with (bottom) disorder. Region of delocalized states are shaded. The dashed line indicates the Fermi level (Taken from ref. [69])

loop. This may be obtained through a "Gedanken" experiment, by adding a flux φ in the middle of the loop, and increasing it slowly in such a way that the system be constantly in equilibrium. If the additional magnetic field $\delta\mathbf{B}$ created by the flux φ, vanishes on the loop, we can describe it with a uniform vector potential $\delta A = \varphi/L$ pointing around the loop. Let x be the coordinate around the loop and y the coordinate transverse to it. We also assume that a transverse electric field $\mathbf{E}$ exists due to the Hall effect. The one-electron hamiltonian is given in the Landau gauge by :

$$(12) \qquad H(\delta A) = [(p_x - q\delta A/c + qBy/c)^2 + p_y{}^2]/2m^* + W(x,y) + e\mathbf{E}y$$

with periodic boundary condition with respect to x and Dirichlet boundary conditions at y=0 and y=δ. We can change this operator through a gauge transformation multiplying the wave functions by the phase factor $e^{i2\pi\alpha x/L}$ where $\alpha = \varphi/\Phi$ is the ratio between the additional flux and the quantum of flux. If α is not an integer, the new wave functions have a discontinuity at x=0. This is a unitary transformation which suppresses the δA dependence in the first term of (12) but modify the domain of H(δA). The advantage of this representation is that in this frame, H(δA) becomes a period one function of the parameter α.

If we now consider the situation where the potential W can be neglected, the one-electron wave function has the form :

$$(13) \qquad \psi(x,y) = e^{i2\pi m x/L}\ h(y-a)\ e^{-[(y-a)/r]^2}$$

where h is some Hermite polynomial properly normalized. The corresponding energy is linear in a. Changing α from zero to one, namely changing the flux through the loop by one quantum unit, has the effect of changing a into $a - \Delta\delta A/B$. Since this transformation maps the system back to itself, the energy increase due to it results from a net transfer of n electrons from one edge to the other [69]. The energy increase is then given by neV where V is the

Hall voltage between the edges. On the other hand the current is given by $I = c\Delta U/\Delta\varphi = ne^2V/h$. Hence the Hall conductance (in two dimension it equals the conductivity) is an integer multiple of e^2/h.

If the system is now dirty we must assume that **the Fermi level lies in a gap of extended states**. In this case the Hamiltonian is still periodic in α and the same argument holds even though now we have no excitation of quasiparticle across the gap, and the charge is transferred only through the extended states with energy close to the Landau level as in (13). The number of electrons transferred in this way may be different but it is still an integer.

Clearly the argument must be supplemented by some more rigorous study. More recently Avron and Seiler [10] produced an argument, using [9,99], which made the Laughlin argument rigorous under the condition that the Fermi level lies in a true gap. Again in this latter work, the topology of the sample seems essential. We should be able to avoid such a constraint which does not fit with the experiment.

Several works were useful to investigate each part of the argument. In his early work, using weak-scattering calculation, Ando & Aoki [3] showed that the presence of an isolated impurity does not affect the Hall current. A similar result was obtained by Prange [83] for a δ-functions impurity, to the leading order in the drift velocity $c\mathbf{E}/B$. Later on Thouless [98], showed indeed that such an invariance holds at least for weak disorder, when a true gap occurs between Landau levels.

An important step was performed by Den Nijs, Kohmoto, Nightingale and Thouless [99] (N_2KT) who recognized that the Hall conductance for a perfect crystal is given by a homotopy invariant quantity over the two dimensional torus given by the Brillouin zone when the Fermi level lies in a gap. Avron, Seiler and Simon [9] showed that it is actually the Chern character of some fibre bundle over this torus. This torus sits in the momentum space rather than in the real space, and is independent of the topology of the sample. The very high stability of the result under perturbation is explained by the homotopy invariance of the Chern character. However, they required the magnetic field to satisfy a rationality condition, namely that the flux through one unit cell be a rational multiple of the quantum of flux. Up to now, no rigorous proof is available yet, when the Fermi level lies in a gap of extended states, and the topology of the sample is arbitrary.

8)- The quantum Hall effect and Connes's cohomology :

In this section, we intend to describe a mathematical framework using C*-algebras in order to give a local proof of the quantum Hall effect, independent of the topology of the sample. We follow the idea of the N_2KT paper to express the Hall conductivity in term of a non commutative Chern character. The main approximations are the following :

- we consider only the one electron theory.
- we work with a 2D disordered sample of infinite size.

In this approach however, we shall miss one point, essential for a physicist, namely, that the result will be valid only if the Fermi level belongs to a gap of the spectrum of the one body hamiltonian. We shall indicate what should happen if the Fermi level belongs only to a gap of extended states.

The Schrödinger operator for a charge carrier in the sample submitted to a uniform magnetic field perpendicular to it is given by :

$$(1) \qquad H_\omega = 1/2m[(p_1-eA_1/c)^2 + (p_2-eA_2/c)^2)] + V_\omega(x_1,x_2) = H_0 + V_\omega$$

where m is the effective mass of the charge carrier, e its electric charge, and ω represents the effect of the impurities on the potential. As previously, ω will be taken in a compact metrisable space Ω with :

$$(2) \qquad V_\omega(x_1,x_2) = v(T_{-\mathbf{x}}\omega) \qquad v \in C(\Omega) \qquad \mathbf{x} = (x_1,x_2) \in \mathbf{R}^2$$

Let us introduce the vector valued operator $\mathbf{K}=\mathbf{p}-e\mathbf{A}/c$ and for $\xi \in \mathbf{R}^2$ and f a function in the Schwartz space $S(\mathbf{R}^2)$, we set :

$$(3) \qquad W(\xi) = \exp(\mathbf{i}\xi\mathbf{K}/\hbar) \qquad W(f) = \int_{\mathbf{R}^2} d^2\xi\ f(\xi)\ W(\xi)$$

These "Weyl operators" fulfill the following commutation relations :

$$(4) \qquad W(\xi)\, W(\xi') = W(\xi+\xi')\, e^{i\pi\alpha\xi\wedge\xi'} \qquad \xi\wedge\xi' = (\xi_1\xi'_2 - \xi'_1\xi_2)$$

$$\alpha = B/\Phi = eB/hc$$

It is a well known fact that the C*-algebra generated by the W(f)'s is isomorphic to the algebra of compact operators $\mathcal{K}$ on a separable Hilbert space. Moreover the resolvent of the free part of the hamiltonian belongs to this algebra. We shall denote by $\mathcal{A}$ the C*-algebra generated by operators of the form $V_\omega W(f)$ with V as in (2) and f in the Schwartz space.

This algebra can be abstractly reconstructed as follows [see M23] : we consider the space of continuous functions with compact support on $\Omega \times \mathbf{R}^2$ endowed with the following structure of *algebra :

$$(5) \qquad ab(\omega,\xi) = \int_{R^2} d^2\xi'\ a(\omega,\xi')b(T_{-\xi'}\omega,\xi-\xi')\ e^{i\pi\alpha\xi.\xi'}$$

$$a^*(\omega,\xi) = a(T_{-\xi}\omega,-\xi)^*$$

We built $\mathcal{A}$ as in the section 2 (equ.5) with the representations π_ω acting on $L^2(\mathbf{R}^2)$:

$$(6) \qquad \pi_\omega(a)\psi(\mathbf{x}) = \int_{R^2} d^2\xi\ a(T_{-\mathbf{x}}\omega,\xi-\mathbf{x})\ e^{i\pi\alpha\xi.\mathbf{x}}\ \psi(\xi)$$

One can check that the resolvent $R_\omega(z)=(z-H_\omega)^{-1}$ of the hamiltonian H_ω has the form $\pi_\omega(r(z))$ for some $r(z)$ in $\mathcal{A}$.

If now μ is a T-invariant ergodic probability measure on Ω, one can define a trace τ_μ on $\mathcal{A}$ as before by the formula (3) of the section 4. In addition we have a differential structure [28,32,33] through the data of the following two derivations of $\mathcal{A}$:

$$(7) \qquad \delta_j a(\omega,\xi) = 2i\pi\xi_j a(\omega,\xi) \qquad j=1,2$$

or

$$\pi_\omega(\delta_j a) = 2i\pi[\pi_\omega(a),Q_j]$$

where the Q_j's are the operators of multiplication by x_j in $L^2(\mathbf{R}^2)$.

The main remark of Thouless et al. in [99] consists in writing the Hall conductivity by mean of the Kubo formula, in a way which identifies it with a Chern character. In our notation this gives the following :

<u>Proposition 11</u> : The expression of the Hall conductivity according to the Kubo formula at zero temperature is given by :

$$(8) \qquad \sigma_H = e^2/h\,(1/2i\pi)\ \tau_\mu(P_F[\delta_1 P_F,\delta_2 P_F])$$

where $\pi_\omega(P_F)$ is the eigenprojection of H_ω on energies smaller than or equal to the Fermi energy E_F. This formula is valid provided the Fermi energy lies in a gap of H_ω (μ-almost surely). ◇

Before using this result, let us comment about the way this formula is derived.

1)-The proof of it is just a matter of calculation once the Kubo formula is accepted. To compute the Kubo formula however, one should use the following steps : (i) as in the section 7 (eq. 7-10) one must compute the current at time t after turning on an electric field ε in the x_1 direction ; (ii) at zero temperature, the Fermi Dirac function reduces to the projection P_F ; (iii) one computes the time average of the statistical average of the current, and (iv) we compute the first order term as the electric field ε goes to zero. Actually, in practice one usually exchanges the time averaging and the limit $\varepsilon \to 0$. The control of this exchange has been performed in this particular case by R. Seiler [92], for a finite size sample. Since there is no thermal dissipation in this problem, the difficulties in doing this seems purely technical. One does not know how to extend the Seiler proof in our framework case.

2)-In the course of the calculation, one uses the fact that the Green function of H_ω, decreases exponentially fast at infinity when the Fermi energy lies in a gap. Following the arguments given by Prange [83], Thouless [98], and Halperin [50], it is likely that the calculation extends to the case for which the Fermi level lies in the pure point spectrum, for in this case, since the states of energy close to E_F are localized, the Green function still decreases fast enough at infinity to insure the convergence of the integrals. However, as shown by Fröhlich and Spencer [44], one must be careful with "almost sure" properties.

3)-It has been proved rigorously recently [20] that in 2D, in the framework of a tight-binding approximation, all states are localized at high disorder, and the spectrum is pure point. At lower disorder, only the states corresponding to energies in the band edges are localized.

Once we are ready to admit the previous formula, one remarks with A. Connes that the expression :

$$(9) \qquad Ch(P) = (1/2i\pi)\, \tau_\mu(P[\delta_1 P, \delta_2 P])$$

where P is a projection in $\mathcal{A}$, satisfies the following properties :

$$(10) \quad \begin{array}{llll} (i) & \text{if } P \approx Q & \text{then} & Ch(P) = Ch(Q) \\ (ii) & \text{if } PQ = QP = 0 & \text{then} & Ch(P \oplus Q) = Ch(P) + Ch(Q) \end{array}$$

This shows that Ch actually defines a mapping on the K-group of $\mathcal{A}$, and that it is a homomorphism. It is therefore sufficient to know a set of generators of the K-group in order to know what are the possible values of Ch(P). Thank to the proposition 11, the Quantum Hall effect will be established in this set up once we are able to answer yes to the following question :

<u>Question 5</u> : Let $(\Omega, T, \mathbf{R}^2, \mu)$ be a dynamical system as before, and let $\mathcal{A}(\Omega \times_T \mathbf{R}^2, \alpha)$ be the corresponding algebra constructed according to eqs. (5-6). Then is the Chern character Ch(P) of any projection in this algebra an integer ?

The answer to this question is yes in a certain number of cases. First of all, if Ω is reduced to a point the algebra is nothing but the algebra of compact operators and the answer is trivially yes. More involved is the case for which V is periodic. In this case, Ω is a two dimensional torus $\mathbf{T}^2$ and the algebra is isomorphic to the stabilized the rotation algebra $\mathcal{A}_\alpha$. The answer in this case is still yes thanks to the Pimsner & Voiculescu [82,86], Rieffel [] who computed the two generators of the K-group, and to an explicit calculation of A. Connes [30], who computed their Chern character. Note that in this previous case, the spectrum is likely to be a Cantor set [18,23,53], and the gaps are not necessarily associated with Landau levels. At last, the answer is still yes at small disorder, for in this case, H_ω is close to an operator (in the norm sense) for which the Chern character can be explicitly computed. Using the homotopy invariance of Ch(P), the value of Ch(P_F) is still an integer as far as E_F stay in a gap while turning on the disorder.

Let us note at last the two homotopy invariance results :

Proposition 12 : (i) Let us assume that the potential V in (1) depends continuously (in the norm sense) upon a parameter λ. Then the Chern character Ch($P_F(\lambda)$) is independent of λ as far as the Fermi level stay in a gap.
(ii) The Chern character Ch(P_F) is continuous in α (i.e. of the magnetic field) as far as the Fermi level stay in a gap. ◊

The first result is just the usual homotopy invariance of the class of K-theory of a projection. The second one can be shown by considering the universal algebra obtained as the union over α of the previous ones [41]. Then the trace τ_μ can be shown to be a pointwise continuous function of α.

APPENDIX

Shubin's formula for a discrete action

Let G be a discrete, countable amenable group. Let Ω be a compact space on which G acts by a family of homeomorphisms such that the map $(\omega,x) \to x^{-1}\omega$ be continuous. Let μ be a G-invariant ergodic probability measure on Ω and as in the §4, let τ_μ be the corresponding trace on $C^*(\Omega,G)$. Let H be a bounded self adjoint operator on $l^2(G)$ and ω_0 in Ω, h in the C*algebra of the dynamical system (Ω,G) such that :

(1) $$H = \pi_{\omega_0}(h)$$

For Λ a finite subset of G let χ_Λ be the projection onto $l^2(\Lambda)$ in $l^2(G)$, and let H_Λ be the restriction of H on Λ namely

(2) $$H_{\omega,\Lambda} = \chi_\Lambda \, \pi_\omega(h) \, \chi_\Lambda$$

Since Λ is finite, H_Λ is finite dimensional, and its dimension is $|\Lambda|$, the cardinality of Λ.

The density of states was defined in the §4 eq.2, and gives rise to a probability measure defined by :

(3) $$\int d\mathcal{N}_\omega(E)\, f(E) = \lim_{n\to\infty} |\Lambda_n|^{-1} \operatorname{Tr}(f(H_{\omega,\Lambda_n}))$$

where Λ_n is a Følner sequence in G. In the sequel we will drop the index n. The first result we need is :

<u>Lemma 1</u> : for μ almost every ω, and all $f \in C(\Omega)$, the limit (3) exists and is equal to :

(4) $$\int d\mathcal{N}_\omega(E)\, f(E) = \tau_\mu(f(h))$$

◊

<u>Proof</u> : For any finite Λ the right hand side of (3) defines a probability measure (Riesz's theorem) on the convex hull of the spectrum of h, for if we denote it by $\mathcal{N}_\Lambda(f)$, we get

(i) $\mathcal{N}_\Lambda(f)$ is linear in f

(ii) $\mathcal{N}_\Lambda(1) = 1$ and $\mathcal{N}_\Lambda(f) \geq 0$ if $f \geq 0$

Consequently we have

(iii) $|\mathbf{N}_\Lambda(f)| \leq \| f \|$ for all Λ

On the other hand from the definition of the trace given in §4 eq.5 we have

(5) $\tau_\mu(f(h)) = \lim_{|\Lambda| \to \infty} |\Lambda|^{-1} \mathrm{Tr}(f(H_\omega)\chi_\Lambda)$ for μ almost all ω

In much thé same way, for each Λ, the right hand side of (5) defines another probability measure on the convex hull of the spectrum of h which we denote by $\tau_\Lambda(f)$. We know from the Birkhoff theorem [M13], that the limit exists μ almost surely. The result will be achieved if we prove that the difference $\mathbf{N}_\Lambda(f) - \tau_\Lambda(f)$ converges to zero as $|\Lambda|$ goes to infinity.

Using the Stone-Weïerstrass theorem it is sufficient to prove the result when f is a monomial. Let k be a positive integer. Then dropping the index ω (for k=1 we get zero) :

(6) $|\Lambda| \{\mathbf{N}_\Lambda(X^k) - \tau_\Lambda(X^k)\} = \mathrm{Tr}\{\chi_\Lambda(H^k - (\chi_\Lambda H \chi_\Lambda)^k)\} =$

$= \Sigma_{1 \leq j \leq k-1} \mathrm{Tr}\{\chi_\Lambda H^j (1-\chi_\Lambda)(H\chi_\Lambda)^{k-j}\} \leq (k-1) \| H \|^{k-2} \mathrm{Tr}\{\chi_\Lambda H(1-\chi_\Lambda) H \chi_\Lambda\}$

Since $H = \pi(h)$ with $h \in C^*(\Omega, G)$, for any $\varepsilon > 0$, there is $h(\varepsilon)$ in $C_c(\Omega x G)$ such that $\| H - H(\varepsilon) \| < \varepsilon$, if $H(\varepsilon) = \pi(h(\varepsilon))$. Replacing H by $H(\varepsilon)$ in the right hand side produces an error $\delta = (k-1) \| H \|^k \varepsilon |\Lambda|$. On the other hand we get :

(7) $\mathrm{Tr}\{\chi_\Lambda H(\varepsilon)(1-\chi_\Lambda)H(\varepsilon)\chi_\Lambda\} = \Sigma_{x \in \Lambda} \Sigma_{y \notin \Lambda} |h(\varepsilon; x^{-1}.\omega, x^{-1}y)|^2$

Now since G is discrete and $h(\varepsilon)$ has a compact support, there is Γ finite in G such that $x^{-1}y \notin \Gamma$ implies $h(\varepsilon; x^{-1}.\omega, x^{-1}y) = 0$ for all $\omega \in \Omega$. Let C be the maximum of $h(\varepsilon; \omega, x)$ over $\Omega x G$, then :

(8) $\mathrm{Tr}\{\chi_\Lambda H(\varepsilon)(1-\chi_\Lambda)H(\varepsilon)\chi_\Lambda\} \leq C \, | \Lambda \Delta (\Lambda a) |$

Therefore, for any $\varepsilon > 0$, there is $C(\varepsilon) > 0$, such that :

(9) $\{\mathbf{N}_\Lambda(X^k) - \tau_\Lambda(X^k)\} \leq (k-1) \| H \|^k \varepsilon + C(\varepsilon)(k-1) \| H \|^{k-2} | \Lambda \Delta (\Lambda \Gamma) | / |\Lambda|$

Since the group G is amenable, Γ is finite and Λ is a member of a Følner sequence, the last term of the right hand side converges to zero as $|\Lambda| \to \infty$. Since ε is arbitrary, we achieve the result.

◇

<u>Lemma 2</u> : The function $E \rightarrow \tau_\mu(\chi(h \leq E))$ is increasing from 0 to 1, and is locally constant outside the spectrum of h. If E is a point of continuity of this function, then :

$$(10) \qquad \mathbf{N}_\omega(E) = \tau_\mu(\chi(h \leq E)) \qquad \text{for } \mu \text{ almost all } \omega$$

◇

<u>Proof</u> : The first assertion is obvious. The left hand side of (10) is defined through the weak the limit of a sequence of probability measures on the convex hull of the spectrum of h, as was proved in lemma1. The result is a consequence of a well known result on weak limits of probability measures on a a complete metric space. ◇

<u>Remark</u> : In general the function $\mathbf{N}(E)$ is not continuous. For if h is a projection in $C^*(\Omega,G)$, then clearly, $\mathbf{N}(E)$ is a step function with a discontinuity at E=1. Conversely, if it is discontinuous at E, then E is an eigenvalue of H of infinite multiplicity and the eigenprojection is actually in the C* algebra. The question is whether such projections exist. The answer depends upon the C*algebra. For an abelian virtual group $(\mathbf{T}^\nu, \mathbf{Z}^D)$ the answer is yes. This was proved firstly by M. Rieffel [86] on the irrational rotation algebra, and his argument extends in the general case. He even exhibited a projection given by a continuous function with compact support. However, there are examples of dynamical systems $(\Omega,\mathbf{Z})$ such that the corresponding C*algebra has no projection and is simple [29] (see also §5)

REFERENCES

MONOGRAPHS:

[M1] S. AGMON, Lectures on Exponential Decay of Solutions of Second-Order Elliptic Equations, Princeton University Press, (1982), Princeton

[M2] T. ANDO, A.B. FOWLER, F STERN, Electronic properties of two-dimensional systems, Rev. Mod. Phys., 54, 437-672, (1982).

[M3] M. ATIYAH, K-Theory, Benjamin, New-York Amsterdam, (1967)

[M4] H. BOHR, Almost Periodic functions, Chelsea Publishing Company, New-York, (1947)

[M5] I.P.CORNFELD, S.V.FORMIN, Ya.G.SINAI, Ergodic Theory, Grundlerhen Bd245, (1982), Springer Verlag, Berlin, Heidelberg, New York.

[M6] J. DIXMIER, Les algèbres d'opérateurs dans l'espace Hilbertien (Algèbres de Von Neumann), Gauthiers-Villars, Paris, (1969).

[M7] M. EASTHAM, The Spectral Theory of Periodic Differential Equations, Scottish Academic Press, Edimburgh, (1973).

[M8] L. FORD, Automorphic Functions, Chelsea Publishing Company, New-York, (1951).

[M9] G. FREUD, Orthogonal Polynomials, Pergamon Press, (1971).

[M10] J. GLIMM, A. JAFFE, Quantum Physics, a functional integral point of view, Springer-Verlag, Berlin Heidelberg New-York, (1981).

[M11] W. GOTTSCHALK, G.A. HEDLUND, Topological Dynamics, A.M.S. Coll. Publ., vol. 36, Providence, (1955).

[M12] F.P. GREENLEAF, Invariant means on topological groups, Van Nostrand, New-York Toronto London Melbourne, (1969).

[M13] P.R. HALMOS, Lectures on Ergodic Theory, Chelsea Publishing Company, New-York, (1956).

[M14] M.R. HERMAN, Sur la conjugaison différentiable des difféomorphismes du cercle à des rotations, Pub. Math. I.H.E.S., 49, 5-234, (1979).

[M15] "Recent developments in gauge theory", G. 't HOOFT, C. ITZYKSON, A. JAFFE, R. STORA, Eds., Plenum Press, New York, (1980)

[M16] R.A. JOHNSON, A Review of Recent Works on Almost Periodic Differential and Difference Operators, Acta Appl. Math., 1, (1983), 161-241.

[M17] Operator Algebras and Applications, Proceedings of symposia in Pure Mathematics, D. KADISON Ed., 38, Part I&II, A.M.S., Providence, Rhode Island, (1982).

[M18] T. KATO, Perturbation theory for linear operators, Die Grundlehren der Math. Wiss., vol. 132, Springer-Verlag, New-York, (1965).

[M19] G. PEDERSEN, C*-algebras and their automorphism groups", Academic Press, London New-York, (1979).

[M20] R.S.PEIERLS, Quantum Theory of Solids, Oxford Clarendon Press, (1955)

[M21] H. POINCARÉ, Les Nouvelles Méthodes de la Mécanique Céleste, vol. I,II,III, Gauthiers-Villars (1892-94-99), Reprinted by Dover, New-York (1957).

[M22] M. REED, B. SIMON, Methods of Modern Mathematical Physics, Vol.IV, Academic Press, New-York London, (1972).

[M23] J. RENAULT, A Groupoid Approach to C*-Algebras, Lecture Notes in Mathematics, 793, (1980), Springer-Verlag, Berlin Heidelberg New-York.

[M24] D. RUELLE, Statistical Mechanics. Rigorous Results, Benjamin, Amsterdam, (1969)

[M25] S. SAKAI, C*-Algebras and W*-Algebras, Ergebnisse der Mathematik und ihrer Grenzgebiete, Band 60, Springer-Verlag, Berlin Heidelberg New-York, (1971).

[M26] C. SIEGEL, J. MOSER, Lectures on Celestial Mechanics, Grundlehren, Band 187, Springer-Verlag, Berlin Heidelberg New-York, (1971).

[M27] B. SIMON, Functional integration and quantum physics, Academic Press, New-York, (1979).

[M28] J.B. SOKOLOFF, Unusual Band Structure Wave Functions and Electrical Conductance in Crystals with Incommensurate Periodic Potentials, Preprint Northeastern Univ., Boston Ma, (1984)

[M29] M. TAKESAKI, Tomita's theory of modular Hilbert algebras and its applications, Lecture Notes in Mathematics, 128 , (1970), Springer-Verlag, Berlin Heidelberg New-York.

ARTICLES :

[1] S.ALEXANDER, Some Properties of the Spectrum of the Serpinski Gasket in a Magnetic Field, Preprint 1984.

[2] T. ANDO, Y. MATSUMOTO, Y. UEMURA, Theory of Hall effect in a two dimensional electron system, J. Phys. Soc. Japan, 39, (1975), 279

[3] H. AOKI, T. ANDO, Effect of localization on the Hall conductivity in the two dimensional system in a strong magnetic field, Solid State Commun., 38, (1981), 1079-1082.

[4] W. ARVESON, The harmonic analysis of automorphism groups, in [M17], (1982), 199-269.

[5] S. AUBRY, The new concept of transition by breaking of analyticity in a crystallographic model, Solid State Sci., 8, (1978), 264.

[6] S. AUBRY, G. ANDRE, Analyticity breaking and Anderson localization in incommensurate lattices, Ann. Israel Phys. Soc., 3, (1980), 133.

[7] S.AUBRY, P.Y. Le DAERON, The Discrete Frenkel-Kontorova Model and its Applications, I, Exact Results for the Ground State, Physica 8D, (1983),381-422.

[8] S.AUBRY, P.Y.Le DAERON, G.ANDRË, Classical Ground State of a one dimensional Model for Incommensurate structures, Preprint Saclay, France, (1982).

[9] J. AVRON, R. SEILER, B. SIMON, Homotopy and quantization in condensed matter physics, Phys. Rev. Lett., 51, (1983), 51-53.

[10] J. AVRON, R. SEILER, Quantization of the Hall Conductance for General, Multiparticle Schrödinger Hamiltonians, Preprint Berlin 84,

[11] J. AVRON, B. SIMON, Transient and recurrent spectrum, J. Func. Anal., 43, (1981), 1-31.

[12] M.F. BARNSLEY, J.S. GERONIMO, A.N. HARRINGTON, Infinite dimensional Jacobi matrices associated with Julia sets, Proc. A.M.S., 88, (1983), 625-630.

[13] M.F. BARNSLEY, J.S. GERONIMO, A.N. HARRINGTON, Almost periodic Jacobi matrices associated with Julia sets for polynomials, Commun. Math. Phys., 99, (1985), 303-317.

[14] J. BELLISSARD, Schrödinger Operators with an Almost Periodic Potential, In "Mathematical Problems in Theoretical Physics", R. SCHRADER, R. SEILER Eds., Lecture Notes in Physics, 153, (1982), 356-359, Springer-Verlag, Berlin Heidelberg New-York.

[15] J. BELLISSARD, D. BESSIS, P. MOUSSA, Chaotic states of almost periodic Schrödinger operators, Phys. Rev. Lett., 49, (1982), 701-704.

[16] J. BELLISSARD, R. LIMA, A. FORMOSO, D. TESTARD, Quasi Periodic Interactions with a Metal Insulator Transition, Phys. Rev., B26, (1982), 3024-3030.

[17] J. BELLISSARD, E. SCOPPOLA, The Density of States of Almost Periodic Hamiltonians and the Frequency Module : a counter example, Commun. Math. Phys., 85, (1982), 301-308.

[18] J. BELLISSARD, B. SIMON, Cantor spectrum for the Almost Mathieu equation, J. Funct. Anal., 48, (1982), 408-419.

[19] J. BELLISSARD, R. LIMA, D. TESTARD, Almost Periodic Schrödinger Operators, in "Mathematics + Physics, Lectures on Recent Results", Vol. 1, L. STREIT ed., World Scientific, Singapore Philadelphia, (1985),1-64.

[20] J. BELLISSARD, D.R. GREMPEL, F. MARTINELLI, E. SCOPPOLA, Localization of electrons with spin-orbit or magnetic interactions in a two dimensional crystal, to appear in Phys. Rev. Rapid Communication B, (1986).

[21] L. BOLTZMANN, a) Wien Ans. 17, (1880), 12 ; b) Phil. Mag., 9,(1880), 307.

[22] M. CASDAGLI, Symbolic Dynamics for the Renormalization Map of a Quasiperiodic Schrödinger Equation, Preprint Univ. of Warwick, July 1985, to appear in Commun. Math. Phys.

[23] F.H. CLARO, G.H. WANNIER, Magnetic Subband Structure of Electrons in Hexagonal Lattices, Phys. Rev., B19, (1979), 6068-6074.

[24] L.A. COBURN, R.D. MOYER, I.M. SINGER, C*-algebras of almost periodic pseudo-differential operators, , Acta Math., 139, (1973), 279-307.

[25] A. CONNES, Une classification des facteurs de type III, Ann. Sci. Ecole Norm. Sup., 4º série, 6, (1973), 133-252.

[26] A. CONNES, The Von Neumann algebra of a foliation, Lecture Notes in Physics, 80, (1978), 145-151, Springer-Verlag, Berlin Heidelberg New-York.

[27] A. CONNES, Sur la théorie non commutative de l'intégration, in "Algèbres d'opérateurs", P. de la HARPE ed., Lecture Notes in Mathematics, 725, (1980), 19-143, Springer-Verlag, Berlin Heidelberg New-York.

[28] A. CONNES, C* algèbres et géométrie différentielle, C.R.A.S., 290, (1980)

[29] A. CONNES, An analog of the Thom homomorphism for crossed products of a C*-algebra by an action of R, Adv. Math., 39, (1981), 31-55.

[30] A. CONNES, A survey of foliations and operator algebras, in [M17], (1982), 521-628.

[31] A. CONNES, G. SKANDALIS, Le théorème de l'index pour les feuilletages, C.R.A.S. I, 292, (1981), 871-876.

[32] A. CONNES, Non Commutative Differential Geometry, Part I : The Chern Character in K-Homology, Preprint IHES, (1982)

[33] A. CONNES, Non Commutative Differential Geometry, Part II : De Rham Cohomology and Non Commutative Algebra, Preprint IHES, (1983)

[34] A. CONNES, Cohomologie cyclique et foncteur Ext^n, C.R.A.S. I, 296, (1983), 953-958.

[35] F. DELYON, B. SOUILLARD, Remark on the continuity of the density of states of ergodic finite difference operators, Commun. Math. Phys., 94, ,(1984), 289-291.

[36] F. DELYON, B. SOUILLARD, The rotation number for finite difference operators and its properties, Commun. Math. Phys., 89, (1983), 415.

[37] A. DENJOY, Sur les courbes définies par les équations différentielles à la surface du tore, J. de Math. Pures et Appl., (9)11, (1932), 333-375.

[38] B.A. DOBRUVIN, V.B. MATVEEV, S.P. NOVIKOV, Non linear equations of Korteweg-de Vries type, finite zone linear operators and abelian varieties, Russ. Math. Surveys, 31, (1976), 59-146.

[39] M. DUNEAU, A. KATZ, Quasiperiodic Patterns, Phys. Rev. Letters, 54, (1985), 2688.

[40] M. DUNEAU, A. KATZ, Quasiperiodic Patterns and Icosahedral Symmetry, Ecole Polytechnique Preprint, Palaiseau, (France) (1985) .

[41] G. ELLIOTT, Gaps in the spectrum of an almost periodic Schrödinger operator, C.R. Math. Ref. Acad. Sci. Canada, 4, (1982), 255.

[42] G. ELLIOTT, on the K-Theory of the C*-algebra generated by a projective representation of a torsion-free discrete abelian group, in Proc. of the Conf. on "Operator Algebras and Group Representations", G. ARSENE Ed., Pitman, London, (1983)

[43] T. FACK, G. SKANDALIS, Structure des idéaux de la C*-algèbre associée à un feuilletage, Preprint Univ. Pierre et Marie Curie, Paris (1980).

[44] J. FRÖHLICH, T. SPENCER, Absence of diffusion for the Anderson tight binding model for large disorder or low energy, Commun. Math. Phys., 35, (1983), 203-245.

[45] J. GLIMM, A. JAFFE, Quantum field theory models, in "Statistical Mechanics and Quantum Field Theory", C. de WITT & R. STORA, Eds., Les Houches summer school 1970, Gordon and Breach Science Pub., New York-London-Paris, (1971).

[46] A. GROSSMAN, Momentum-Like Constants of Motion, 1971 Europhysics Conference on Statistical Mechanics and Field Theory, Haifa, R.N. SEN & C. WEIL Eds. (1972)

[47] R. HAAG, H. HUGENHOLTZ, M. WINNINK, On the equilibrium states in quantum statistical mechanics, Commun. Math. Phys., 5, (1967), 215.

[48] R. HAAG, D. KASTLER, An algebraic approach to quantum field theory, J. Math. Phys., 5, (1964), 848-851.

[49] E.H. HALL, On a New Action of the Magnet on Electric Currents, Amer. J. of Math., 2, (1879), 287, and Phil. Mag., 9, (1880), 225

[50] B.I. HALPERIN, Quantized Hall conductance, current-carrying edge states, and the existence of extended states in a two dimensional disordered potential, Phys. Rev. B25, (1982), 2185

[51] P.G. HARPER, Single Band Motion of Conduction Electrons in a Uniform Magnetic Field, Proc. Phys. Soc. London , A68, (1955), 874,

[52] M.R. HERMAN, Une méthode pour minorer les exposants de Lyapounov, et quelques exemples montrant le caractère local d'un théorème d'Arnold et Moser sur un tore de dimension 2, Commun. Math. Helv., 58, (1983), 453-502.

[53] D.R. HOFSTADTER, Energy levels and wave functions of Bloch electrons in a rational or irrational magnetic field, Phys. Rev., B14, (1976), 2239.

[54] R. JOHNSON, The Recurrent Hill's Equation, J. Diff. Eqns, 46, (1982), 165-194

[55] R. JOHNSON, J. MOSER, The rotation number for almost periodic potentials, Commun. Math. Phys., 84, (1982), 403.

[56] R.A.JOHNSON, Exponential dichotomy, rotation number, and linear differential equations, to appear in J. Diff. Eqns., (1985).

[57] A.KATOK, Some Remarks on Birkhoff and Mather Twist Map Theorem, Erg. Theo. and Dyn. Syst., 2, (1982), 185-194.

[58] A.KATOK, More about Birkhoff Periodic Orbits and Mather Sets for Twist Maps, Preprint I,IHES, (1982), Preprint II, Univ. Maryland, (1982).

[59] S. KAWAJI, J. WAKABAYASHI, Hall conductivity in n-type silicon inversion layers under strong magnetic fields, Surf. Science, 98, (1980), 299- 307.

[60] S. KAWAJI, J. WAKABAYASHI, Hall current measurement under a strong magnetic field for silicon MOS inversion layers, J. Phys. Soc. Japan, 48, (1980), 333- 334.

[61] M. KLEMAN, DONNADIEU, Use of tessellations in the study of properties of covalent glasses, to appear in Phil. Mag. (1985).

[62] K. von KLITZING, G. DORDA, M. PEPPER, Realization of a resistance standard based on fundamental constants, Phys. Rev. Letters, 45, (1980), 494-497.

[63] M. KOHMOTO, L. KADANOFF, C. TANG, Localization problem in one dimension : mapping and escape, Phys. Rev. Lett., 50, (1983), 1870-1872.

[64] P. KRAMER, R. NERI, Acta Crystallogr., A40, (1984), 580-587.

[65] P. KRAMER, On the theory of a non periodic quasilattice associated with the icosahedral group, Preprint Tübingen, May 1985.

[66] H. KUNZ, B. SOUILLARD, Sur le spectre des opérateurs aux différences finies aléatoires, Commun. Math. Phys., 78, (1980), 201-246.

[67] H. KUNZ, B. SOUILLARD, The localization transition on the Bethe lattice, J. Physique Lett.(Paris), 44, (1983), L411.

[68] L. LANDAU, Z. für Phys., 64, (1930), 629

[69] R.B. LAUGHLIN, Quantized Hall conductivity in two dimensions, Phys. Rev. , B23, (1981), 5652-565

[70] D. LEVINE, P. STEINHARDT, Quasicrystals: A New Class of Ordered Structures, Phys. Rev. Letters, 53, (1984), 2477.

[71] G. MACKEY, Ergodic theory, group theory and differential geometry, Proc. Nat. Acad. Sci. U.S.A., 50, (1963), 1184-1191.

[72] G. MACKEY, Ergodic Theory and Virtual Subgroups, Math. Ann., 166, (1966), 187-207, see especially section 4, p. 190 and section 6, p.195.

[73] J.N.MATHER, Existence of Quasi Periodic Orbits for Twist Homeomorphisms of the Annulus, Topology, 21, (1982), 457-467.

[74] J. MATHER, Amount of Rotation About a Point and the Morse Index, Commun. Math. Phys., 94, (1984), 141-153.

[75] J. MOSER, An example of a Schrödinger equation with an almost periodic potential and a nowhere dense spectrum, Commun. Math. Helv., 56, (1981), 198.

[76] R. MOSSERI, J.F. SADOC, in "Structure of non crystalline materials 1982", Taylor & Francis Ed. (1983), London.

[77] P.MOUSSA, Un Opérateur de Schrödinger Presque Périodique à Spectre Singulier Associé aux Itérations d'un Polynome, Proc. RCP 25, CNRS nº34, (1984), 43, Strasbourg.

[78] D. NELSON, S. SACHDEV, Statistical Mechanics of pentagonal and icosahedral order in dense liquids, Phys. Rev., B32, (1985), 1480-1502.

[79] S. OSTLUND, R. PANDIT, D. RAND, H. SCHNELLHUBER, E. SIGGIA, One dimensional Schrödinger equation with an almost periodic potential, Phys. Rev. Lett., 50, (1983), 1873-1875.

[80] S.OSTLUNDT, R.PANDIT, Renormalization Group Analysis of a Discrete Quasi Periodic Schrödinger Equation, Preprint (1984).

[81] M.PIMSNER, D. VOICULESCU, Exact Sequence for K groups and Ext groups of certain cross-product C*-algebras, J. Operator Theory, 4, (1980), 93-118.

[82] M.PIMSNER, D. VOICULESCU, Imbedding the irrational rotation C*-algebra into an AF algebra, J. Operator Theory, 4, (1980), 211-218.

[83] R.PRANGE, Quantized Hall resistance and the measurement of the fine structure constant, Phys. Rev., B23, (1981), 4802-4805.

[84] I. PUTNAM, K. SCHMIDT, C. SKAU, C*-algebras associated with Denjoy homeomorphisms of the circle, Preprint Trondheim, (Norway), (1985).

[85] R. RAMMAL, Spectrum of Harmonic Excitations on Fractals, J. Physique, 45, (1984), 191-206.

[86] M.A. RIEFFEL, Irrational rotation C*-algebra, Short communication at the International Congress of Mathematicians, (1978).

[87] M.A. RIEFFEL,a) Morita equivalence for operator algebras, in [M17], p.285-298; b) Applications of strong-Morita equivalence to transformation group C*-algebras, in [M17], p.299-310; see also reference therein.

[88] D. RUELLE, D. SULLIVAN, Cycles and the dynamical study of foliated manifolds, Invent. Math., 36, (1976), 225-255.

[89] R. SACKEL, G. SELL, Existence of dichotomies and invariant splittings for differential systems I) J. Diff. Eqns, 15, (1974), 429-458. II) J. Diff. Eqns, 22, (1976), 478-496, III) J. Diff. Eqns, 22, (1976), 497-522, IV) J. Diff. Eqns, 27, (1974), 106-137.

[90] R. SACKEL, G. SELL, A spectral theory for linear differential systems, J. Diff. Eqns, 27, (1978), 320-358.

[91] D. SCHECHTMAN, I. BLECH, D. GRATIAS, J.V. CAHN, Metallic Phase with Long-Range Orientational Order and No Translation Symmetry, Phys. Rev. Lett., 53, (1984), 1951.

[92] R. SEILER, Private communication.

[93] M.A. SHUBIN, The spectral theory and the index of elliptic operators with almost periodic coefficients, Russ. Math. surveys, 34, (1979), 109-157.

[94] B. SIMOŃ, Almost Periodic Schrödinger Operators, Adv. Appl. Math., 3, (1982), 463-490.

[95] M.M. SKRIGANOV, Proof of the Bethe-Sommerfeld conjecture in dimension two, Sov. Math. Dokl., 20, (1979), 956-959.

[96] S.SMALE, Differential Dynamical Systems, Bull. A.M.S., 73, (1967), 747-897, see especially the appendix II, p. 797.

[97] B. SOUILLARD, G. TOULOUSE, M. VOOS, L'effet Hall en quatre actes, Preprint Palaiseau (France), (1984).

[98] D. THOULESS, Localization and the two-dimensional Hall effect, J. of Phys., C14, (1982), 3475-3480.

[99] D. THOULESS, M. KOHMOTO, M. NIGHTINGALE, M. den NIJS, Quantized Hall conductance in two dimensional periodic potential, Phys. Rev. Lett., 49, 405, (1982).

[100] C.H. WILCOX, Theory of Bloch waves, J. Analyse Math., 33, (1978), 146-167.

[101] This method is well known from the experts. I thank Michael Barnsley for mentioning the Schur complement, and Pierre Duclos for the reference to Feschback. Some informations can be found in : P. LOWDIN, "The calculation of upper and lower bounds of energy eigenvalues in perturbation theory by mean of partitioning technics" in Int. Summer School on " Quantum theory of polyatomic molecules", Menton, (1965).

QUANTUM FIELD THEORY WITHOUT CUTOFFS: RENORMALIZABLE AND NONRENORMALIZABLE

A. Kupiainen
Research Institute for Theoretical Physics, Helsinki University
Helsinki, Finland

1. INTRODUCTION

Constructive quantum field theory set as its goal to construct models of quantum fields in the continuum with nontrivial interactions. In the 70's, this goal was fullfilled in the context of superrenormalizable models [see [1] for references], whereas the renormalizable case remained an enigma. In fact, some doubts whether renormalizability is a sufficient condition for the existence of nontrivial theories were raised by the work of Aizenmann and Fröhlich [2], [3], who presented "almost proofs" of the conjecture that the lattice regulated ϕ_4^4 theory with positive coupling has only a trivial (gaussian) continuum limit.

In these notes I would like to present some work done jointly with K. Gawedzki on the question of necessity/suffieciency of renormalizability for the existence of these theories. What will be presented is a rigorous construction of nontrivial renormalizable theories, the Gross-Neveu model in two dimensions and the ϕ_4^4 model with negative coupling constant as well as a nonrenormalizable version of the Gross-Neveu model. We should mention that by different methods, Feldman, Magnen, Rivasseu and Seneor have succeeded constructing the Gross-Neveu model [4], 't Hooft and Rivasseu [5], [6] the planar negative coupling ϕ_4^4 model and Felder [7] a nonrenormalizable version of it.

2. EFFECTIVE ACTIONS

The problem of continuum limit in (euclidean) QFT is to try to establish the limit of the Green functions

$$<\pi\phi(x_i)>_{S_\Lambda} = \int D\phi e^{-S_\Lambda} \pi\phi(x_i) / \int D\phi e^{-S_\Lambda} \tag{1}$$

as the ultraviolet cutoff Λ is taken to infinity. S_Λ in (1) is some action involving the cutoff Λ both explicitly (in the form of lattice spacing or momentum cutoff) and implicitly in the various couplings $g_i(\Lambda)$ occurring in it (see below).

It is advantageous to look at this problem from a slightly different point of view, that is, to study the continuum limit of *the effective actions*. Vaguely, the effective action for momenta $\lesssim \tilde{\Lambda}$, $S^{EFF}_{\tilde{\Lambda}}$ describes the theory (its Green functions) for momenta $\lesssim \tilde{\Lambda}$ and is produced from S_Λ by performing a partial functional integral: we integrate out fluctuations ψ of the fields having momenta $\tilde{\Lambda} \lesssim |p| \lesssim \Lambda$, schematically,

$$\exp\left[-S^{EFF\Lambda}_{\tilde{\Lambda}}(\phi)\right] = \int D\,\psi \exp\left[-S_\Lambda(\phi+\psi)\right]. \tag{2}$$

Concrete ways of realizing (2) will be sketched below. Continuum limit means now that we are to find the bare couplings $g_i(\Lambda)$ s.t. the limit

$$\lim_{\Lambda\to\infty} S^{EFF\Lambda}_{\tilde{\Lambda}} = S^{EFF}_{\tilde{\Lambda}} \tag{3}$$

exists in some strong enough sense. The Green functions of the theory are determined by the family $\left\{S^{EFF}_{\tilde{\Lambda}}\right\}^{\infty}_{\tilde{\Lambda}=0}$, provided we gain sufficient control of it (as will be the case below).

3. THE OLD PERTURBATIVE APPROACH

To contrast with the rigorous analysis sketched below, let us recall the standard perturbative approach to the continuum limit. This amounts to computing (2) by means of a formal perturbation expansion in the couplings $g_i(\Lambda)$. One obtains a formal sum

$$S^{EFF\Lambda}_{\tilde{\Lambda}}(\phi) = \sum_N \int dx_1 \ldots dx_N \, \Gamma^N_{\tilde{\Lambda}}(x_1 \ldots x_N, \Lambda) \pi \, \phi(x_i) \tag{4}$$

where the vertex functions $\Gamma^N_{\tilde{\Lambda}}$ are given in terms of a formal power series in the coupling(s)

$$\Gamma^N_{\tilde{\Lambda}}(x_i, \Lambda) = \sum_n g(\Lambda)^n \, \Gamma^N_{\tilde{\Lambda}n}(x_i, \Lambda) \, . \tag{5}$$

As is well known, the coefficients $\Gamma^N_{\tilde{\Lambda}n}$ in (5) diverge (in most models) as $\Lambda \to \infty$. This should come as no surprise; after all (5) is a relation between the data $(\Gamma^N_{\tilde{\Lambda}})$ describing the theory at the scale $\tilde{\Lambda}$ and the data $(g(\Lambda))$ at scale Λ. Such a relation in most interesting cases should be singular.

The old renormalization theory can now be formulated as an attempt to save (5) by expressing $g_i(\Lambda)$ as functions (formal power series) of renormalized couplings g_{iR} describing the theory at some fixed scale (the renormalization point μ)

$$g(\Lambda) = \sum_n g_R^n \, C_n(\Lambda, \mu) \, . \tag{6}$$

(6) and (5) combine to a series

$$\Gamma^N_{\tilde{\Lambda}}(x_i, \Lambda) = \sum_n g_R^n \, \bar{\Gamma}^N_{\tilde{\Lambda}n}(x_i, \Lambda, \mu) \tag{7}$$

and the theory is called *renormalizable* if the coefficients $\bar{\Gamma}^N_{\tilde{\Lambda}n}$ have a limit as $\Lambda \to \infty$. In such a case, all the effective actions $S^{EFF}_{\tilde{\Lambda}}$ have a computable expansion in terms of a finite amount of data g_{iR} at scale μ.

The theory is *nonrenormalizable* in case the above procedure doesn't work: typically in such a case an infinite number of couplings g_i^R $(g_i(\Lambda))$ are needed, invalidating the whole approach.

Finally, the work of Aizenman and Fröhlich teaches us that a theory may be perturbatively renormalizable, i.e. $\lim_{\Lambda\to\infty} \bar{\Gamma}^N_{\tilde{\Lambda}n}(\Lambda)$ exist, but still be trivial for a nonperturbative reason: for lattice ϕ^4_4 with positive coupling the only value of g_R that may be achieved is zero.

4. THE WILSON APPROACH

In the late 60's and early 70's a new approach to the continuum limit was developed by K. Wilson (see [8] for references). The idea is, that rather than trying to establish a direct relation between S_Λ and $S^{EFF}_{\tilde{\Lambda}}$ (as in (5)) we reduce the cutoff little by little, establishing a *flow* of effective actions

$$S_\Lambda \to S^{EFF}_{\Lambda/L} \to S^{EFF}_{\Lambda/L^2} \to \dots \to S^{EFF}_{\tilde{\Lambda}} \to \dots \tag{8}$$

in some "space of actions". Here L is some number of order unity. To bridge a connection to dynamical systems, it is advantageous to make a trivial change of variables (to dimensionless quantities) so that the cutoffs Λ/L^n are all the same. We define a *"Hamiltonian"*

$$S_\Lambda(\phi) = H_\Lambda(\Lambda^{-\mathcal{D}}\phi(\tfrac{\cdot}{\Lambda})) \tag{9}$$

and similarly for $S^{EFF}_{\tilde{\Lambda}}$. $\mathcal{D}$ in (9) is the scale dimension of the field, which we have to determine in each model separately (see below).

(9) amounts to the formulation of the continuum limit as *a scaling limit*; we are to find a sequence of Hamiltonians H_Λ and to look at larger and larger distances as $\Lambda\to\infty$:

$$\langle \prod^{N} \phi(x_i)\rangle_{S_\Lambda} = \Lambda^{N\mathcal{D}}\langle \prod^{N} \phi(\Lambda x_i)\rangle_{H_\Lambda} . \tag{10}$$

Thus we study the flow of Hamiltonians

$$H_\Lambda \overset{R}{\to} H^{EFF}_{\Lambda/L} \to \ldots \tag{11}$$

each of cutoff 1 and related to each other by the *Renormalization Group Transformation* (RGT) R

$$e^{-RH(\phi)} = \int D\,\psi\; e^{-H(L^{-\mathcal{D}}\,\phi(\frac{\cdot}{L}) + \psi)} \tag{12}$$

which is a *nonlinear* map in some "space of Hamiltonians" $\mathcal{H}$. In terms of R the continuum limit is thus expressed as

$$H^{EFF}_{\tilde{\Lambda}} = \lim_{N\to\infty} R^N\, H_{L^N\tilde{\Lambda}} . \tag{13}$$

The canonical example of (13) is obtained if R has some fixed point H^*. Then, if H_Λ is made to approach the stable manifold of H^* in a suitable way, the $H^{EFF}_{\tilde{\Lambda}}$ will lie on the unstable manifold of H^*, which will then parametrize the continuum limit. Obviously in our case the space $\mathcal{H}$ is complicated and a very good knowledge of R is needed to establish such a picture. We now turn to concrete examples where such an analysis in fact may be carried out.

5. THE RENORMALIZABLE CASE: FERMIONS

We take S_Λ the Gross-Neveu model of two dimensional Dirac fermions with a four-fermion coupling:

$$S_\Lambda = \int d^2x\, [\bar{\psi} i \not{\partial}_\Lambda \psi - g(\Lambda)(\bar{\psi}\psi)^2] . \tag{14}$$

Here $\psi = (\psi^i_\alpha)$ are Grassmann variables (they belong to an infinite dimensional Grassmann algebra; see below and [9] for a rigorous formulation) where $\alpha = 1,2$ are the Lorentz indices and $i = 1,\ldots,N$, $N > 1$ are vector internal symmetry indices, suppressed below. The cutoff Λ is put to the propagator in momentum space:

$$(i\not{\partial}_\Lambda)^{-1}(p) = \frac{\not{p}}{p^2}\, e^{-p^2/\Lambda^2} . \tag{15}$$

The theory (15) is renormalizable but not superrenormalizable. Thus $g(\Lambda)$ is dimensionless, which is reflected in the corresponding Hamiltonian

$$H_\Lambda = \int [\bar{\psi} i \not{\partial}_1 \psi - g(\Lambda)(\bar{\psi}\psi)^2] . \tag{16}$$

We took $\mathcal{D} = \frac{1}{2}$ so as to keep the coefficient of the gaussian part unity.

The RGT is now explicitly given as

$$\exp[-RH_\Lambda(\psi)] = e^{-\int \bar{\psi} i \not{\partial}_1 \psi} \int D\chi \exp[-(\bar{\chi},\Gamma^{-1}\chi) - g(\Lambda)V(L^{-\frac{1}{2}}\psi(\tfrac{\cdot}{L})+\chi)] \tag{17}$$

($V = (\bar{\psi}\psi)^2$) where the fluctuation covariance is

$$\Gamma(p) = \frac{\not p}{p^2}(e^{-p^2} - e^{-L^2p^2}). \tag{18}$$

Note how Γ has mass of O(1) and UV cutoff too.

The point of studying a fermionic theory is, that a rigorous analysis of R is especially simple. In fact it turns out, we may use standard perturbation theory to evaluate (17)! Just as in (4) and (5) expand (17) in powers of $g(\Lambda)$ ($\psi^{\#} = \psi$ or $\bar{\psi}$):

$$H^{EFF}_{\Lambda/L} = \sum_N \int dx_1 \ldots dx_N \, \Gamma^N_{\Lambda/L}(x_1,\Lambda) \, \pi \, \psi^{\#}(x_i) \tag{19}$$

with again

$$\Gamma^N_{\Lambda/L} = \sum_n g(\Lambda)^n \, \Gamma^N_{\Lambda/Ln} \, . \tag{20}$$

However, this time both the sums over N and n *converge*. To understand this, note that $\Gamma^N_{\Lambda/Ln}$ is given in terms of all connected graphs with n 4-point vertices and N legs with lines carrying Γ of (18). There are $\sim n!$ such graphs which in the case of bosonic theories leads to the divergence of (20) in n. However, in the case of fermions minus signs occur saving us. More explicitly, the free expectation

$$\langle \overset{N}{\pi} \, \bar{\psi}(x_i) \psi(y_i) \rangle_\Gamma \tag{21}$$

in the covariance Γ is given by determinant $\det \Gamma(x_i - y_i)$ of a $N \times N$ matrix. Since Γ has both UV and infrared (IR) cutoffs, (21) is bounded by $(\text{const})^N$. This observation leads to the convergence of (20) in quite a straightforward way.

As for (19), recall that the graphs entering (20) are local, since Γ is. Thus convergence of the integrals in (19) follows, for $\psi^{\#} = O(1)$, as well as the sum over N since the graphs carry explicit $g^{O(N)}$ factors. In fact we may now set up the space $\mathcal{H}$ where R acts as a nice Banach space. We take $H \in \mathcal{H}$ of the form

$$H = \int [\bar{\psi} i \not{\partial}_1 \psi + \delta Z \bar{\psi} i \not{\partial} \psi - g(\bar{\psi}\psi)^2] + \sum_N \int \Gamma^N(x_i) \pi \Psi(x_i) \quad (22)$$

$$\equiv (\delta Z, g, \Gamma^N)$$

Here $\Psi = (\psi, \partial\psi)$, i.e. we allow also derivatives of ψ. These appear, since the local parts δZ and g are extracted from Γ^2 and Γ^4 leaving extra derivatives on the fields. The structure of the local parts follows from symmetries. Introduce next the norm (for some small g_0)

$$||\Gamma^N|| = g_0^{-N} \int dx_2 \ldots dx_N \Gamma^N(x_1 \ldots x_N) e^{\mathcal{L}(x_1 \ldots x_N)} \quad (23)$$

with $\mathcal{L}$ the shortest connected graph on $\{x_1 \ldots x_N\}$, and for H:

$$||H|| = |\delta Z| + |g| + \sum_N ||\Gamma^N|| \quad (24)$$

In this way $\mathcal{H}$ gets a Banach space strucure. We may again evaluate RH perturbatively, the boundedness of fermionic expectations leading to the cenvergence of the expansion. The coordinates $(\delta Z', g', \Gamma'^N)$ of RH are analytic functions of those of H and we get

$$\begin{aligned} g' &= g - \bar{\beta}_2 g^2 - \bar{\beta}_3 g^3 - \bar{\beta}(g, \delta Z, \Gamma^N) \\ \Gamma'^N &= L^{\mathcal{D}_N} \Gamma^N(L\cdot) + \gamma_N(g, \delta Z, \Gamma^N) \end{aligned} \quad (25)$$

where some leading terms were separated. The coefficient $\bar{\beta}_2$ is negative; thus the g direction is marginally unstable. $\mathcal{D}_N$ is the scale dimension of Γ^N. Since we separated the marginal parts in (22), the Γ^N directions are unstable. Indeed, in terms of the norm (23)

$$\| L^{\mathcal{D}_N} \Gamma^N (L\cdot) \| \leqslant L^{-d_N} \|\Gamma^N\| , d_N > 0 \tag{26}$$

It is now a straightforward matter [9] to establish the continuum limit of $H^{EFF\Lambda}_{\tilde{\Lambda}}$ in the norm (24). Only a fine tuning of $g(\Lambda)$ is needed, it turns out, that only the three first perturbative terms in (25) are needed, we take

$$g(\Lambda)^{-1} = g_R^{-1} - \beta_2 \log \Lambda/\mu + \frac{\beta_3}{\beta_2} \log(1 - g_R \beta_2 \log \Lambda/\mu) \tag{27}$$

Here β_i are related to $\bar{\beta}_i$; they are the first two coefficients of the perturbative β-function. With the choice (27)
$\| \lim_{\Lambda} H^{EFF\Lambda}_{\tilde{\Lambda}} \| = O(g(\tilde{\Lambda}))$.
How about the Green functions? For them we need $H^{EFF}_{\tilde{\Lambda}}$ for all $\tilde{\Lambda}$. However, as $\tilde{\Lambda}$ decreases, $g_{\tilde{\Lambda}}$ increases, eventually leaving the region where perturbation converges. Thus we cannot study this way the infrared behaviour of our model. Nevertheless, adding an explicit mass term to (14) the full set of $H^{EFF}_{\tilde{\Lambda}}$ may be constructed as well as the Green function which satisfy Euclidean axioms guaranteeing the existence of Wightman QFT. We expect to control the "massless" theory for N, the number of components, large (but finite) when the model is expected to show dynamic mass generation and symmetry breaking [10].

6. THE NONRENORMALIZABLE CASE

By a slight modification of the free propagator, we may make the Gross-Neveu model in d=2 nonrenormalizable. Namely, consider again (14), this time with

$$(i\not{\partial}_{\Lambda})^{-1}(p) = \not{p}\int_{\Lambda^{-2}}^{\infty} dx\, e^{-\alpha p^2}\alpha^{-\varepsilon/2} \qquad (28)$$

Note that for $\Lambda = \infty$ (28) becomes $\alpha\not{p}/p^{2-\varepsilon}$ and the perturbation series nonrenormalizable. The Hamiltonian now is

$$H_{\Lambda} = \int(\bar{\psi}\, i\,\not{\partial}_1\,\psi - g(\Lambda)\,\Lambda^{2\varepsilon}(\bar{\psi}\psi)^2). \qquad (29)$$

Note, how the coupling has a non-zero "dimension" just as formally in $d=2+\varepsilon$. All the analysis of the previous section goes through; the new Γ is again massive. We get the recursion for

$$\lambda_{\tilde{\Lambda}} = \tilde{\Lambda}^{2\varepsilon} g_{\tilde{\Lambda}} \quad \text{as}$$

$$\lambda' = L^{-2\varepsilon}\lambda - \bar{\beta}_2\lambda^2 + \ldots \quad . \qquad (30)$$

This time λ is stable at zero. However a non-gaussian fixed point is seen to emerge for $\lambda = O(\varepsilon)$. It is a standard matter of non-linear analysis to construct [11] a $H^* = (O(\varepsilon), O(\varepsilon), O(\varepsilon^2))$ which turns out to have one unstable direction, roughly corresponding to λ. The sequence of bare theories (29) is shown to intersect the stable manifold of H^* at a point $(0, \lambda_c, 0)$, Then the continuum limit exists, provided we choose

$$\lambda(\Lambda) = g(\Lambda)\,\Lambda^{2\varepsilon} = \lambda_c + \left(\frac{\Lambda}{\mu}\right)^{-\nu^*}(\lambda - \lambda_c) \qquad (31)$$

where ν^* is the largest exponent at H^*.

The massless theory exists now as well, provided $\lambda \leqslant \lambda_c$. The IR is then governed by the gaussian fixed point. For $\lambda > \lambda_c$ again a dynamic mass generation is expected and the present method only yields the massive theory with explicit mass term. Needless to say, the resulting theory is not reflection positive since even the free theory us not. However we feel, that this example shows that nonrenormalizable theories indeed should make sense.

7. BOSONIC THEORIES

The approach sketched above is applicable to the ϕ_4^4 model with some modifications [12], [13] . Now we have

$$H_\Lambda = \sum_{x \in \mathbb{Z}^4} \tfrac{1}{2}(\nabla \phi(x))^2 + \tfrac{1}{2}\mu(\Lambda)\,\phi(x)^2 + \tfrac{1}{4}\lambda(\Lambda)\,\phi(x)^4 \tag{32}$$

(Here it is more advantageous to work with lattice cutoff, the RGT is then given by the block spin transformation). The main modification to the fermionic case comes from the unboundedness of ϕ: The perturbation expansion for RH diverges as n!. However, the main point of R is, that the fluctuation integral is massive. The role of the perturbation expansion will now be played by a convergent high-temperature expansion. The analysis will be more complicated since one has to keep separately track of the small and large field properties of $H^{EFF}_{\tilde{\Lambda}}$ (for a pedagogical exposition, see [14]).

In the region of field space where ϕ is not too large a representation (19) again emerges with a convergent N-sum. The $\Gamma^N_{\tilde{\Lambda}}$ have perturbative and small non-perturbative contributions, again leading to a recursion of the type (25) where β and γ_N are under good control. Summarizing, for $\lambda(\Lambda) > 0$ in (25) the λ recursion is

$$\lambda' = \lambda - \beta_2 \lambda^2 + \ldots \tag{33}$$

with $\beta_2 > 0$ this time. The gaussian fixed point has only a gaussian unstable manifold for $\lambda(\Lambda)$ small: all such continuum limits are trivial. This proves the conjecture of Aizenmann and Fröhlich in the case of small bare soupling. It should be noted that this is in no contradiction with the renormalizability of the model, i.e. the existence of the limit $\Lambda \to \infty$ of $\Gamma^{-N}_{\tilde{\Lambda}n}(\Lambda)$ in (7). Indeed renormalizability manifests itself in the fact that if we keep the cutoff very high $(O(e^{-c/g_R}))$, then for $\tilde{\Lambda} \ll \Lambda$ the cutoff-dependence of $H^{EFF(\Lambda)}_{\tilde{\Lambda}}$ is very small. Thus positive coupling ϕ^4_4 is a perfectly good *effective field theory* describing "physics" for a wide range of energy-scales.

From the recursion (33) we see, that formally the theory for negative λ has an unstable direction, and thus is a candidate for continuum theory. Of course the problem then is the stability of the theory. In fact it is very easy to produce a stable cutoff theory by analytic continuation from positive λ[13]. In practical terms this is achieved by taking (32) with λ negative and rotating the fields in complex plane to $e^{i(\pm\frac{\pi}{4} - \varepsilon)}\mathbb{R} \equiv C^{\pm}$.

Then in the functional integral (in a finite box V)

$$\int_{C^V} D\phi\, e^{-H_\Lambda} \tag{34}$$

both the ϕ^4 part and the kinetic part have positive real parts:

$$\begin{aligned} \mathrm{Re}\,\lambda\,\phi^4 &= |\lambda\phi^4|\,\cos\varepsilon \\ \mathrm{Re}(\nabla\phi)^2 &= |\nabla\phi|^2\,\sin\varepsilon\,. \end{aligned} \tag{35}$$

(35) allow us to carry out the RG analysis of this theory as in the positive coupling case. The upshot is, that one may prove the existence of the continuum limit of the Euclidean Green functions of the negative coupling theory. These functions are non-trivial, they have the formal renormalized perturbation series as an asymptotic expansion and satisfying the axions *except* reflection positivity. In fact we expect a violation of this property. Namely, the two different theories with $C = C^{\pm}$ in (34) correspond to two different analytic continuations in the upper and lower complex λ-plane. The difference of these theories is the imaginary part of the Green functions and is given formally by an instanton contribution. Thus heuristically we expect the Green functions to have a nonperturbative $O(e^{-c/g_R})$ imaginary part violating reflection positivity. One might still ask, whether this theory at least formally would represent any Minkowski theory with a decaying vacuum. A standard instanton computation gives a surprising answer to this question. Namely, formally the imaginary part of the ground state energy density, i.e. the decay rate per unit volume of the false vacuum is given by

$$\varepsilon \approx \lim_{\Lambda\to\infty} \int_{\Lambda^{-1}}^{m^{-1}} \frac{d\rho}{\rho}\,\rho^{-4}\, e^{-S(\rho)/\lambda(\rho)} \tag{36}$$

where ρ is the scale of the scale of the instanton, ρ^{-4} the dimensional factor for energy density, $S(\rho)$ the action for instantons of scale ρ and $\lambda(\rho)$ the running coupling at length scale ρ. The numbers are

$$S(\rho) = 16\pi^2 \quad , \quad \lambda(\rho)^{-1} \approx \frac{1}{\lambda_R} - \frac{3}{16\pi^2} \log \rho . \tag{37}$$

Thus (36) behaves as Λ as $\Lambda \to \infty$. The lifetime of the false vacuum tends to zero as the cutoff is removed, and this unstability is a UV effect, even though the theory is asymptotically free! (For a contrast, in QCD m = 0 and (36) diverges in the IR).

REFERENCES

[1] J. Glimm, A. Jaffe, Quantum Physics, Springer (1981)

[2] M. Aizenman, Phys. Rev. Lett. 47 (1981)

[3] J. Fröhlich, Nucl. Phys. B200, 281 (1982)

[4] J. Feldman, J. Magnen, V. Rivasseu, R. Seneor, Phys. Rev. Lett. 54, 1479 (1985)

[5] G. 't Hooft, Commun. Math. Phys. 86, 449 (1982)

[6] V. Rivasseu, Preprint, Ecole Polytechnique (1984)

[7] G. Felder, Preprint ETH-Zürich (1985)

[8] K. Wilson, Rev. Mod. Phys. 55, 583 (1983)

[9] K. Gawedzki, A. Kupiainen, Phys. Rev. Lett. 54, 2191 (1985)

[10] D. Gross, A. Neveu, Phys. Rev. D10, 3235 (1974)

[11] K. Gawedzki, A. Kupiainen, Helsinki Preprint (1985) and Phys. Rev. Lett. 55, 363 (1985)

[12] K. Gawedzki, A. Kupiainen, Commun. Math. Physics 100 (1985)

[13] K. Gawedzki, A. Kupiainen, Nuclear Physics B257, 474 (1985)

[14] K. Gawedzki, A. Kupiainen, Harvard Preprint (1985), to appear in Proceedings of Les Houches Ecole de l'été 1984.

RENORMALIZATION GROUP METHODS IN RIGOROUS QUANTUM FIELD THEORY

K. Gawędzki
CNRS, IHES
91440 Bures-sur-Yvette
France

One of the fundamental problems of Quantum Field Theory (QFT) is the existence of non-trivial models of local quantum fields. For many years the opinion dominated that only perturbatively renormalizable (or super-renormalizable) models have a chance to exist. However, in the course of time, especially with the development of the Renormalization Group (RG) ideas, the evidence began to accumulate that perturbative renormalizability is neither a sufficient nor a necessary condition for the existence of a quantum field. The key to this realization lies in K.G. Wilson's description [1] of (euclidean) QFT through the sequence of effective statistical-mechanical Hamiltonians $H(\mu)$ corresponding to the momentum scales $\leq \mu$. For easy comparison of the theory on various scales, it is convenient to rescale each $H(\mu)$ to the unit cut-off regime. The Hamiltonians $H(\mu)$ are related by the RG transformations

$$H(\mu) \xrightarrow{R_L} H(\mu/L) \tag{1}$$

for $L > 1$, consisting (on the level of the Gibbs states associated to the Hamiltonians) of integrating out the degrees of freedom with momenta between μ/L and μ (and of rescaling the result to the unit cut-off regime). Wilson's description allows to translate the properties of the field theory into the properties of the flow of the semi-group R_L in a space of the unit cut-off Hamiltonians. For example, the (non-perturbative) renormalizability of a family $H(g_i)$ of interactions described by local (cut-off) Hamiltonians becomes equivalent to the existence of functions $g_i^{bare}(g^{ren},\Lambda)$ such that

$$H^{eff}(g^{ren},\mu) = \lim_{\Lambda\to\infty} R_{\Lambda/\mu}\, H(g^{bare}(g^{ren},\Lambda)) \tag{2}$$

exists for each μ and is a non-degenerate function of g_i^{ren}. This notion should be distinguished form the perturbative renormalizability where the existence of the limit in (2) is required only order by order in the formal expansion in powers of g_i^{ren}. A plausible scenario for renormalizability of a family of interactions has been put forward by Wilson : In the space of Hamiltonians we have a fixed point H_* of the RG transformations with finite dimensional expanding and infinite dimensional contracting manifolds, see Fig.1. The family $H(g_i)$ crosses transversally the contracting manifold. It should be then possible to choose $H(g_i^{bare}(\Lambda))$ converging to the contracting manifold in such a way that $R_{\Lambda/\mu}H(g_i^{bare}(\Lambda))$ converge to a Hamiltonian on the expanding manifold of H_*. The continuum limit $(\Lambda \to \infty)$ theory would be then described by effective Hamiltonians $H^{eff}(\mu)$ lying on the finite dimensional manifold independent of the detailed form of $H(g_i)$ (universality in field theory). One would also have

(3) $$\lim_{\mu\to\infty} H^{eff}(\mu) = H_* .$$

We would say that the fixed point H_* governs the continuum limit and the ultra-violet behavior of the field theory.

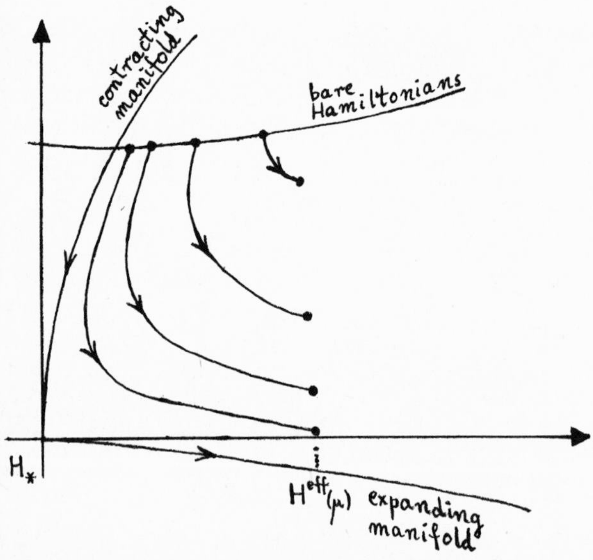

A possible stategy for the demonstration of renormalizability of an interaction may consist of establishing the above scenario. This can be approached by approximate analysis (perturbative RG, numerical analysis of the approximate RG recursions, Monte-Carlo RG) or done with full rigour. Only the latter case will be discussed here.

Exact analysis of Wilson's RG is a formidable task. The simplest situation occurs when the continuum limit is governed by the gaussian fixed point i.e. by H_* quadratic in fields. Such field theories are called asymptotically free. One may achieve then that the whole relevant flow of the RG takes place in the vicinity of H_* where the perturbative expansion around the gaussian theory provides a reliable tool for the control of the flow. In fact all the models of quantum fields constructed until recently like $P(\varphi)_2$, $(\varphi^4)_3$, Y_2, $(\text{sine-Gordon})_2$, $(\text{abelian Higgs})_{2,3}$ (see e.g. [2]) were asymptotically free (or could be solved exactly by mapping into free fields as the Thirring and Schwinger models). Indeed, the methods developed by the so-called constructive QFT seemed suitable for any asymptotically free super-renormalizable theory i.e. one with a finite number of divergent Feynman graphs. It was generally expected, however, that treatment of the renormalizable case would be much more difficult. As the first constructions [3] of the asymptotically free renormalizable models obtained recently have shown this was an excessively pessimistic point of view. In particular, for the purely fermionic Gross-Neveu model in 2 dimensions (the $\bar\psi(i\not\partial+m)\psi-g(\bar\psi\psi)^2$ theory), the RG transformations are given by convergent perturbation expansions, which largely simplifies the analysis, see A. Kupianen's contribution to this volume. However, we are also able to show the existence of the continuum limit of the bosonic negative coupling φ^4 theory in four euclidean dimensions [4]. This theory is renormalizable and asymptotically free, although it describes a metastable system and lacks as a result the so-called Osterwalder-Schrader positivity. On the other hand, as many rigorous and numerical studies suggest [5], the positive coupling φ_4^4 theory, lacking asymptotic freedom, seems to have no continuum limit. Although the main problem in the field, the construction of non-abelian four-dimensional asymptotically free and renormalizable gauge theories has not been solved yet, it seems now within reach, see [6] and references therein.

The case of φ_4^4 raises the question whether the asymptotic freedom is a necessary condition for the non-perturbative renormalizability. There are strong rigorous arguments to the contrary. If in the Gross-Neveu model we replace the free fermionic propagator $\frac{1}{-\not{p}}$ by $\frac{|p|^{\varepsilon}}{-\not{p}}$ we render the theory perturbatively non-renormalizable. Nevertheless, for small $\varepsilon > 0$, it is still possible to renormalize the theory non-perturbatively by constructing the continuum limit governed by a non-gaussian fixed point [7], see also A. Kupianinen's lecture in this volume. For small ε the new fixed point is close to the gaussian one, so that again the whole flow of the RG can be studied by the convergent perturbation expansion. The Green functions of the model are not infinitely differentiable in the weak physical coupling what explains the difficulties of the perturbative renormalization. Similar analysis works for the planar $\varphi_{4+\varepsilon}^4$ theory [8]. Also for the Gross-Neveu model in 3 dimensions in the limit when the number of fermionic species (flavors) goes to infinity one can establish the Wilson's scenario with the similar RG flow as in the $\frac{|p|^{\varepsilon}}{-\not{p}}$ case. Although genuine non-renormalizable field theories remain to be constructed, it seems that they may be consistent in the presence of non-gaussian fixed points of the RG as argued in [9]. This opens the possibility of occurrence of a realistic non-renormalizable sector at ultra high energies which should perhaps be given more thought in model building and future Monte-Carlo RG simulations.

References.

[1] K.G. Wilson, J. Kogut, Phys. Rep. 12C (1974), 75.

[2] J. Glimm, A. Jaffe, Quantum Physics, A functional integral point of view, Springer, New York 1981.

[3] J. Feldman, J. Magnen, V. Rivasseau, R. Sénéor, Phys. Rev. Lett. 54 (1985), 1479 and Ecole Polytechnique preprint.

K. Gawędzki, A. Kupiainen, Phys. Rev. Lett. 54 (1985), 2191 and IHES preprint.

[4] K. Gawędzki, A. Kupiainen, Commun. Math. Phys. 99 (1985), 197 and Nucl. Phys. B2 (1985).

[5] J. Fröhlich, Nucl. Phys. B200 FS4 (1982), 281,

M. Aizenman, R. Graham, Nucl. Phys. B225 FS9 (1983), 261,

D. Callaway, R. Petronzio, Nucl. Phys. B240 FS12 (1984), 577.

[6] T. Bałaban, Commun. Math. Phys. 99 (1985), 389 and references therein.

[7] K. Gawędzki, A. Kupiainen, Phys. Rev. Lett. 55 (1985), 363 and IHES preprint.

[8] G. Felder, ETH preprint.

[9] K. Symanzik, Commun. Math. Phys. 45 (1975), 79,

G. Parisi, Nucl. Phys. B100 (1975), 368.

Renormalization group methods for circle mappings

Oscar E. Lanford III
IHES
91440 Bures-sur-Yvette
France

I review here a number of recent results all having as central theme the application of renormalization group methods to the study of the iteration of circle mappings. I will concentrate on mappings which are smooth but which have critical points and hence are not diffeomorphisms. This review will be divided into two quite distinct parts. The first part is devoted to the theory of circle mappings with a special rotation number, the *golden ratio*. The theory in this case is by now relatively complete. None of the results described in this section are my own; they are due independently to Feigenbaum, Kadanoff, and Shenker[3] and to Ostlund, Rand, Sethna, and Siggia[7].

A second section discusses how to extend this analysis to general rotation numbers. In contrast to the first section, the analysis in this part focuses on the variation of rotation number with parameter rather than on the detailed dynamics of individual circle mappings. This part of the subject is much less developed than is the study of special rotation numbers. The main new idea which will be presented in this section is an extension of the standard renormalization group analysis—which studies the consequences of the existence of a hyperbolic fixed point for a renormalization operator—to a situation where a renormalization operator leaves invariant a hyperbolic set more complicated than a fixed point or periodic cycle. This extended renormalization group analysis appears to apply to the theory of circle mappings with general rotation numbers and makes strikingly strong predictions about the dependence of rotation number on parameter in one-parameter families.

Golden ratio rotation number.

We will use the term *circle mapping* to denote a continuous (usually in fact analytic) strictly increasing mapping f of $\mathbf{R}$ to itself satisfying the *circle mapping identity*

$$f(x+1) = f(x) + 1.$$

The reason for calling such an f a circle mapping is that it induces, by passage to quotients, a continuous one-one mapping of the circle $\mathbf{R}/\mathbf{Z}$ to itself, and, conversely, any continuous one-one mapping of the circle to itself which is orientation-preserving ("increasing") is induced by an f satisfying these conditions and unique up to an additive integer.

The circle mapping identity can equivalently be expressed as the statement that $f(x) - x$ is periodic with period one, or that f commutes with $x \mapsto x+1$. A *critical* circle mapping will mean one satisfying $f'(0) = 0$; i.e., one having a critical point which, for convenience, we situate at the origin. Since circle mappings are by definition increasing, this implies $f''(0) = 0$ also. The *rotation number* of a circle mapping f, denoted by $\rho(f)$, is defined by

$$\rho(f) \equiv \lim_{n\to\infty} \frac{f^n(x) - x}{n}.$$

The definition makes sense because, by a simple argument due to Poincaré, the limit on the right exists and is independent of x. The *golden ratio* will mean $(\sqrt{5}-1)/2$. In this section, we reserve the symbol σ to denote this number. What we are going to be studying is *analytic critical circle mappings with rotation number* σ.

The *Fibonacci sequence* Q_n is defined by the recursion relation

$$Q_{n+1} = Q_n + Q_{n-1}$$

with initial conditions $Q_1 = Q_2 = 1$. The Fibonacci sequence is closely related to the golden ratio; one connection, of which we will make heavy use, is the identity

$$\sigma Q_n - Q_{n-1} = (-1)^{n+1}\sigma^n$$

which can easily be proved by solving explicitly the recursion relation defining the Fibonacci sequence. From this identity, if f is a circle mapping with rotation number σ, then

$$f_n(x) \equiv f^{Q_n}(x) - Q_{n-1}$$

(again a circle mapping) has rotation number $(-1)^{n+1}\sigma^n$, which goes rapidly to zero as n goes to infinity. This suggests that the f_n should converge (in some sense) to the identity mapping. We will look at the limiting behavior, in the immediate vicinity of the critical point 0.

For that purpose, we define

$$\lambda^{(n)} = f_n(0) \quad \text{and} \quad \eta_n(x) = \frac{1}{\lambda^{(n-1)}} f_n(\lambda^{(n-1)}x)$$

Numerical experiments strongly indicate that the $\lambda^{(n)}$ converge to zero (consistent with the guess that the f_n converge on a coarse scale to the identity) and that the η_n (which describe the fine-scale behavior of the f_n) converge to a limit η^* which doesn't depend on which f we start from provided always that f is an analytic critical circle mapping with rotation number σ.

A completely elementary argument shows that, if the η_n constructed from some f converge to a limit η^*, then η^* must be quite a remarkable function: It satisfies a *functional equation*

$$\eta^*(x) = \frac{1}{\lambda^*}\eta^*(\frac{1}{\lambda^*}\eta^*(\lambda^*\lambda^*x)$$

(where λ^* denotes $\eta^*(0)$). Furthermore, writing ξ^* for η^* rescaled by λ^*:

$$\xi^*(x) \equiv \frac{1}{\lambda^*}\eta^*(\lambda^*x)$$

we find that η^* and ξ^* commute , i.e., η^* has the unusual property of commuting with a non-trivial rescaling of itself. On the other hand, neither η^* nor ξ^* has any reason to satisfy the circle mapping identity, and, in fact, the limits found in numerical experiments do not satisfy it.

The proof of the functional equation for η^* goes as follows: We begin by observing that the recursion relation for the Fibonacci sequence, the circle mapping identity, and the definition of the f_n imply

$$f_{n+1} = f_n \circ f_{n-1}.$$

Rescaling this equation by $\lambda^{(n)}$ and reorganizing in a straightforward way gives:

$$\eta_{n+1}(x) = \frac{1}{\lambda_n}\eta_n(\frac{1}{\lambda_{n-1}}\eta_{n-1}(\lambda_n\lambda_{n-1}x), \qquad \text{where} \quad \lambda_n \equiv \frac{\lambda^{(n)}}{\lambda^{(n-1)}} = \eta_n(0).$$

Since, by assumption, η_n converges to a limit η^*, λ_n converges to a limit λ^*, and we can thus take the limit of the above formula for η_{n+1} to get the functional equation.

We can now easily put the above into a renormalization group setting: We introduce, along with η_n, the function ξ_n defined by

$$\xi_n(x) \equiv \frac{1}{\lambda^{(n-1)}} f_{n-1}(\lambda^{(n-1)}x).$$

i.e., by rescaling f_{n-1} by the same factor as f_n was rescaled by to produce η_n. It should be noted for later use that f_n and f_{n-1} commute (since they are both essentially iterates of the same mapping f), and thus ξ_n and η_n (for any given n) commute. The calculation sketched in the preceding paragraph shows that

$$\eta_{n+1}(x) = \frac{1}{\lambda_n}\eta_n(\xi_n(\lambda_n x))$$

and it is immediate that

$$\xi_{n+1}(x) = \frac{1}{\lambda_n}\eta_n(\lambda_n x).$$

This leads us to introduce a "renormalization operator" $\mathcal{T}$ acting on appropriate pairs (ξ, η) and producing new pairs $(\bar{\xi}, \bar{\eta})$ by

$$\bar{\eta}(x) = \frac{1}{\lambda}\eta \circ \xi(\lambda x)), \quad \bar{\xi}(x) = \frac{1}{\lambda}\eta(\lambda x) \qquad \text{where } \lambda \equiv \eta(0).$$

By what we have just shown

$$\mathcal{T}(\xi_n, \eta_n) = (\xi_{n+1}, \eta_{n+1})$$

(if the sequence of η's and ξ's is generated starting from a circle mapping f), although $\mathcal{T}$ "doesn't depend on f". In our formalism, the η_n and ξ_n are defined only starting at $n = 2$, but we can clean up the notation slightly by observing that if we introduce $\varsigma_1(f) = (\xi_1, \eta_1)$ by

$$\eta_1(x) = -f(-x); \qquad \xi_1(x) = x + 1$$

then we get

$$(\xi_n, \eta_n) = \mathcal{T}^{n-1}\varsigma_1(f) \qquad \text{for all } n \geq 1.$$

Various choices are possible for the space on which the renormalization operator is to act. We will describe one of the possibilities which seems to work reasonably well. The description will be semi-formal in that we will suppose we are dealing with functions defined and well-behaved on the whole real axis, although for serious applications of renormalization group ideas it is necessary to select bounded domains in the complex plane in such a way that the operator $\mathcal{T}$, acting on functions defined on those domains, gives functions with strictly larger domains.

Our space of mappings will consist of pairs (ξ, η) of real valued functions which are defined, analytic, and strictly increasing on **R**, with $\eta'(0) = 0$, normalized by $\xi(0) = 1$, and satisfying

$$\eta(x) < x; \quad \xi(x) > x; \quad \eta \circ \xi(x) > x \qquad \text{for all } x.$$

These latter conditions can be interpreted as follows: The first says that η moves points to the left; the second (which actually follows from the first and third) says that ξ moves points to the right; and the third says that the action of ξ dominates that of η.

The pairs (ξ_n, η_n) constructed from circle mappings have a further property which is extremely important for the analysis—they commute. Formally, we would like to restrict ourselves to working in the space of *commuting* pairs. For technical purposes, this may not be convenient—the set of commuting pairs may not form a space with good properties—and so, instead of exact commutation, it may be better to require only that $\xi \circ \eta$ and $\eta \circ \xi$ agree up to some finite order at the origin. Either exact commutativity or approximate commutativity to any given order is preserved by the renormalization operator. In our semi-formal discussion, we may as well consider the space of exactly commuting pairs. We will use $\mathcal{M}$ to denote the space of all pairs satisfying the above conditions.

It is easy to check that the formally defined operator $\mathcal{T}$, acting on a pair (ξ, η) in $\mathcal{M}$, gives a new pair again in $\mathcal{M}$ if and only if the original pair satisfies

$$\eta^2 \xi(x) < x \qquad \text{for all } x,$$

i.e., if and only if the action of η is more than half as strong as that of ξ, (in addition to the conditions defining $\mathcal{M}$.) Thus, if we let $\mathcal{D}$ denote the subspace of $\mathcal{M}$ satisfying this condition, $\mathcal{T}$ becomes a well-defined operator mapping $\mathcal{D}$ into $\mathcal{M}$.

In terms of this machinery, we reconsider the construction of a sequence (ξ_n, η_n) in $\mathcal{M}$ starting from a circle mapping f, dropping the requirement that the rotation number of f be exactly σ. If f is a critical circle mapping, it is easy to check that the pair $\varsigma_1(f)$ defined above is in $\mathcal{M}$ if and only if the rotation number of f is (strictly) between 0 and 1, and is in $\mathcal{D}$ if and only if it is between 1/2 and 1. More generally, it is not difficult to show that, for $n = 2, 3, \ldots$

$\rho(f)$ is (strictly) between Q_n/Q_{n+1} and Q_{n-1}/Q_n if and only if $\varsigma_1(f)$ is in $\mathcal{D}(\mathcal{T}^{n-1})$.

Hence, in particular, $\varsigma_1(f)$ is in $\mathcal{D}(\mathcal{T}^n)$ for all n if and only if f has rotation number σ. I stress that this is *not* a difficult or deep result; its proof uses only some very elementary theory of continued fractions (and a standard trick about rotation numbers to make sure the borderline cases come out as asserted).

The renormalization group picture now goes as usual: $\mathcal{T}$ has a fixed point (the pair (ξ^*, η^*) introduced above as the limit of rescaled f_n's; the functional equation satisfied by η^* just expresses the fixed point property.) This fixed point is hyperbolic, with one-dimensional unstable manifold and, hence, stable manifold (critical surface) of codimension one. Since anything in the stable manifold is in the domain of all powers of $\mathcal{T}$, we expect that the stable manifold will be, locally at least, the set of pairs with rotation number σ. Motion in $\mathcal{M}$ transverse to the stable manifold—in particular, motion along the unstable manifold—means changing the rotation number.

The validity of this picture can be proved, using estimates proved by computer. This has been done by B. Mestel[6], whose results are described in detail in a preprint of

Rand [8]. Partial results (existence of the fixed point) have been obtained by de la Llave and the author. I have been told by K. M. Khanin that Sinai and his collaborators in Moscow also have some results along these lines, but I do not know any details.

Rand's preprint extends Mestel's work considerably, using traditional analytic (non-computer) methods (but starting from Mestel's computer-verified estimates). In particular, Rand shows that Mestel's fixed point really can be obtained as a limit of rescaled f_n for an analytic circle mapping f, and that the stable manifold really is locally the pairs with rotation number σ.

As usual, the renormalization group picture leads to many precise quantitative statements about "universal rates". Here is one sample: Consider a one-parameter family of critical circle mappings f_μ which crosses the stable manifold transversally; for notational simplicity, assume that the crossing occurs for μ equal to zero. Then

$$\rho(f_\mu) - \rho(f_0) \sim \mu^{.92403\cdots},$$

where the "$\sim$" means that the ratio of the two sides is bounded and bounded away from zero for small enough μ. The exponent, which is "universal", i.e., is the same for all families which cross the stable manifold transversally, is given by $-2\log(\sigma)/\log(|\delta|)$, where $\delta = -2.83361\cdots$ is the expanding eigenvalue of the linearization of $\mathcal{T}$ at the fixed point.

General rotation number.

We will now rework the above construction of a renormalization operator to make a more general operator which can be used to study critical circle mappings with arbitrary rotation numbers. This will involve a certain amount of repetition in a more general framework of things done in a concrete way in the preceding section. The construction we describe is due to Ostlund, Rand, Sethna, and Siggia[7].

We begin by recalling some elementary facts about continued fractions. First, as a typographical convenience, we will write $[r_1, r_2, \ldots, r_n]$ for

$$\cfrac{1}{r_1 + \cfrac{1}{r_2 + \cfrac{1}{\ddots \, \cfrac{}{r_{n-1} + \cfrac{1}{r_n}}}}}$$

If ρ is a number in $(0,1)$, we put

$$r_1 = \mathrm{int}(\frac{1}{\rho}) \quad \text{and} \quad \rho_1 = \mathrm{frac}(\frac{1}{\rho}) \quad \text{so that} \quad \rho = \frac{1}{r_1 + \rho_1}.$$

We repeat this process as long as we can: If $\rho_n \neq 0$, we put

$$r_{n+1} = \mathrm{int}(\frac{1}{\rho_n}) \quad \text{and} \quad \rho_{n+1} = \mathrm{frac}(\frac{1}{\rho_n})$$

It is well-known that:

- The process terminates in a finite number of steps if and only if ρ is rational. If the process terminates at the n-th step, then $\rho = [r_1, \ldots, r_n]$, $r_n \geq 2$, and there is no other representation of ρ in the form $[s_1, \ldots, s_m]$ with the s_i strictly positive integers and $s_m \geq 2$. Without this last condition, there is a trivial ambiguity in the continued fraction representation of rational numbers, illustrated by

$$\frac{1}{2} = \frac{1}{1 + \frac{1}{1}}.$$

  When we speak of the continued fraction representation of a rational number, we always mean the unique representation whose last entry is at least two.
- If ρ is irrational,

$$\rho = \lim_{n \to \infty} [r_1, r_2, \ldots, r_n],$$

  and no other sequence of strictly positive integers has this property In other words, an irrational number has a unique continued fraction representation, and this representation can be generated by the iterative procedure defined above. We write $\rho = [r_1, r_2, \ldots]$ as an abbreviation for the above limiting relation.

In the preceding section, we constructed the renormalization operator as acting on a space of commuting pairs of mappings of the line, rather than as acting on circle mappings themselves; circle mappings f were identified with commuting pairs via $f \mapsto (x+1, -f(-x))$. We will proceed in the same way here, but we formalize these considerations with a little more generality. When we speak of a *commuting pair* in what follows, we mean a pair of continuous strictly increasing mappings (ξ, η) of $\mathbf{R}$ to itself which commute with each other and which satisfy in addition

$$\xi(x) > x \qquad \text{for all } x.$$

We will say that a commuting pair is *normalized* if $\xi(0) = 1$ and *critical* if $\eta'(0) = 0$. A non-normalized commuting pair (ξ, η) can be normalized by a rescaling, i.e., by replacing it by the essentially equivalent pair

$$(\gamma^{-1}\xi(\gamma x), \gamma^{-1}\eta(\gamma x)) \qquad \text{where } \gamma = \xi(0).$$

All the commuting pairs we encountered in the preceding section were normalized and critical.

The notion of rotation number generalizes from circle mappings to commuting pairs: If $\varsigma = (\xi, \eta)$ is a commuting pair, we define

$$\rho(\varsigma) = \sup\{\frac{p}{q} : p, q \in \mathbf{Z}, q > 0, \xi^p \circ \eta^q(x) < x \quad \text{for all } x\}.$$

It is easy to see that when a commuting pair is associated as above with a circle mapping, the rotation number for the commuting pair defined in this way agrees with the usual rotation number of the circle mapping. It is also easy to extend many of the elementary properties of the ordinary rotation number to this more general situation. Thus, for example, a commuting pair (ξ, η) has a rational rotation number p/q if and only if $\xi^p \circ \eta^q$ has a fixed point.

To define a renormalization operator, we work, as in the first section, in the space $\mathcal{M}$ of normalized critical commuting pairs with rotation number strictly between zero and one. The operator will be defined for all pairs ς in this space for which the rotation number is *not* the reciprocal of an integer. The construction will be guided by the objective of making a "canonical" (i.e., coordinate independent) operator $\mathcal{T}$ with the property that

$$\rho(\mathcal{T}\varsigma) = \mathrm{frac}(\frac{1}{\rho(\varsigma)}).$$

This property will imply immediately that the action of $\mathcal{T}$ on commuting pairs induces an action on rotation numbers which, in the continued fraction representation, is simply the left shift. More explicitly, a ς in $\mathcal{M}$ is in the domain ot $\mathcal{T}$ if and only the continued fraction representation of $\rho(\varsigma)$ has length at least two, and, if we write $\rho(\varsigma)$ as $[r_1, r_2, \ldots]$, then $\rho(\mathcal{T}\varsigma) = [r_2, \ldots]$.

The idea of the construction is as follows: Let (ξ, η) be a normalized commuting pair with rotation number $\rho > 0$. Loosely, the reversed pair (η, ξ) ought to have rotation number $1/\rho$. With the conventions we have adopted, this doesn't make sense, since $\rho > 0$ implies $\eta(x) < x$ for all x, whereas the first elements of our pairs are defined to move points to the right. We can deal with this, and also produce a normalized pair, by making a negative rescaling. Thus, we form the pair

$$(\frac{1}{\lambda}\eta(\lambda x), \frac{1}{\lambda}\xi(\lambda x)) \qquad \text{with } \lambda = \eta(0)(< 0).$$

It is easy to see that this pair satisfies all the conditions, is normalized, and does indeed have rotation number $1/\rho$. It also follows easily from the definitions that, if $(\bar{\xi}, \bar{\eta})$ is a normalized commuting pair with rotation number $\bar{\rho}$, then $(\bar{\xi}, \bar{\xi} \circ \bar{\eta})$ is also a normalized commuting pair but has rotation number $\bar{\rho} - 1$.

With all this in mind, we are ready to give the formal definition of the renormalization operator. For $r = 1, 2, \ldots$, we let $\mathcal{D}_r$ denote the set of normalized critical commuting pairs (ξ, η) with rotation number strictly between $(r+1)^{-1}$ and r^{-1}. We then define $\mathcal{T}_r$ on $\mathcal{D}_r$ by

$$\mathcal{T}_r(\xi, \eta)(x) \equiv (\frac{1}{\lambda}\eta(\lambda x), \frac{1}{\lambda}\eta^r \circ \xi(\lambda x)),$$

and it follows easily from the above that what is produced is a normalized critical commuting pair with rotation number

$$\frac{1}{\rho((\xi, \eta))} - r \in (0, 1),$$

i.e., is a member of $\mathcal{M}$. We have thus constructed a renormalization operator $\mathcal{T}_r$ for each strictly positive integer r. The operator $\mathcal{T}_1$ is what we called simply $\mathcal{T}$ in the preceding section. The domains of the operators for different r's are disjoint, so we can, as a notational convenience, think of all these operators taken together as a single operator $\mathcal{T}$, i.e., we define an operator $\mathcal{T}$ with domain $\mathcal{D} \equiv \cup_{r=1}^{\infty} \mathcal{D}_r$ by

$$\mathcal{T}\varsigma = \mathcal{T}_r\varsigma \qquad \text{when } \varsigma \in \mathcal{D}_r.$$

As already asserted, the domain $\mathcal{D}$ is exactly the set of all pairs in $\mathcal{M}$ with rotation number which is not the reciprocal of an integer. If, for $r = 2, 3, \ldots$, we let $\mathcal{V}_r$ denote the set of ς in $\mathcal{M}$ with rotation number r^{-1}, we get a disjoint decomposition

$$\mathcal{M} = \left(\cup_{r=1}^{\infty} \mathcal{D}_r\right) \cup \left(\cup_{r=2}^{\infty} \mathcal{V}_r\right).$$

The renormalization group picture developed in the first section can be expressed in the present language as the assertion that $\mathcal{T}_1$ has a fixed point in the neighborhood of which it is expansive in one direction and contractive in all others. It is natural to ask whether the same is true for $\mathcal{T}_r$ for general r. Numerical studies for a few other values suggest that it is; if so, then there is a theory paralleling that developed in the first section for all rotation numbers of the form $[r, r, \ldots]$. It even seems that any finite product of $\mathcal{T}_r$'s has such a fixed point; if this is so, there is a theory for each ***quadratic irrational*** rotation number.

It appears, in fact, that something considerably stronger is true—that there is a large open domain in $\mathcal{M}$ on which $\mathcal{T}$ acts expansively in one direction and contractively in all the others. This expansive-contractive character of $\mathcal{T}$ is referred to loosely as ***hyperbolicity***. The kind of hyperbolicity property which seems to hold is analogous—although with some important technical differences—to that of the Smale horseshoe map. In the remainder of this article we will explore the consequences of $\mathcal{T}$'s having such a property. This exploration will be semi-heuristic; the objective is to develop a picture rather than to make mathematically precise conjectures about exactly what hyperbolicity properties $\mathcal{T}$ is likely to have. Indeed, as will be seen, there is an important point on which it isn't yet clear exactly what one should expect. The exploration will be guided by the well-known theory of the Smale horseshoe (for an exposition in the spirit of this review, see [5]) and by some standard or almost-standard results from the general theory of hyperbolic sets which it should be straightforward to extend to the situation at hand once precise hyperbolicity properties have been established. I expect that (if careful numerical studies confirm the validity of this picture), it should eventually be possible to derive it from a precise set of statements about the action of $\mathcal{T}$ which could, in principle and perhaps also in practice, be proved rigorously by computer. I stress, however, that this is still far in the future. At this time, the principal justification for the picture is "physical"—it provides an economical explanation for some non-trivial phenomena which aren't understood otherwise. There is also a certain amount of indirect evidence for it (see, for example, Farmer and Satija[2] and Cvitanović, Jensen, Kadanoff, and Procaccia[1]), and, so far as I know, no evidence against it.

For simplicity of exposition, we will assume the domain on which $\mathcal{T}$ acts hyperbolically is all of $\mathcal{M}$. The sets $\mathcal{D}_i$, $\mathcal{V}_i$ defined above organize $\mathcal{M}$ into "layers", as illustrated in Fig 1. In this figure, rotation number increases upwards; $\mathcal{D}_1$ is on top, $\mathcal{V}_2$ immediately below it, $\mathcal{D}_2$ below that, and so on. Each $\mathcal{T}_r$ maps the corresponding $\mathcal{D}_r$ (a short wide rectangle) into a long thin rectangle running the full height of $\mathcal{M}$. The assertion that the image of $\mathcal{D}_r$ runs the full height of $\mathcal{M}$ is nothing mysterious; it is simply a geometrical translation of the statement that the image of the interval $((r+1)^{-1}, r^{-1})$ under $\rho \mapsto \mathrm{frac}(1/\rho)$ is all of $(0,1)$).

Consider a sequence $r_0, r_1, r_2, \ldots$ and ask what the sets $\mathcal{D}(\mathcal{T}_{r_n}\mathcal{T}_{r_{n-1}}\cdots\mathcal{T}_{r_0})$ look like. Take first $n = 1$. Then

$$\mathcal{D}(\mathcal{T}_{r_1}\mathcal{T}_{r_0}) = (\mathcal{T}_{r_0})^{-1}(\mathcal{D}_{r_1} \cap \mathcal{T}_{r_0}\mathcal{D}_{r_0}))$$

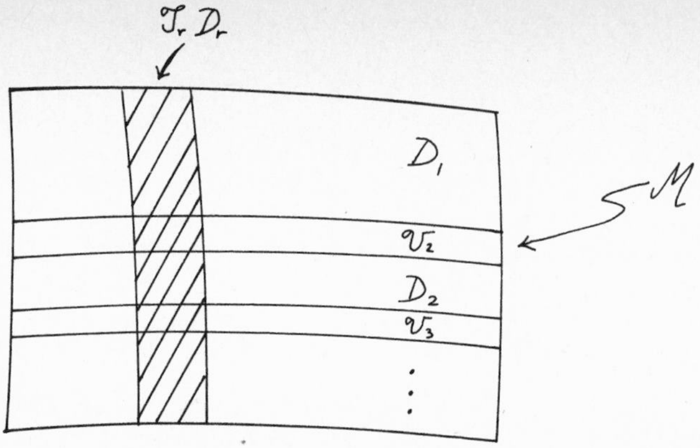

Figure 1.

As shown in Fig. 2, $D_{r_1} \cap (T_{r_0} D_{r_0})$ is a short rectangle running across $T_{r_0} D_{r_0}$, and since we are supposing that T_{r_0} is expansive vertically and contractive horizontally, its pre-image will be a shorter rectangle running all the way across M. Repeating the argument for larger values of n, we see that the domains $D(T_{r_n} T_{r_{n-1}} \cdots T_{r_0})$ are a sequence of shorter and shorter rectangles running all the way across M. It is a routine application of standard techniques from the theory of hyperbolic sets to show that what they converge to is a smooth codimension-one hypersurface, which we will denote by $W^s(r_0, r_1, \ldots)$.

Note, however, that from the definition of the T_r's it follows immediately that

$$\varsigma \in D(T_{r_n} T_{r_{n-1}} \cdots T_{r_0}) \quad \Longleftrightarrow \quad \rho(\varsigma) \text{ lies between } [r_0, r_1, \ldots, r_n] \text{ and } [r_0, r_1, \ldots, r_n + 1],$$

and hence that

$$\varsigma \in D(T_{r_n} T_{r_{n-1}} \cdots T_{r_0}) \text{ for all } n \quad \Longleftrightarrow \quad \rho(\varsigma) = [r_0, r_1, \ldots]$$

In other words, $W^s(r_0, r_1, \ldots)$ is exactly the set of ς's with rotation number $[r_0, r_1, \ldots]$.

We now need express a technical reservation. The operators T_r become more and more singular as $r \longrightarrow \infty$. It is not evident, heuristically, that we should expect the hyperbolicity of the action of T to hold uniformly in r (i.e., all the way down to the bottom of the figures we have been drawing). It may turn out that T is not uniformly hyperbolic on all of $\cup_{r=1}^{\infty} D_r$ but that it is uniformly hyperbolic on $\cup_{r=1}^{R} D_r$ for each R, without any uniformity in R. Under this weaker assumption, the above analysis

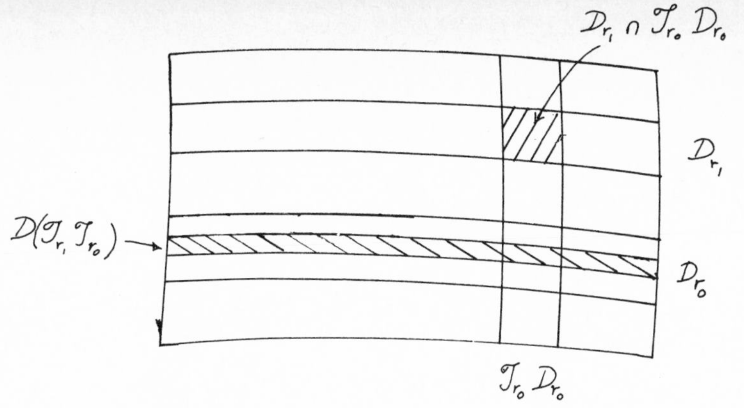

Figure 2.

of $W^s(r_0, r_1, \ldots)$ works only for *bounded* sequences. The behavior of $\mathcal{T}_r$ for large r is interesting and apparently non-trivial. At this time it is only partially understood, so it is not possible to say with any confidence whether hyperbolicity uniform in r is likely to be true. For the remainder of this exposition, we will describe results which follow from the assumption of uniform hyperbolicity; there are corresponding (slightly weaker, and more complicated) result which follow from the weakened hyperbolicity assumption.

Having described the structure of the domains of products of $\mathcal{T}_r$'s, we next describe the images of these products. We have already noted that $\mathcal{T}_{r_1} \mathcal{D}_{r_1}$ is a thin rectangle running the full height of $\mathcal{M}$. Since $\mathcal{T}_{r_0}$ contracts horizontally, it must map $\mathcal{D}_{r_0} \cap \mathcal{T}_{r_1} \mathcal{D}_{r_1}$ into a still thinner rectangle, again running the full height of $\mathcal{M}$ and contained, of course, in $\mathcal{T}_{r_0} \mathcal{D}_{r_0}$. (See Fig. 3.) Continuing in this way we see that, given any sequence $r_0, r_1, \ldots$, the images of the products $\mathcal{T}_{r_0} \mathcal{T}_{r_1} \cdots \mathcal{T}_{r_n}$ form a decreasing sequence of thinner and thinner rectangles of full height collapsing down to a smooth curve which we will denote by $W^u(r_0, r_1, \ldots)$.

For any doubly infinite sequence $\mathbf{r} = (\ldots, r_{-1}, r_0, r_1, \ldots)$, the curve $W^u(r_{-1}, r_{-2}, \ldots)$ and the hypersurface $W^s(r_0, r_1, \ldots)$ intersect in a single point which we will denote by $\phi(\mathbf{r})$. The mapping ϕ is continuous and

$$\mathcal{T} \phi(\mathbf{r}) = \phi(\tau \mathbf{r}),$$

where τ is the left shift mapping. The image of ϕ, which we denote by $\mathcal{X}$, is mapped to itself by $\mathcal{T}$; it is the analogue, in the present analysis, of a fixed point for the renormalization operator in standard renormalization group analyses. Numerical studies suggest

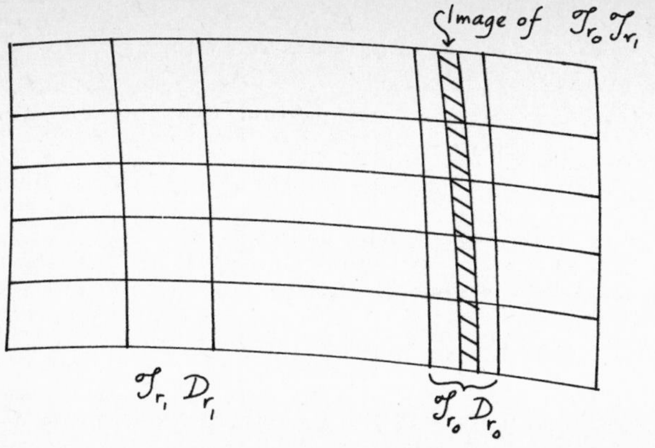

Figure 3.

that ϕ is one-to-one (There is, nevertheless, no convincing general reason why it must be, and the analysis we are going to present does not require it.) If it is, then $\mathcal{X}$ is an ***imbedded*** ∞***-shift***, with a Cantor-set-like structure. In any event $\mathcal{X}$ is something only slightly more general than a ***hyperbolic set*** in the classical sense of Anosov-Smale with one-dimensional unstable manifolds and codimension-one stable manifolds. It may be worth noting explicitly that the set $\mathcal{X}$ is ***most definitely not*** a strange attractor. It might reasonably be described, however, as a ***strange saddle***.

What we have been doing so far is just elaborating on what we really mean when we say that $\mathcal{T}$ acts expansively in one direction and contractively in all the others, on an adequately large region of the space of commuting pairs. We now want to extract some consequences from this picture. These consequences concern one-parameter families in $\mathcal{M}$, and it will be convenient to restrict the class of one-parameter families we consider and to impose some normalizations on how they are parametrized. From now on, when we speak of a one-parameter family ς_μ in $\mathcal{M}$, we always mean:

- The rotation number is a non-decreasing function of μ.
- The parameter μ runs from 0 to 1.
- The rotation number is 0 for $\mu = 0$, 1 for $\mu = 1$, and strictly between 0 and 1 for μ strictly between 0 and 1.

Given such a family, and a rational number p/q between zero and one, the set of μ's where $\rho(\varsigma_\mu) = p/q$ is an interval of non-zero length; we will refer to this interval as the ***phase-locking*** interval with rotation number p/q.

Consider, now, such a normalized one-parameter family ς_μ. For any $r = 1, 2, \ldots$, the set of μ's where $\varsigma_\mu \in \mathcal{D}_r$ is an interval (μ_r^-, μ_r^+). As μ runs over this interval, $\rho(\mathcal{T}_r \varsigma_\mu)$ runs from 1 to 0. By making an affine (decreasing) change of parametrization, we replace the interval (μ_r^-, μ_r^+) by $(0, 1)$ and recover the normalization. This gives a new one-parameter family, which we will denote by $\varsigma_\mu^{[r]}$. Explicitly:

$$\varsigma_\mu^{[r]} = \mathcal{T}_r\left(\varsigma_{\mu_r^+ + \mu\cdot(\mu_r^- - \mu_r^+)}\right).$$

Repeating this process, we define recursively

$$\varsigma_\mu^{[r_1, r_2, \ldots, r_n]} = \left(\varsigma_\mu^{[r_2, \ldots, r_n]}\right)^{[r_1]}.$$

(The order here is chosen so that $\varsigma_\mu^{[r_1, r_2 \ldots, r_n]}$ is obtained by applying $\mathcal{T}_{r_1} \mathcal{T}_{r_2} \cdots \mathcal{T}_{r_n}$ *in that order* to ς_μ for μ in the appropriate subinterval and reparametrizing.)

With this notation, we can formulate two principal technical results on which our analysis of the length of phase locking intervals rests. The following two propositions, while not quite standard, are not difficult to prove, using standard methods from the theory of hyperbolic sets, given appropriate hypotheses on the global hyperbolicity of $\mathcal{T}$. The first result concerns ways of parametrizing the curves W^u. For each $r_0, r_1, r_2, \ldots$, $\mathcal{T}_{r_0}$ maps part of $W^u(r_1, r_2, \ldots)$ onto all of $W^u(r_0, r_1, r_2, \ldots)$. What is claimed, in essence, is that there is a unique way to parametrize all the curves W^u in such a way as to make this action of $\mathcal{T}$ affine in the respective parametrizations.

Proposition 1. *There is a unique continuous way of writing all the curves $W^u(r_1, r_2, \ldots)$ as normalized one parameter families $W^u_\mu(r_1, r_2, \ldots)$ so that, for all $r_0, r_1, r_2, \ldots$,*

$$\left(W^u_\mu(r_1, r_2, \ldots)\right)^{[r_0]} = W^u_\mu(r_0, r_1, r_2, \ldots).$$

The second proposition says that applying $\mathcal{T}$ repeatedly to an "arbitrary" arc gives a sequence of arcs converging to one of the W^u.

Proposition 2. *There is an open set $\mathcal{U}$ of normalized one-parameter families (containing in particular all the $W^u_\mu(r_1, r_2, \ldots)$ such that, for any $\varsigma_\mu \in \mathcal{U}$ and any $r_1, r_2, \ldots$, the sequence of one-parameter families*

$$\varsigma_\mu^{[r_1]}, \varsigma_\mu^{[r_1, r_2]}, \ldots, \varsigma_\mu^{[r_1, r_2, \ldots, r_n]}, \ldots$$

converges to $W^u_\mu(r_1, r_2, \ldots)$. For fixed ς_μ, the convergence is uniform in $r_1, r_2, \ldots$ and exponentially fast in n.

We will refer to the set $\mathcal{U}$ of the above proposition as the ***universality class***. As noted, $\mathcal{U}$ is non-empty, since it contains the parametrized unstable manifolds. There are other ways of constructing many elements of $\mathcal{U}$. For example, if ς_μ is any one-parameter family crossing the ***stable*** manifold $W^s(r_0, r_1, \ldots)$ transversally, it is easy to see that $\varsigma^{[r_n, \ldots, r_0]}$ is in $\mathcal{U}$ for sufficiently large n. Roughly, $\mathcal{U}$ is the set of all one-parameter families which cross all the stable manifolds transversally and for which this transversality is sufficiently uniform.

We will write $\sigma(r_0|r_1, r_2, \ldots)$ and $\gamma(r_0|r_1, r_2, \ldots)$ for the lengths of the parameter intervals where $W^u_\mu(r_1, r_2, \ldots)$ is in $\mathcal{V}_{r_0}$ and $\mathcal{D}_{r_0}$ respectively. The functions σ and γ

are a convenient way of representing the metric structure of the hyperbolic invariant set $\mathcal{X}$. We are now going to argue that this metric structure determines a large number of "universal properties" of one-parameter families of mappings.

Let ζ_μ be in the universality class, and for any $r_0, \ldots, r_n$ let

$S(r_0, \ldots, r_n)$ denote the length of the parameter interval where ζ_μ is phase locked with rotation number $[r_0, \ldots, r_n]$.

$G(r_0, \ldots, r_n)$ denote the length of the parameter interval where $\zeta_\mu \in \mathcal{D}(\mathcal{T}_{r_n} \cdots \mathcal{T}_{r_0})$.

$$s(r_n | r_{n-1}, \ldots, r_0) \text{ denote } \frac{S(r_0, \ldots, r_n)}{G(r_0, \ldots, r_{n-1})}$$

$$g(r_n | r_{n-1}, \ldots, r_0) \text{ denote } \frac{G(r_0, \ldots, r_n)}{G(r_0, \ldots, r_{n-1})}$$

It follows readily from the definitions that we can also identify $s(r_n | r_{n-1}, \ldots, r_0)$ and $g(r_n | r_{n-1}, \ldots, r_0)$ with the lengths of the parameter intervals where $\zeta_u^{[r_{n-1}, \ldots, r_0]}$ is in $\mathcal{V}_{r_n}$ and $\mathcal{D}_{r_n}$ respectively. Thus, from Proposition 2, for large n,

$$s(r_n | r_{n-1}, \ldots, r_0) \approx \sigma(r_n | r_{n-1}, \ldots, r_0, r_{-1}, \ldots)$$
$$g(r_n | r_{n-1}, \ldots, r_0) \approx \gamma(r_n | r_{n-1}, \ldots, r_0, r_{-1}, \ldots)$$

Here, $r_{-1}, r_{-2}, \ldots$ can be chosen arbitrarily. These approximations hold in the precise sense that the ratio of the two sides differs from one by a quantity which is exponentially small in n, uniformly in $r_n, r_{n-1}, \ldots$. Note that the right-hand sides do not depend on the particular one-parameter family ζ_μ from which we started.

With these observation in mind, we write

$$\begin{aligned} S(r_0, r_1, \ldots, r_n) &= \frac{S(r_0, r_1, \ldots, r_n)}{G(r_0, r_1, \ldots, r_{n-1})} \frac{G(r_0, r_1, \ldots, r_{n-1})}{G(r_0, r_1, \ldots, r_{n-2})} \cdots \frac{G(r_1, r_0)}{G(r_0)} G(r_0) \\ &= s(r_n | r_{n-1}, \ldots, r_0) g(r_{n-1} | r_{n-2}, \ldots, r_0) \cdots g(r_1 | r_0) G(r_0) \end{aligned}.$$

This differs from

$$\sigma(r_n | r_{n-1}, r_{n-2}, \ldots) \gamma(r_{n-1} | r_{n-2}, \ldots) \cdots \gamma(r_1 | r_0, r_{-1} \ldots) \gamma(r_0 | r_{-1} \ldots)$$

by a factor which is bounded, and bounded away from zero, uniformly in $(n, r_n, r_{n-1}, \ldots)$. Since this latter expression does not depend on which one-parameter family ζ_μ we started from, we see in particular that

If $\zeta_\mu^{(1)}$ and $\zeta_\mu^{(2)}$ are any two members of the universality class, and if $S^{(1)}(r_0, \ldots, r_n)$ and $S^{(2)}(r_0, \ldots, r_n)$ are the corresponding lengths of phase-locking intervals, then the ratios

$$\frac{S^{(1)}(r_0, \ldots, r_n)}{S^{(2)}(r_0, \ldots, r_n)}$$

are bounded, and bounded away from zero, uniformly in $(n, r_n, \ldots, r_0)$

More loosely formulated: Up to bounded corrections, the lengths of all phase locking intervals are universal, i.e., the same for all members of the universality class.

This is only one of a large set of consequences which follow from this approach to the study of the lengths of phase-locking intervals. For example, it is a very small step to extend the above analysis to show that the Hausdorff dimension of the set of μ's where $\rho(\varsigma_\mu)$ is irrational is the same for all ς_μ's in the universality class. This universality was discovered numerically by Jensen, Bak, and Bohr[4]. Moreover, using methods coming from classical statistical mechanics, one can derive a prescription for determining this universal Hausdorff dimension from the quantities $\gamma(r_0|r_1\ldots)$.

References.

1. P. Cvitanović, M. H. Jensen, L. P. Kadanoff, and I. Procaccia, Renormalization, unstable manifolds, and the fractal structure of mode locking, *Phys. Rev. Lett.* **55** (1985) 343–346.
2. J. D. Farmer and I. I. Satija, Renormalization of the quasiperiodic transition to chaos for arbitrary winding numbers, *Phys. Rev. A.* **5** (1985) 3520–3522.
3. M. J. Feigenbaum, L. P. Kadanoff, and S. J. Shenker, Quasi-periodicity in dissipative systems: a renormalization group analysis, *Physica* **5D** (1982) 370–386.
4. M. H. Jensen, P. Bak, and T. Bohr, Complete devil's staircase, fractal dimension, and universality of mode-locking structure in the circle map, *Phys. Rev. Lett.* **50** (1983) 1637–1639.
5. O. E. Lanford, Introduction to the mathematical theory of dynamical systems, in *Chaotic Behaviour of Deterministic Systems, Les Houches Session XXXVI (1981)*, G. Iooss, R. H. G. Helleman, and R. Stora eds. (North Holland, Amsterdam, 1983) 5–51.
6. B. Mestel, Ph. D. Dissertation, Department of Mathematics, Warwick University (1985).
7. S. Ostlund, D. Rand, J. Sethna, and E. Siggia, Universal properties of the transition from quasi-periodicity to chaos in dissipative systems, *Physica* **8D** (1983) 303–342.
8. D. Rand, Universality for critical golden circle maps and the breakdown of dissipative golden invariant tori, Cornell University Laboratory of Atomic and Solid State Physics preprint, September 1984.

CORRELATIONS AND FLUCTUATIONS IN CHARGED FLUIDS

Ph. A. Martin
Institut de Physique Théorique
Ecole Polytechnique Fédérale de Lausanne
PHB-Ecublens
CH-1015 Lausanne, Switzerland

I. INTRODUCTION

We describe in this lecture specific properties of Gibbs states of charged particles which are not present in equilibrium states of particles interacting with short range forces. These new properties, multipolar sum rules, complete shielding, reduced fluctuations..., express the total or partial screening of the Coulomb force. Using field theoretic methods (Sine-Gordon formalism), the thermodynamic limit of the state can be constructed in some cases by means of correlation inequalities (charge symmetric systems [1]) or cluster expansions [2,3] , see also [4] for a review. Then, most of the above properties follow naturally from this field theoretical analysis.

Here we adopt a non contructive and general view point : we explore the constraints imposed by the long range Coulomb potential, assuming that the infinitely extended state exists and that it obeys appropriate equilibrium equations and cluster conditions. We treat general multicomponent systems, as well as the jellium model of point charges moving on a uniformly charged background of density ρ_B (OCP). We consider only homogeneous fluid phases in $\mathbb{R}^\nu$ in dimension $\nu = 2,3$, and have no results for the crystal (the one dimensional Coulomb system is solvable and well understood [5, 6]). Inhomogeneous fluid phases obtained by submitting the system to external fields or confining it by hard walls are of considerable interest, some recent results can be found in [7,8,9,10,11].

In this text, we state the relevant propositions and sketch the main arguments in their simplest form. We refer to the quoted references for details and completeness.

II. MULTIPOLAR SUM RULES

The screening properties of the Coulomb force are conveniently expressed in terms of the excess particle density at q , $\hat{\rho}(q|q_1 \cdots q_n)$, when particles are fixed at $q_1 \cdots q_n$

$$\hat{\rho}(q|q_1 \ldots q_n) = \frac{\rho(q q_1 \ldots q_n)}{\rho(q_1 \ldots q_n)} + \sum_{i=1}^{n} \delta_{qq_i} - \rho(q) \qquad (1)$$

$q = (\alpha, x)$ denotes the position x of a particle of species α and charge e_α, $\delta_{qq_i} = \delta_{\alpha\alpha_i}\,\delta(x - x_i)$ and $\rho(q_1 \ldots q_n)$ are the correlation functions of the infinitely extended state.

A phase with good screening properties (plasma or conducting phase) is characterized by the fact that the excess charge density carries no multipole of any order

$$\int dx\, \mathcal{Y}_\ell(x) \sum_\alpha e_\alpha \hat{\rho}(\alpha, x | q_1 \ldots q_n) = 0 \qquad (2)$$

where $\mathcal{Y}_\ell(x)$ is an harmonic polynomial on $\mathbb{R}^\nu$. This set of constraints for $\ell = 0,1,2\ldots$, $n = 1,2,\ldots$ (ℓ, n sum rules), which do not exist for short range forces, express that typical configurations in a plasma phase shield the multipoles induced by specifying any arrangement of the system's charges. The case $\ell = 0$ (resp. $\ell = 1$) are called the charge (resp. dipole) sum rules. The general (ℓ, n) sum rules can be established by two different methods.

Proposition 1 [12]

The states constructed by Brydges and Federbush [2], and Imbrie [3] in the Debye screening regime $\beta e^2 / \ell_D \ll 1$ satisfy the (ℓ, n) sum rules for all $\ell = 0, 1 \ldots$, $n = 1,2,\ldots$ ($\ell_D = (\beta \sum_\alpha e_\alpha^2 z_\alpha)^{-\frac{1}{2}}$ Debye length).

In the Sine-Gordon formalism used here, the sum rules follow from the fact that the Gaussian measure with covariance given by the inverse of the Coulomb potential (the Laplacian $-\Delta$) is formally invariant under translations by harmonic functions $\mathcal{Y}(x)$ with $\Delta \mathcal{Y}(x) = 0$. Technically, one has first to approximate $\mathcal{Y}(x)$ by a smooth function with compact support, the resulting boundary terms do not contribute as a consequence of exponential clustering which is known to hold in the Debye regime [12].

Alternatively, the (ℓ, n) sum rules can be deduced from the BGY equilibrium equations (assuming now that the state exists in the thermodynamic limit and verifies the BGY hierarchy) whenever the rate of clustering is sufficiently fast [13, 8, 14]. For an homogeneous neutral state, the BGY hierarchy has the form

$$\beta^{-1}\nabla_1\rho(q_1 q_2) = \sum_{j=2}^{n} F(q_1 q_j)\rho(q_1 \dots q_n) + \int dq\, F(q_1 q)\big(\rho(q_1 \dots q_n q) - \rho(q)\rho(q_1 \dots q_n)\big) \quad (3)$$

$F(q_1 q_2) = -e_{\alpha_1} e_{\alpha_2} \nabla \phi_1(x_1 - x_2)$ is the force, and $\phi(x)$ is the Coulomb potential regularized at the origin. Introducing the fully truncated correlations $\rho_T(q_1 \dots q_n)$ defined in the usual way, we have :

<u>Proposition 2</u> [8 ,14]

If the correlations satisfy Eq. (3) and

$$|r^{\ell_0 + \nu + \varepsilon} \rho_T(q_1 \dots q_n)| < M \quad , \quad \nu \geq 2$$

for some $\varepsilon > 0$, $r = \sup_{ij} |x_i - x_j|$, $i j = 1, \dots n$, $n = 2 \dots n_0 + 2$ then the (ℓ, n) sum rules hold for $\ell \leq \ell_0$, $n \leq n_0$.

The main argument is simple : written in terms of truncated functions (3) is equivalent to

$$\beta^{-1}\nabla_1 \rho_T(q_1 Q) = e_{\alpha_1}\rho(q_1)\rho(Q)E(x_1) + \sum_{j=2}^{n} F(q_1 q_j)\rho_T(q_1 Q) + \int dq\, F(q_1 q)\rho_T(q_1 q\, Q) \quad (4)$$

with the notation $Q = (q_2 \dots q_n)$, $\int dq = \int dx \sum_\alpha$, and

$$\begin{aligned}\rho_T(q_1 Q) &= \rho(q_1 Q) - \rho(q_1)\rho(Q)\\ \rho_T(q_1 q Q) &= \rho(q_1 q Q) - \rho(q)\rho(q_1 Q) - \rho(q_1)\big(\rho(q\, Q) - \rho(q)\rho(Q)\big)\\ &\quad - \rho(Q)\big(\rho(q_1 q) - \rho(q_1)\rho(q)\big)\end{aligned} \quad (5)$$

In Eq. (4) $E(x_1) = \int dx\, F(x_1 - x) \sum_\alpha e_\alpha \hat\rho(\alpha x | Q)$ is precisely the electric field at x_1 generated by the excess charge density $\sum_\alpha e_\alpha \hat\rho(\alpha x | Q)$. We integrate Eq. (4) on a ball of fixed radius centered around x_1 and let $|x_1| \to \infty$. Then, the cluster assumption implies that all terms except the first one in the r.h.s. are $o(|x_1|^{-\nu+1+\ell_0})$. Hence, $E(x_1)$ also decays faster than $|x_1|^{-\nu+1+\ell_0}$. Introducing its multipolar expansion, we conclude that $\sum_\alpha e_\alpha \hat\rho(\alpha x | Q)$ has no multipoles up to order ℓ_0.

Let $C(x) = \sum_j e_{\alpha_j}\delta(x - x_j) + \rho_B$ be the charge density and

$$\begin{aligned} S(x_1, x_2) &= \langle (C(x_1) - \langle C(x_1)\rangle)(C(x_2) - \langle C(x_2)\rangle)\rangle \\ &= \sum_{\alpha_1 \alpha_2} e_{\alpha_1} e_{\alpha_2}\big(\rho(q_1 q_2) - \rho(q_1)\rho(q_2) + \delta_{q_1 q_2}\rho(q_1)\big)\end{aligned} \quad (6)$$

be truncated charge charge correlation function .

In an homogeneous neutral state $\langle C(x)\rangle = \sum_\alpha e_\alpha \rho_\alpha + \rho_B = 0$

and $S(x_1,x_2) = S(x_1-x_2,0) \equiv S(x_1-x_2)$.

Then, with (6), the (0,1)-sum rule has the important consequence that the total integral of $S(x)$ vanishes

$$\int dx\, S(x) = 0 \tag{7}$$

Since $S(x)$ is spherically symmetric, the first non trivial higher order sum rule is the dipole sum rule when two particles are fixed (ℓ=1, n = 2)

$$\int dq\, e_\alpha x\, \hat{\rho}(q|q_1q_2) = 0 \tag{8}$$

III. COMPLETE SHIELDING

The (ℓ=0,n) charge sum rule, which holds under the condition of integrable clustering, can be interpreted as the shielding of test particles of the same species as those which constitute the system itself.

The situation is however not the same when we introduce test charges which are different from the system's charges : they may be shielded or not according to the plasma or dielectric nature of the phase [15,4]. We say that the system has the complete shielding property if any external distribution is screened by the system's charges.

In this case, the charge-charge correlation has to satisfy the additional constraint

$$\int dy \int dx\, \frac{1}{|x|}\, S(x-y) = \beta^{-1} \tag{9}$$

which can easily be derived by expressing the shielding of an infinitesimal charge in the linear response theory.

An alternative form of (9) is the second moment Stillinger-Lovett condition [16]

$$\lim_{k\to 0} \varepsilon^{-1}(k) = \lim_{k\to 0}\left(1 - \frac{\omega_\nu\beta}{|k|^2}\tilde{S}(k)\right)$$

$$= 1 + \frac{\omega_\nu\beta}{2\nu}\int dx\, |x|^2 S(x) = 0 \tag{10}$$

where $\tilde{S}(k)$ is the Fourier transform of $S(x)$, $\omega_2 = 2\pi$, $\omega_3 = 4\pi$ and $\varepsilon(k)$ is the dielectric function. The property $\lim_{k\to 0} \varepsilon^{-1}(k) = 0$ characterizes a perfect conductor. The equivalence of (9) and (10) result of an integration by parts and of the Poisson equation $\Delta \frac{1}{|x|} = -\omega_\nu \delta(x)$ writing

$$\int dy \int dx \frac{1}{|x-y|} S(x) = \frac{1}{2\nu} \int dy\, |y|^2 \Delta_y \int \frac{1}{|x-y|} S(x)\, dx = -\frac{\omega_\nu}{2\nu} \int dy\, |y|^2 S(y)$$

We show in the next proposition that the complete shielding relation (9) occurs under slightly stronger clustering conditions than those needed for the simple electroneutrality (ℓ=0,n) sum rule : one needs also that the dipole sum rule (8) is satisfied.

Proposition 3 [17]
Assume that

$$\int dx\, |x|^2 |S(x)| < \infty, \quad \int dx \int dx_2\, |x_2|\, |\rho_T(q_1 q q_2)| < \infty$$

If the (0,1) charge and (1,2) dipole sum rules are verified, then the complete shielding relation (9) holds.

In the case n=2 ($Q = q_2$), Eq. (4) can be written, setting $q_1 = (\alpha_1, 0)$ and using translation invariance, as

$$-\beta^{-1} \nabla_2 \rho_T(q_1 q_2) = e_{\alpha_1} \rho_{\alpha_1} \rho_{\alpha_2} \int dx\, F(-x) \sum_\alpha e_\alpha \hat{\rho}(\alpha x | q_2)$$

$$+ \int dq\, F(q_1 q) \left(\rho_T(q_1 q_2 q) + (\delta_{q_2 q} + \delta_{q_2 q_1}) \rho_T(q_1 q) \right) \tag{11}$$

(Notice that $\int dx\, F(-x) \rho_T(\alpha_1 0, \alpha x) = 0$ by the invariance of the state under rotations and the antisymmetry of the force). We take now the scalar product of (11) with $e_{\alpha_2} x_2$ and sum over α_2 . After an integration by part the l.h.s. of (11) equals

$$\nu \beta^{-1} \int dq_2\, e_{\alpha_2} \rho_T(q_1 q_2) = -\nu \beta^{-1} e_{\alpha_1} \rho_{\alpha_1} \tag{12}$$

where the charge sum rule has been used. It is easy to check that the dipole sum rules (8) implies also for the truncated functions

$$\int dq_2\, e_{\alpha_2} x_2 \left[\rho_T(q_1 q_2 q) + (\delta_{q_2 q} + \delta_{q_2 q_1}) \rho_T(q_1 q) \right] = 0 \tag{13}$$

Hence, exchanging the q and q_2 integrals the contribution of the last term of (11) vanishes. Finally, introducing the definition (6) in the first term of the r.h.s. of (11) gives with (12), (13)

$$-\nu\beta^{-1}e_{\alpha_1}\rho_{\alpha_1} = e_{\alpha_1}\rho_{\alpha_1}\int dx_2\, x_2 \cdot \int dx\, F(-x)\, S(x-x_2)$$

$$= e_{\alpha_1}\rho_{\alpha_1}\int dx_2\, x_2 \cdot \nabla_2 \int dx\, \frac{1}{|x_2-x|}\, S(x)$$

After an integration by part, this gives (9).

This result implies that in a phase where arbitrary test charges are not screened and (9) does not hold (as in the Kosterlitz-Thouless phase [15, 4]), the dipole sum rule (8) cannot be true. It then follows from Prop.2 that the clustering cannot be faster than $|x|^{-(\nu+1)}$ in a dielectric phase.

IV. CHARGE FLUCTUATIONS

A direct consequence of screening is the reduced growth of the charge fluctuations [18]. Heuristically, a free charge carries always its neutralizing cloud (of radius of the order ℓ_D), constituting with it a neutral entity. In a macroscopic region Λ (of volume $|\Lambda| \gg \ell_D$) only those entities cutting the boundary $\partial\Lambda$ of Λ at random contribute to the net charge C_Λ in Λ. The fluctuations $\langle (C_\Lambda^2 - \langle C_\Lambda \rangle^2) \rangle$ may then be expected to be proportional to the surface $|\partial\Lambda|$ of Λ, and not to its volume as are the usual fluctuations of extensive quantities.

The correlations between two subdomains Λ_1 and Λ_2 with characteristic functions $\chi_{\Lambda_1}(x)$ and $\chi_{\Lambda_2}(x)$ are

$$\langle C_{\Lambda_1} C_{\Lambda_2} \rangle = \int_{\Lambda_1} dx_1 \int_{\Lambda_2} dx_2\, S(x_1-x_2) = \int dx\, S(x)\, \gamma_{\Lambda_1\Lambda_2}(x) \tag{14}$$

where $\gamma_{\Lambda_1\Lambda_2}(x) = \int dy\, \chi_{\Lambda_1}(y)\, \chi_{\Lambda_2}(x+y)$ is the volume of the intersection of Λ_1 with the x-translate of Λ_2.

Choosing first $\Lambda_1 = \Lambda_2 = \Lambda$ = cube of side L, we have

$$\gamma_{\Lambda\Lambda}(x) = L^\nu - \sum_{r=1}^{\nu} |x^r|\, L^{\nu-1} + O(L^{\nu-2})$$

With (7), the volumic contribution vanishes, and we find that $\langle C_\Lambda^2 \rangle$ is of the order of the surface

$$\lim_{L\to\infty} \frac{\langle C_\Lambda^2 \rangle}{2\nu L^{\nu-1}} = -\frac{1}{2\nu}\int dx \sum_{r=1}^{\nu} |x^r|\, S(x) \equiv K > 0 \tag{15}$$

We consider now the charge fluctuations in two adjacents cubes of volume L^ν centered at (0,0...0) and (L, 0...0). In this case

$$\gamma_{\Lambda_1\Lambda_2}(x) = |x^1| L^{\nu-1} + \mathcal{O}(L^{\nu-2}) \;, \quad -L \leq x^1 \leq 0 \;,$$

and thus

$$\lim_{L\to\infty} \frac{\langle C_{\Lambda_1} C_{\Lambda_2}\rangle}{2\nu L^{\nu-1}} = \frac{1}{4\nu}\int dx\, |x^1|\, S(x) = -\frac{1}{2\nu} K < 0 \qquad (16)$$

The negative correlations of adjacent regions reflects again the fact that the system behaves as built of neutral entities.

More generally, let $\mathbb{R}^\nu = \bigcup_j \Lambda_j$ be divided into disjoint cubes of volume L^ν centered on a simple cubic lattice, and let $c_{\Lambda_j} = (2\nu L^{\nu-1})^{-1/2} C_{\Lambda_j}$ be the appropriately normalized charges in Λ_j. Then, as stated in the next proposition, these random variables are jointly Gaussian with covariance given by the finite difference Laplacien $-\Delta_{ij}$ on $\mathbb{Z}^\nu$.

Proposition 4 [19]

Assume that

(i) $\int dx_2 \ldots dx_n \,|\rho_T(\alpha_1 0, q_2 \ldots q_n)| < \infty$ ($\mathcal{L}^1$-clustering)

(ii) $\int dx_2 \ldots dx_n \,|x_2|\,|\rho_T(\alpha_1 0, q_2 \ldots q_n)| < \infty$ for n = 2,3

(iii) the (ℓ=0,n) sum rules hold for n = 1, 2, 3.

Then the joint distribution of the c_{Λ_j} is Gaussian as $L\to\infty$ with covariance $K\left(\delta_{ij} - \frac{1}{2\nu}\delta_{|i-j|,1}\right) = \frac{K}{2\nu}(-\Delta)_{ij} \equiv d_{ij}$.

To obtain the covariance d_{ij}, one notes that (in addition to (15) and (16)) when the cubes are not adjacents, one has always $\gamma_{\Lambda_1\Lambda_2}(x) = \mathcal{O}(L^{\nu-2})$; thus non adjacent cubes are not correlated. The Gaussian distribution results from the fact that, with (i), (ii) and the sum rules (iii) all n^{the} order cumulants vanish except when n = 2.

V. POTENTIAL AND FIELD FLUCTUATIONS

For particles with short range interaction (for instance with finite range d), the potential at x is a local quantity, i.e. it depends only on the part of the particle configuration which is in a sphere of radius d around x. Then the potential and force fluctuations are well defined as soon as the state has a thermodynamic limit. This is not the case for the Coulomb potential : here $V(x) = \int dy \frac{1}{|x-y|} C(y)$ is genuinely non local and particles far away will contribute to the fluctuations at x . Our main point is that whenever the potential fluctuations exist in the thermodynamic limit, they can as well be computed as limit of averages of strictly local functions in the infinitely extended state (we consider here the three dimensional situation $\nu = 3$).

We first split the potential at x into a local part $V_R^{in}(x)$ due to the particles inside a sphere Σ_R of radius R centered at x , and a global outside part $V_R^{out}(x)$ due to the particles in the exterior of this sphere

$$V(x) = V_R^{in}(x) + V_R^{out}(x) \tag{17}$$

with $V_R^{in}(x) = \int\limits_{|y-x| \leq R} dy \, \frac{1}{|x-y|} \, \zeta(y)$, $V_R^{out}(x) = \int\limits_{|y-x| \geq R} dy \, \frac{1}{|x-y|} \, \zeta(y)$.

We introduce moreover the spatial average of the total potential on this sphere

$$\overline{V}_R(x) = \frac{1}{|\Sigma_R|} \int\limits_{|z-x| \leq R} dz \, V(z) \, , \qquad |\Sigma_R| = \frac{4\pi R^3}{3} \, .$$

Then the formula

$$\int\limits_{|y| \leq R} dy \, \frac{1}{|x-y|} = \begin{cases} \frac{4\pi R^3}{3} \, \frac{1}{|x|} \, , & |x| \geq R \\ -\frac{2\pi}{3} |x|^2 + 2\pi R^2 \, , & |x| \leq R \end{cases}$$

yields the identity

$$V_R^{out}(x) = -\hat{V}_R(x) + \overline{V}_R(x) \tag{18}$$

with

$$\hat{V}_R(x) = \int_{|y-x| \leq R} dy \left(-\frac{1}{2R^3} |y-x|^2 + \frac{3}{2R} \right) \zeta(y) \, . \tag{19}$$

Proposition 5 [20]

Let $W_\Lambda(x,y) = \int_\Lambda dx_1 \int_\Lambda dy_1 \, \frac{1}{|x_1 - x|} \, \frac{1}{|y_1 - y|} \, S_\Lambda(x_1, y_1)$

be the potential fluctuations in the finite volume system. If (i) the thermodynamic limit $W(x-y) = \lim\limits_{\Lambda \to \mathbb{R}^3} W_\Lambda(x,y)$ exists and (ii) $\lim\limits_{|x| \to \infty} W(x) = 0$, then the potential fluctuations can be computed locally in the infinite state by

$$W(x-y) = \lim_{R\to\infty} \langle (\hat{V}_R(x) - \langle \hat{V}_R(x)\rangle)(\hat{V}_R(y) - \langle \hat{V}_R(y)\rangle)\rangle \tag{20}$$

with $\hat{V}_R(x) = V_R^{in}(x) - \breve{V}_R(x)$.

The proposition results from the fact that the spatially averaged potential $\overline{V}_R(x)$ does not contribute to the fluctuations. Indeed, the assumption (i) implies that for each fixed R

$$\lim_{\Lambda\to\mathbb{R}^3} \langle (\overline{V}_R(x) - \langle \overline{V}_R(x)\rangle_\Lambda)^2\rangle_\Lambda = \frac{1}{|Z_R|^2} \int\limits_{|x_1-x|\le R} dx_1 \int\limits_{|y_1-x|\le R} dy_1 \; W(x_1-y_1)$$

$$= \left(\frac{3}{4\pi}\right)^3 \int\limits_{|x_1|\le 1} dx_1 \int\limits_{|y_1|\le 1} dy_1 \; W(R(x_1-y_1))$$

and by (ii) this expression tends to zero as $R\to\infty$.

We therefore conclude that the contribution $V_R^{out}(x)$ of the particles at infinity to the fluctuations can be calculated from the local function $-\hat{V}_R(x)$, and the decomposition (17) provides a natural distinction between the contribution of the nearby and far away particles. The next proposition expresses these fluctuations with the help of the charge-charge correlation and shows that the potential from distant regions has a Gaussian distribution.

<u>Proposition 6</u> [20]
If $\int dx_2 \, |x_2| \, |\rho_T(\alpha_1 0, \alpha_2 x_2)|$ exists and the (0, 0)-charge sum rule holds, one has

$$W(x) = -2\pi \int dy \, |x+y| \, S(y)$$

and

$$W(x) = \frac{1}{|x|}\left(-\frac{2\pi}{3}\int dy \, |y|^2 S(y)\right) + o\left(\frac{1}{|x|}\right) \quad \text{as } |x|\to\infty . \tag{21}$$

Moreover, if the state is $\mathcal{L}^1$-clustering, $V_R^{out}(x)$ are jointly Gaussian as $R\to\infty$ with covariance $W^{out}(x) = -\pi \int dy \, |x+y| \, S(y)$.

It is interesting to note that particles at infinity contribute in $W^{out}(x)$ for half of the total potential fluctuations.

To obtain the asymptotic behaviour (21) we use the charge sum rule and the spherical symmetry to write

$$W(x) = -2\pi \int dy \left(|x+y| - |x| - y \cdot \frac{x}{|x|} \right) S(y)$$
$$= -\frac{\pi}{|x|} \sum_{r,s}^{3} \left(\delta_{rs} - \frac{x^r x^s}{|x|^2} \right) \int dy \, y^r y^s S(y) + o\left(\frac{1}{|x|}\right)$$

Since $\int dy \, y^r y^s S(y) = \frac{1}{3} \delta_{rs} \int dy \, |y|^2 S(y)$, we obtain (21). We see that the decay of the potential fluctuations is always slow, even if the clustering is exponentially fast. Moreover, when the complete shielding relation (10) holds, the asymptotic behaviour of $W(x)$ is $\beta^{-1} \frac{1}{|x|}$, i.e. universal and independent of the short range part of the interaction. The electric field fluctuations can be treated in the same way, replacing everywhere $V(x)$ by $-\nabla V(x)$, see [20].

In two dimensions, the fluctuations of the potential at a point are infinite; however, the potential differences have finite fluctuations similar to the ones which occur in three dimensions [21].

VI. DECAY OF CORRELATIONS

Not much is known on the asymptotic form of the correlations in charged systems, except in the Debye regime where it is bounded by an exponential [2]. However, because of the local relation between the charge density and the field generated by it (due to the harmonicity of the Coulomb potential), all types of asymptotic behaviour are not allowed : if the correlations are integrable and monotonous at infinity, they have to decay faster than any inverse power. This is in sharp contrast with the behaviour of a neutral fluid of particles interacting with an integrable potential $\phi(x) = b|x|^{-s}$, $s > \nu$: here, at low activity, the correlation decay monotonously, exactly as $|x|^{-s}$ [22].

We formulate our result for the case of the two point function of the homogeneous OCP in three dimensions. Denoting $\rho_T(r) = \rho(x,0) - \rho^2$, $(|x| = r)$, the spherically symmetric truncated two point function, Eq. (4) reduces in this case to

$$\beta^{-1} \frac{d}{dr} \rho_T(r) = e E(r) + \frac{e^2}{r^2} \rho_T(r) + e^2 \frac{x}{|x|} \cdot \int dy \, F(x-y) \, \rho_T(x,y,0) \qquad (22)$$

where $E(r)$ is the radial component of the electric field determined by the Poisson equation

$$\frac{1}{4\pi r^2}\frac{d}{dr}\left(r^2 E(r)\right) = e\left(\frac{\rho_T(r)}{\rho} + \delta(x)\right) \tag{23}$$

(In the OCP a short range regularization of the Coulomb potential is not needed).

Proposition 7 [23]
Assume that $\rho_T(x,0)$ is integrable and $\rho_T(r)$ tends monotonously to zero as $r\to\infty$. Moreover, for $|x|$ large enough, $|\rho_T(x,y,0)| \le M(t)\,|\rho_T(x,0)|$, $t = \min(|x-y|, |y|)$ and $\lim_{t\to\infty} M(t) = 0$, then
$\lim_{r\to\infty} r^p \rho_T(r) = 0$ for all $p>0$.

The assumed bound on the three point truncated function is slightly stronger than in Prop. 2 in the sense that some joint decay is required as a second particle is sent to infinity. This bound is compatible with known exact results and bounds indicated by perturbative expansions [23].

Let us suppose first that $\lim_{r\to\infty} r^p \rho_T(r) = A$ for some $p>3$ (24)

Integrating the Poisson equation (23) gives for $r\neq 0$

$$E(r) = -\frac{4\pi e}{\rho r^2}\int_r^\infty dr'\, r'^2 \rho_T(r') = -\frac{4\pi e}{\rho(p-3)}\frac{A}{r^{p-1}} + o\left(\frac{1}{r^{p-1}}\right) \tag{25}$$

The dominant contribution to $E(r)$, which would be $\frac{e}{r^2}\int dx\left(\frac{\rho_T(x,0)}{\rho} + \delta(x)\right)$, vanishes because of the charge sum rule. Moreover, the condition on the three point function implies that (lemma 1 of [23])

$$\int dy\, F(x-y)\,\rho_T(x,y,0) = o\left(\frac{1}{r^{p-1}}\right) \tag{26}$$

Hence, inserting (25) and (26) in (22) gives

$$\beta^{-1}\frac{d}{dr}\rho_T(r) = -\frac{4\pi e^2}{\rho(p-3)}\frac{A}{r^{p-1}} + o\left(\frac{1}{r^{p-1}}\right) \tag{27}$$

and therefore, by integration

$$\beta^{-1}\rho_T(r) = \frac{4\pi e^2}{\rho(p-3)(p-2)}\frac{A}{r^{p-2}} + o\left(\frac{1}{r^{p-2}}\right) \tag{28}$$

Thus, we conclude from (24) that $A=0$.

The a priori assumption (24) can be removed to exclude all kinds of monotonous non algebraic decays which are not faster than inverse powers [23]. However, the monotonicity hypothesis plays an essential role, and we cannot exclude an oscillatory behaviour as $\frac{\cos \lambda r}{r^{\rho}}$ because then the field and the charge density are of the same order at infinity. These arguments can be extended to higher order correlations and to multicomponent systems.

VII. TIME DISPLACED CORRELATIONS

It is of interest to investigate to what extent the screening sum rules remain valid in a dynamical situation. We consider the time dependent correlations of the homogeneous OCP involving k particles at time t and n particles at time $t=0$, formally defined by

$$\rho(U,t|V)=\Big\langle\Big[\sum_{i_1\neq\dots\neq i_k}\delta(u_1-\hat{u}_{i_1}(t))\dots\delta(u_k-\hat{u}_{i_k}(t))\Big]\Big[\sum_{j_1\neq\dots\neq j_n}\delta(v_1-\hat{u}_{j_1}(0))\dots\delta(v_k-\hat{u}_{j_n}(0))\Big]\Big\rangle \quad (29)$$

$u=(x,p)$ denotes the position and momentum of a particle, and $U=(u_1\dots u_k)$ $V=(v_1\dots v_n)$ are sets of coordinates; $\hat{u}_i(t)=(x_i(t),p_i(t))$ are the coordinate of the ith particle at time t under the time evolution and $\langle\dots\rangle$ is the thermal average on the initial conditions $\hat{u}_i(0)=\hat{u}_i$. When the set V is empty, the correlations reduce to their equilibrium value

$$\rho(U,t)=\rho(U)=\rho(x_1\dots x_k)\prod_{i=1}^{k}\left(\frac{\beta}{2m}\right)^{3/2}\exp\left(-\beta\frac{p_i^2}{2m}\right).$$

The existence of the dynamics in the infinite system has so far not been proven : we assume here that the correlations (29) exist in the thermodynamic limit and obey the BBGKY equation (for $k=1$)

$$\frac{\partial}{\partial t}\rho(u_1,t|V)=-\frac{p_1}{m}\cdot\nabla_{x_1}\rho(u_1,t|V)$$

$$-e^2\int dx_2 F(x_1-x_2)\cdot\nabla_{p_1}\left(\rho(u_1,x_2,t|V)-\rho\,\rho(u_1,t|V)\right) \quad (30)$$

The time dependent generalization of the static excess particle density (1) is*)

$$\hat{\rho}(x,t|V)=\frac{1}{\rho(V)}\int dp\,\rho(x,p,t|V)-\rho \quad (31)$$

---

*)In the sequel, we simply suppress momentum arguments in the correlation functions when they have been integrated out.

and reduces to (1) at time $t=0$ (notice that self correlations are included in the definition (29), but not in the definition of the static correlations). We introduce the truncated correlations $\rho_T(u_1,t|V)$ and $\rho_T(u_1,u_2,t|V)$ defined as in (5). Then, under some reasonable clustering assumptions, the charge and dipole sum rule still hold for all times in the homogeneous OCP.

Proposition 8 [11]
Assume that (i) $|\rho(x,p,t|V)| \leq \frac{M}{p^{5+\varepsilon}}$, $|p|\to\infty$, (insuring the existence of the kinetic energy density), (ii) $\rho_T(x,p,t|V) \leq \frac{M}{|x|^3}$, $|x|\to\infty$, and (iii) $\int dx_1 \int dx_2\, |x_2|\, |\rho_T(x_1,x_2,t|V)| < \infty$. If the static $(0,n)$ and $(1,n)$ sum rules hold, then

$$e\int dx\;\hat{\rho}(x,t|V) = 0 \tag{32}$$

$$e\int dx\; x\;\hat{\rho}(x,t|x_1\ldots x_n) = 0 \tag{33}$$

for all t, $x_1\ldots x_n$ and $V=(x_1,p_1\ldots x_n,p_n)$.

Using the definition of the truncated functions as well as (3) to cancel the contribution of time independent terms, we find the equivalent of (30) written for the truncated functions :

$$\frac{\partial}{\partial t}\rho_T(u_1,t|V) = -\frac{p_1}{m}\cdot\nabla_{x_1}\rho_T(u_1,t|V)$$

$$-e^2\nabla_{p_1}\rho(u_1)\cdot\int dx_2\,F(x_1-x_2)\,\rho_T(x_2,t|V) - e^2\int dx_2\,F(x_1-x_2)\cdot\nabla_{p_1}\rho_T(u_1,x_2,t|V) \tag{34}$$

The evolution of a general momentum independent function $f(x)$ is obtained by integrating out the momentum in (34), the ∇_p - terms giving no contribution

$$\frac{\partial}{\partial t}\int dx\,f(x)\,\rho_T(x,t|V) = -\int dx\,f(x)\,\nabla_x\cdot\int dp\,\frac{p}{m}\,\rho_T(x,p,t|V) \tag{35}$$

Setting now $f(x)=1$, the r.h.s. of (35) vanishes in view of (ii). Taking into account the initial condition $\int dx\,\rho_T(x,t|V)\big|_{t=0} = \rho(V)\int dx\,\hat{\rho}(x,t|V)\big|_{t=0} = 0$ (the $(0,n)$ static sum rule) leads to (32).

To derive the dipole sum rule, it is useful to consider the second time derivative obtained from (34), (35) and partial integrations on momentum

$$\frac{\partial^2}{\partial t^2}\int dx_1\, f(x_1)\, \rho_T(x_1,t|V) = -\omega_p^2 \int dx_1\, f(x_1)\, \rho_T(x_1,t|V) \tag{36}$$

$$-\frac{1}{m}\int dx_1 \int dp_1\, (p_1\cdot\nabla_{x_1} f(x_1))\,(p_1\cdot\nabla_{x_1}\rho_T(x_1,p_1,t|V)) \tag{37}$$

$$+\frac{e^2}{m}\int dx_1 \int dx_2\, (\nabla_{x_1} f(x_1))\cdot F(x_1-x_2)\, \rho_T(x_1,x_2,t|V) \tag{38}$$

The first term of the r.h.s. results of Poisson equation $\nabla_x\cdot F(x) = 4\pi\delta(x)$ and $\omega_p = \left(\frac{4\pi e^2\rho}{m}\right)^{1/2}$ is the plasmon frequency. Choosing now $f(x) = x$ the terms (37) and (38) vanish, (37) being the integral of a gradient and (38) because of the antisymmetry of the force. So (36) becomes an ordinary differential equation. From the $(1,n)$-static sum rules and (35) the initial conditions are found to be (with $V = (x_1, p_1 \ldots x_n, p_n)$)

$$\int dx\, x\, \rho_T(x,t|V)\Big|_{t=0} = 0$$

$$\frac{\partial}{\partial t}\int dx\, x\, \rho_T(x,t|V)\Big|_{t=0} = \frac{1}{m}\left(\sum_{j=1}^{n} p_j\right)\rho(V)$$

and the solution is thus

$$\int dx\, x\, \rho_T(x,t|V) = \frac{1}{m\omega_p}\left(\sum_{j=1}^{n} p_j\right)\rho(V)\sin\omega_p t \tag{39}$$

Averaging (39) on initial momenta gives the dipole sum rule (33).

There is also a time dependent generalization of the complete shielding relation (9) :

$$\int dy \int dx\, \frac{1}{|x|}\, S(x-y,t) = -\frac{2\pi}{3}\int dx\, |x|^2\, S(x,t) = \beta^{-1}\cos\omega_p t \tag{40}$$

where $S(x,t) = \langle C(x,t)\, C(0,0)\rangle = e^2(\rho(x,t|0) - \rho^2)$ is the time displaced charge-charge correlation. The relation (40) is well known in the frame work of linear response, our point being here to derive it from the microscopic dynamical equations. For this, we set $f(x) = |x|^2$, $V = (0,p)$ in (36) and integrate over p . Then, under the same hypothesis as in Prop. 8, the terms (37) and (38) vanish again as a consequence of the sum rules (32) and (33). Thus (36) reduces to the differential equation

$$\frac{\partial^2}{\partial t^2}\int dx\, |x|^2\, S(x,t) = -\omega_p^2 \int dx\, |x|^2\, S(x,t) \tag{41}$$

The initial conditions (9) and $\frac{\partial}{\partial t}\int dx\, |x|^2\, S(x,t)\Big|_{t=0} = 0$ (which follows from (35)) gives (40).

After the change of variables $x_1 = y-x$, $x_2 = -x$ and using translation invariance the term (38) becomes (exchanging the x and y integrals)

$$\frac{2e^2}{m}\int dy\, F(y)\cdot\int dx\, (y-x)\, \rho_T(y,0,t|x) =$$

$$\frac{2e^2}{m}\int dy\, F(y)\cdot\int dx\,(y-x)\, \rho_T(x,-t|y,0) \tag{42}$$

We have also used $\rho(V,t|V) = \rho(V,-t|V)$ which follows from the stationarity of the equilibrium state. It is easily checked that (32) and (33) imply the same relations for $\rho_T(x,t|y,0)$ instead of $\hat{\rho}(x,t|y,0)$, thus (42) vanishes. The term (37) is treated in the same way.

It should be stressed that multicomponent systems show a much more complex dynamical behaviour : only the charge sum rule (32) remains true for them. The validity of (33) and (40) for the OCP is due to the special feature here that the electric current is proportional to the total momentum, which is not affected by interparticle collisions. Even in the OCP the higher order ℓ-sum rules are not expected to hold, a fact which is linked to the existence of long range time dependent momentum-momentum correlations of the type [11]

$$\rho(x,p,t|x_1,p_1) = \frac{\beta e^2 t^2}{2m}\, \rho(x,p,x_1,p_1)\,(p\cdot\nabla_x)(p_1\cdot\nabla_{x_1})\,\frac{1}{|x-x_1|} + o(t^2) \tag{43}$$

VIII. QUANTUM SYSTEMS

Some properties of the classical correlations can be extended to the quantum mechanical situation. Let $\rho(Q|Q')$, $Q=(q_1\ldots q_k)$, $Q'=(q'_1\ldots q'_\ell)$, be the reduced density matrices (RDM) formally defined by

$$\rho(Q|Q') = \langle a^*(q'_1)\ldots a^*(q'_\ell)\, a(q_k)\ldots a(q_1)\rangle$$

where $a^*(q')$, $a(q)$ are the creation and annihilation operators satisfying the canonical commutation or anticommutation relations. Gauge invariance of the state (charge conservation) implies $\rho(Q|Q') = 0$ if $\sum_{i=1}^{k} e_{\alpha_i} = \sum_{j=1}^{\ell} e_{\alpha'_j}$. We define the quantum mechanical analog of the excess particle density by

$$\rho(Qq|Q'q) = \rho(Qq|Q'q) - \rho(Q|Q')\,\rho(q|q) + \frac{1}{2}\Big(\sum_{i=1}^{k}\delta_{q_i q} + \sum_{j=1}^{\ell}\delta_{q'_j q}\Big)\,\rho(Q|Q') \tag{44}$$

When $Q = Q'$ (diagonal RDM), (44) is identical to (1) up to the factor $\rho(Q|Q)$ and has the same interpretation. Then the ℓ-sum rules are

$$\int dq\, e_\alpha\, Y_\ell(x)\, \hat{\rho}(Qq|Q'q) = 0 \tag{45}$$

(45) can be derived from the quantum BBGKY hierarchy under appropriate cluster assumptions, following the same method as in Prop. 2 [24]. The cluster properties of the RDM needed to carry out the proof have so far not been rigorously established, even in the Debye-Hückel regime. In fact, different behaviours will occur depending on the type of observables which are considered. In the plasma phase, one may expect that the correlation of the charge with any other local observable has a fast (at least integrable decay). However, up to second order in Plank's constant, the current-current correlations $\langle J^r(x)\, J^s(0)\rangle$ have a non integrable decay, even when the corresponding classical state shows Debye screening e.g. for the OCP [25]

$$\langle J^r(x)\, J^s(0)\rangle = -\frac{1}{12}\hbar^2 \beta \rho^2 \frac{\partial^2}{\partial x^r \partial x^s}\left(\frac{1}{|x|}\right) + o(\hbar^2) \tag{46}$$

It is therefore important to distinguish in (45) between the diagonal sum rules $Q = Q'$, and the off-diagonal sum rules $Q \neq Q'$. The diagonal sum rules will hold for all ℓ, as in the classical case, if the correlation of the charge decays faster than any inverse power. However, because of the long range (46), the off-diagonal sum rules can only be established with the method of [24] for the case $\ell = 0$, and they are presumably not true when $\ell \geq 1$.

A generalization of the complete shielding relation (9)-(10) to the quantum mechanical OCP is

$$\int dx\, |x|^2 S(x) = -\frac{3}{4\pi}\hbar\omega_p \coth\left(\frac{\beta\hbar\omega_p}{2}\right) \tag{47}$$

$S(x) = \langle C(x)\, C(0)\rangle$ is the quantum mechanical charge-charge correlation and $\omega_p = \left(\frac{4\pi e^2 \rho}{m}\right)^{1/2}$ is the plasmon frequency.

(47) can be derived from the energy-entropy balance inequalities [26]

$$\beta \langle A^* [H, A]\rangle \geq \langle A^* A\rangle \ln\left(\frac{\langle A^* A\rangle}{\langle A A^*\rangle}\right) \tag{48}$$

A being a local observable. The proof is based on the remark that in the

OCP, the local dipole D_Λ and current J_Λ (attached to a cylindrical region Λ) behave as a quantum oscillator of frequency ω_p as $\Lambda \to \mathbb{R}^3$. Setting $A_\Lambda^\pm = D_\Lambda \pm i \frac{1}{\omega_p} J_\Lambda$, we write the inequality (48) for the pairs $(A_\Lambda^+, A_\Lambda^-)$ and $(A_\Lambda^-, A_\Lambda^+)$

$$-\beta \frac{\langle A_\Lambda^+ [H, A_\Lambda^-] \rangle}{\langle A_\Lambda^+ A_\Lambda^- \rangle} \leq \ln \left(\frac{\langle A_\Lambda^- A_\Lambda^+ \rangle}{\langle A_\Lambda^+ A_\Lambda^- \rangle} \right) \leq \beta \frac{\langle A_\Lambda^- [H, A_\Lambda^+] \rangle}{\langle A_\Lambda^- A_\Lambda^+ \rangle} \tag{49}$$

Under the assumption that the diagonal sum rules (45)(resp. off-diagonal holds for $\ell = 0, 1$, (resp. for $\ell = 0$) one finds removing the cut off that

$$\lim_{|\Lambda| \to \infty} \frac{\langle A_\Lambda^\pm A_\Lambda^\mp \rangle}{|\Lambda|} = -\frac{1}{3} \int dx \, |x|^2 S(x) \pm \frac{\hbar \omega_p}{4\pi} \tag{50}$$

$$\lim_{|\Lambda| \to \infty} \frac{\langle A_\Lambda^\pm [H, A_\Lambda^\mp] \rangle}{|\Lambda|} = \mp \hbar \omega_p \lim_{|\Lambda| \to \infty} \frac{\langle A_\Lambda^\mp A_\Lambda^\pm \rangle}{|\Lambda|} \tag{51}$$

Inserting this in (49) leads immediately to the result (47).

It should be stressed again that the relation (47) is specific to the OC and does not extends to multicomponent systems. In a general quantum cha ged system, the proper generalization of the Stillinger-Lovett condition (9)-(10) involves the two point Duhamel function

$$\overline{S}(x) = \frac{1}{\beta} \int_0^\beta d\tau \, \langle C_\tau(x) \, C(0) \rangle \tag{52}$$

where $C_\tau(x)$ is the imaginary time displaced charge operator formally gi ven by $C_\tau(x) = \exp\left(-\frac{H\tau}{\hbar}\right) C(x) \exp\left(\frac{H\tau}{\hbar}\right)$. In the linear response theory, the shielding of an infinitesimal test charge leads to the same cons- traints (9) or (10) with $\overline{S}(x)$ in place of $S(x)$, i.e. in three dimensions [27]

$$1 + \frac{2\pi\beta}{3} \int dx \, |x|^2 \, \overline{S}(x) = 0 \tag{53}$$

There is also a non perturbative proof of (53) (the quantum mechanical a nalog of Prop. 3) relying on the assumption that the imaginary time dis- placed charge correlations have suitable spatial decay properties and sa tisfy the $\ell = 0, 1$ sum rules (for details, see [28]).

REFERENCES

1 J. Fröhlich, Y.M. Park : Comm. Math. Phys. 57, 235 (1978)

2 D. Brydges, P. Federbusch : Comm. Math. Phys. 73, 197 (1980)

3 J. Imbrie : Comm. Math. Phys. 87, 515 (1983)

4 See the articles by J. Fröhlich and T. Spencer, D.C. Brydges and P. Federbusch, M. Aizenman, J.L. Lebowitz in the Proceedings of the Erice Summer School, Ed. Velo and Wightman, Plenum Press (1981)

5 S.F. Edwards, A. Lenard : J. Math. Phys. 3, 778 (1962), A. Lenard : J. Math. Phys. 4, 533 (1963), H. Kunz : Ann.Phys. 85, 303 (1974)

6 M. Aizenman, Ph.A. Martin : Comm.Math. Phys. 78, 99 (1980)
Ch. Lugrin, Ph.A. Martin : J. Math. Phys. 23, 2418 (1982)

7 P. Federbush : Surface effects in Debye screening, preprint Michigan University

8 Ch. Gruber, J.L. Lebowitz, Ph.A. Martin : J.Chem.Phys. 75, 944 (1981)

9 B. Jancovici : J. Stat. Phys. 28, 43 (1982), 29, 263 (1982), 34, 803, (1982)

10 J.L. Lebowitz, Ph.A. Martin : Phys. Rev. Lett. 54, 1506 (1985)

11 B. Jancovici, J.L. Lebowitz, Ph.A. Martin : Time-Dependent Correlations in an Inhomogeneous Plasma, to appear in J. Stat. Phys.

12 J.R. Fontaine, Ph.A. Martin : J. Stat. Phys. 36, 163 (1984)

13 Ch. Gruber, Ch. Lugrin, Ph.A. Martin : J. Stat.Phys. 22, 193 (1980)

14 L. Blum, Ch. Gruber, J.L. Lebowitz, Ph.A. Martin : Phys. Rev. Lett. 48, 1769 (1982)

15 J. Fröhlich, T. Spencer : Comm. Math. Phys. 84, 55 (1982)

16 F. Stillinger, R. Lovett : J. Chem.Phys. 49, 1991 (1968)

17 Ph.A. Martin, Ch. Gruber : J. Stat. Phys. 31, 691 (1983)

18 Ph.A. Martin, T. Yalçin : J. Stat. Phys. 22, 435 (1980)

19 J. L. Lebowitz : Phys. Rev. A 27, 1491 (1983)

20 J. L. Lebowitz, Ph.A. Martin : J. Stat.Phys. 34, 287 (1984)

21 A. Alastuey, B. Jancovici : J. Stat. Phys. 34, 557 (1984)

22 J. Benfatto, Ch. Gruber, Ph.A. Martin : Helv.Phys.Acta 57, 63 (1984)

23 A. Alastuey, Ph.A. Martin : J. Stat.Phys. 37, 405 (1985)

24 Ph.A. Martin, Ch. Gruber : Phys. Rev. A 30, 512 (1984)

25 Ph.A. Martin, Ch. Oguey : J. of Physics A 18, (1985)

26 M. Fannes, A. Verbeure : Comm. Math. Phys. 55, 125 (1977)

27 G. Mahan : Many Particle Physics - Plenum New-York (1981)

28 Ch. Oguey : Ph. D. Dissertation, EPF-Lausanne (1985)

MODELS OF STATISTICAL MECHANICS IN ONE DIMENSION ORIGINATING FROM QUANTUM GROUND STATES

Herbert Spohn

Universität München, Theoretische Physik,
Theresienstr. 37, 8000 München 2, FRG

Abstract. We consider the ground state of a few quantum mechanical degrees of freedom coupled to a Bose field. Examples are the spin-boson Hamiltonian, the polaron, an electron in an external potential coupled to the radiation field, and, the recently popular, quantum coherence with dissipation. Feynman's method to integrate over the Bose field yields models of statistical mechanics in one dimension, possibly with several components. They are ferromagnetic and may undergo a phase transition as the coupling is varied. In simple cases we have a fairly complete understanding. Other cases are not yet covered and require a non-trivial extension of the methods available for one-dimensional systems. We summarize recent results and list nine challenging problems.

1. Introduction

Many physical problems can be idealized as a single (or a few) degrees of freedom in interaction with an infinitely extended, ideal (=quasi-free) system. As a standard, initially the ideal system is only slightly perturbed away from global thermal equilibrium or from the vacuum - a situation which therefore persists in the course of time. We have in mind here nonrelativistic either classical or quantum mechanical systems, although in many other dynamical problems a similar structure can be found. For simplicity we often refer to the few degrees of freedom as "particle", although physically it could be something else. The nonlinearity, and hence the difficulty of the problem, originates either in a nonlinear (non-harmonic) interaction between the particle and

and the free system or from a nonlinear external potential acting on the particle (or both). Our hope is to be able to analyze these models is nourished by the fact that the dynamics of the ideal system is so well understood.

Let us list a few examples :

A Brownian particle in a fluid is modelled as a classical particle moving in an infinitely extended system of non-interacting particles. The Brownian particle interacts with the fluid particles via a short range potential. Possibly, the Brownian particle is confined by an external potential acting on it, e.g. by a double well potential if thermally activated jumps are under investigation. In the limit where the mass of the fluid particles is very large they become immobile scatterers. We obtain then the motion of a classical particle in a static random potential. It is known as the Lorentz gas, because it was introduced many years back by H.A.Lorentz as a primitive model for the scattering of conduction electrons in a metal by impurities. If the centers of the scatterers are arranged on a regular lattice, the periodic Lorentz gas, and a hard core interaction potential is assumed, then the system is equivalent to the Sinai billiard.

The motion of a conduction electron in an ideal solid is modelled as a particle coupled to an infinite harmonic crystal. Another system of fundamental physical importance is a charge, better a charge cloud to avoid divergencies, coupled to the electromagnetic field.

The quantized version of these models hardly need any mentioning. The Anderson model describes the motion of an electron in a static random potential. The polaron is the prototype model for an electron in interaction with phonons. In this particular case the electron is coupled to the longitudinal optical mode of an ionic crystal. Absorption and emission of radiation by an atom, the scattering of electromagnetic radiation from atoms and molecules, resonance fluorescence, etc. are all phenomena obtained theoretically from the model of an atom (or a simplified version of it) coupled to the radiation field. The Kondo problem (a spin impurity coupled to the free electron gas) and the spin-boson problem (a spin impurity coupled to an ideal solid) are famous examples from solid state physics. Recently, the quantum tunneling out of a metastable minimum including dissipation became popular. No surprise, the dissipation mechanism is modelled as a free field of bosonic excitations coupled linearly to the quantum particle.

I regard it as a challenge to mathematical physics to elucidate some of the basic features of these models.

The type of questions one asks seem to be fairly standardized. Two cases should be distinguished :

(I) The motion of the particle is unconfined. This simply means that the particle can go anywhere, but whereever it interacts with the ideal system in the same fashion (= translation invariant interaction).

The main question of interest are the transport properties of the particle. This refers to the long time asymptotics of the particle as characterized by the mean square displacement $\langle q(t)^2 \rangle$ for large t. Here (q(t) is the position of the particle at time t and the average is over the thermal equilibrium state for the ideal system and a state localized near the origin for the particle. One distinguishes

$$\langle q(t)^2 \rangle \sim \begin{cases} \text{bounded} & \text{localization, zero diffusion coefficient, insulator} \\ t & \text{regular transport, diffusive behavior, conductor} \\ t^2 & \text{free motion, infinite diffusion coefficient,} \\ & \text{ideal conductor} \end{cases} .$$

Examples of intermediate behavior are known. More generally, of interest are time correlation functions (Green's functions) with such properties as self energy, effective mass, etc. .

The localization for the Anderson model (Schrödinger equation with a static random potential) is rather well understood by now [1]. Despite considerable effort, the diffusive behavior on the basis of a strictly mechanical model can be proved only in a few cases. The "best" result is for the two-dimensional periodic Lorentz gas where convergence to Brownian motion is established [2]. The impurity in a harmonic chain is an explicitely soluble case. A system under attack is a particle in one space dimension interacting through a hard core potential with an ideal gas. If the mass of the particle is the same as the mass of the ideal gas particles, diffusive behavior has been established some time back [3]. If the masses are unequal, upper and lower bounds proportional to t are proved [4].

A number of limits where a physical parameter is changed are fairly well understood : the limit of a large mass of the Brownian particle [5], the limit of a weak interaction between particle and ideal fluid, both classically [6] and quantum mechanically [7], the limit of low density for the ideal fluid [8].

(II) The motion of the particle is confined. Typically there is an external potential which prevents the particle from moving out. In some quantum mechanical models only the spin degrees of freedom of an atom at a fixed location are taken into account. Similar models are obtained by approximating the bound states of the particle by a two- or multilevel system.

In physical applications (I) and (II) are not that cleanly separated. E.g. a hydrogen atom coupled to the radiation field has the center of mass translational degrees of freedom, and bound and scattering states for the relative motion with the possibility of transitions between them. Still the distinction (I) and (II) is useful, in my opinion.

For (II) the main question of interest is the approach to equilibrium as $t \to \infty$. Because the total system is only slightly perturbed from global equilibrium one also speaks of return to equilibrium. Physically a more detailed description of how the system reaches equilibrium is of crucial importance. In simple approximations one obtains exponential decay with some life time possibly superimposed with oscillations. This can be understood as a Markovian time evolution which approximates the true reduced dynamics in a limiting regime. Often, as a physical fact, equilibrium has been reached already and one is interested then in time correlation functions in equilibrium.

For the Brownian particle confined by some potential, in the classical case, techniques have been developed in the recent year which are powerful enough to prove convergence to equilibrium [9]. A detailed proof has been worked out yet only in a specific one-dimensional model. If the potential confines the particle to some compact region, the method in [10] gives the desired result. For quantum models, to my knowledge, only in case the total system is quasi-free the return to equilibrium is proved [11]. Markovian approximations for confined systems have been worked out only for quantum systems, specifically in the limits of weak [12] and singular [13] couplings and of low density [14].

I want to end here my brief summary. It is incomplete and is intended as a rough sketch only of my view of the landscape. I hope it explains parts of my motivation behind the investigation of the models I am going to discuss.

I would like to understand the structure of the ground state time correlations, i.e. time correlations at zero temperature, of some of the quantum models mentioned above. Emphasis is on the cases where the particle's motion is confined. Unfortunately, I am still far from a satisfactory understanding. Instead the more primitive question of the uniqueness of the ground state will be investigated.

This leads to other parts of my motivation. Quantum ground states are in essence equivalent to a statistical mechanics system with one dimension more than the spatial dimension of the quantum system. As observed by Feynman, already in his earliest papers [15], the path integral representation of e^{-tH} for the above mentioned models yields a

one-dimensional statistical mechanics system, possibly with several components, for the particle and some Gaussian measure for the free system. For the usual couplings the integration over the Gaussian variables can be carried out explicitely leading to an effective interaction for the one-dimensional system. Altogether quantum ground state expectations can be expressed then through averages in a one-dimensional statistical mechanics system over the real line, however with a nonlocal interaction. The dictum says that for statistical mechanics in one dimension one knows everything, one wants to know. However, to my own surprise, the statistical mechanics models obtained from the quantum ground states of interest have been hardly touched upon in the mathematical physics literature. Three complicating features appear : the field is over the real line and unbounded; the nonlocal interaction, although attractive, is not quadratic in the field; the nonlocal interaction may be given by stochastic integrals with no way to rewrite them as ordinary integrals. It would be nice to have some machinery available to deal with some of the basic properties of these statistical mechanics models, such as uniqueness of the phase, analyticity of the free energy, decay of correlations.
The paper is organized as follows : Section 2 summarizes the spin-boson Hamiltonian. This has been investigated in [16]. Some natural and physically important extensions of this model are considered in Section 3. Section 4 deals with the unconfined, translation invariant case.

2. The Spin-Boson Hamiltonian

The spin-boson Hamiltonian is

$$H = - \varepsilon \sigma_x \otimes 1 + 1 \otimes \int dk\, \omega(k)\, a^+(k)\, a(k) + \sqrt{\beta}\, \sigma_z \otimes \int dk\, \lambda(k)\, (a^+(k) + a(k)) - h\, \sigma_z \otimes 1 \quad . \tag{2.1}$$

σ_x, σ_y, σ_z are the Pauli spin matrices. $\{a^+(k), a(k)\}$ is a Bose field over the real line with the standard commutation relations $[a(k), a^+(k')]=\delta(k-k')$. More realistically $\{a^\#(k)\}$ should be over R^3. The changes are trivial and are omitted for simplicity. $\omega(\cdot)$ is the dispersion relation of the free Bose field, $\omega(k)\geqq 0$. $\lambda(k)$ are the couplings The uniqueness of the ground state depends only on the coupling strength at given energy. Therefore we introduce

$$\rho(\omega)d\omega = \int dk\ \lambda(k)^2\ \delta(\omega(k)-\omega)\ d\omega \quad . \tag{2.2}$$

In general, λ,ω and ρ are measures.
Let W be the Laplace transform of ρ,

$$W(t) = \int_0^\infty d\omega\ \rho(\omega)\ e^{-\omega|t|} \quad . \tag{2.3}$$

To have a well defined ground state we impose that $W(t)$ is integrable for large t and that $tW(t)$ is integrable for small t. In fact, to simplify somewhat, the latter condition is replaced by the stronger one of $W(t)$ to be bounded.

ω, λ and hence $\rho(\omega)d\omega$ are assumed to be given. We want to understand the ground state properties of H in dependence on the following parameters :

- The level splitting ε. 2ε is the separation of eigenvalues of the uncoupled spin. By symmetry we may assume $\varepsilon \geq 0$.
- The coupling strength $\beta, \beta \geq 0$. To set a scale for β, one possibility is the normalization $2\int_0^\infty d\omega\rho(\omega)\ (1/\omega) = 1 = \int_{-\infty}^{\infty} dt W(t)$ (or some other rule if W is not integrable at $t=0$).
- The asymmetry h. For $t \neq 0$ there is a unique ground state. To investigate the degeneracy at $h=0$ we let the asymmetry $h \to 0_+$, resp. $\to 0_-$.

Ground state properties are governed by the infrared behavior of $\rho(\omega)$. We characterize it by

$$\rho(\omega) = \omega^\gamma \tag{2.4}$$

for small ω, $\gamma > 0$.

The Hamiltonian (2.1) has a rich history and very likely my account is neither accurate nor complete. To my knowledge, (2.1) was first used by Anderson and Yuval [17] as an approximation to the Kondo Hamiltonian, cf. also [18,19]. (2.1) appears as the standard two-level approximation for an atom coupled to the radiation field [20]. The Bose field is then the quantized electromagnetic field and is therefore over R^3 with two transversal components. Also the dipole approximation is invoked. Pfeifer [21] proposed (2.1) as a model in his search for the explanation of the observed chirality (handedness) of molecules. A chiral molecule would have then a degenerate ground state due to the coupling to the electromagnetic field. In the context of solid state physics (2.1) describes a spin impurity coupled to the phonon field [22,23]. Recently, the Hamiltonian (2.1) was taken up again as a model for quantum coherence with dissipation [24-27]. Despite the wide range of physical applications one always finds

$$\gamma = 1 \quad , \tag{2.5}$$

a fact, which I find astonishing.

To handle (2.1) we follow Feynman. The simple observation is that, in the representation where σ_z is diagonal, $\varepsilon(\sigma_x - 1)$ generates a spin flip process $\sigma(t)$ with flip rate ε. This means that $t \to \sigma(t) = \pm 1$ is piecewise constant and that σ flips to $-\sigma$ after an exponentially distributed holding time with average $1/\varepsilon$. $\int_0^\infty dk\omega(k)\, a^+(k)\, a(k)$ generates a collection of oscillator (=Ornstein-Uhlenbeck) processes. In the representation chosen the interaction is a multiplication operator. Therefore we may integrate over the Bose field resulting in an effective interaction for the spin.

Let us describe the Ising model so obtained. We are interested only in the + state. Therefore $\sigma(t) = 1$ for $|t| > T$. The free measure is a spin flip process with rate ε starting at $t = -T$ with $\sigma(-T) = 1$ and conditioned such that $\sigma(T) = 1$. Equivalently, we may place an even number of points in $[-T, T]$ randomly, i.e. according to a Poisson distribution, with density ε and construct the spin configuration $t \to \sigma(t)$ such that it changes sign exactly at the given points. The free measure is denoted by $\mu^{\varepsilon}_{T,+}(d\sigma(\cdot))$. The measure of the Ising model is then given by

$$\begin{aligned} \frac{1}{Z}\, \mu^{\varepsilon}_{+,T}(d\sigma(\cdot)) \exp[& -\beta\frac{1}{4}\int_{-T}^{T} dt \int_{-T}^{T} ds\; W(t-s)(\sigma(t) - \sigma(s))^2 \\ & - \beta\frac{1}{4}\int_{-T}^{T} dt \left(\int_{-\infty}^{-T} ds + \int_{T}^{\infty} ds \right) W(t-s)\; (1-\sigma(t))^2 \;] \;. \end{aligned} \tag{2.6}$$

The second term in the exponential arises from the + boundary conditions. Note that W is even, cf. (2.3). The limit $T \to \infty$ of (2.6) defines the probability measure $\langle \cdot \rangle_+$ on the space of piecewise constant functions, i.e. functions with a finite number of discontinuities in any finite interval, with range $\{-1, 1\}$.

From (2.6) the conditions on $\rho(\omega)\, d\omega$, i.e. on W, are readily understood. At large distances $W(t)$ has to be integrable in order for the free energy to be extensive. At short distances $\int dt \int ds\; W(t-s)(\sigma(t) - \sigma(s))^2$ has to be finite. Exploiting $\sigma(t) = \pm 1$ yields the condition $tW(t)$ to be integrable for small t. The physical $\gamma = 1$ implies

$$W(t) \sim t^{-2} \tag{2.7}$$

for large t. Therefore we are forced to deal with long range interactions.

The quantum mechanical ground state expectations are obtained in the limit of zero temperature followed by the limit $h \to 0_+$. Again, we denote them by $\langle\cdot\rangle_+$. Then the following identities hold :

$$\langle\sigma_z\rangle_+ = \langle\sigma(0)\rangle_+ \quad , \tag{2.8}$$

$$\langle\sigma_y\rangle_+ = 0 \quad , \tag{2.9}$$

$$\langle\sigma_x\rangle_+ = \lim_{t\to 0_+} \frac{1}{\varepsilon} \frac{1}{t} \langle(\sigma(0)-\sigma(t))^2\rangle_+ \quad . \tag{2.10}$$

Also ground state expectations of the Bose field can be obtained from Ising expectations, e.g.

$$\langle a^+(f)a(g)\rangle_+ = \int_0^\infty dt \int dk\ \lambda(k)\ f(k) \int dk'\ \lambda(k')\ g(k')$$

$$\langle\sigma(0)\sigma(t)\rangle_+ \int_0^t d\tau\ e^{-\omega(k)(t-\tau)}\ e^{-\omega(k')\tau} \quad , \tag{2.11}$$

where $a^\#(f)=\int dkf(k)\ a^\#(k)$ with a suitable test function f, and similarly for higher order correlations. As an example for a dynamical quantity let us consider the time-dependent spin-spin correlation $\langle\sigma_z(t)\sigma_z\rangle_+=\langle e^{itH}\sigma_z\ e^{-itH}\sigma_z\rangle_+$. For this one needs the spectral representation of the Ising spin-spin correlation

$$\langle\sigma(t)\sigma(0)\rangle_+ = \int_0^\infty \nu(d\lambda)e^{-\lambda|t|} \tag{2.12}$$

which is analytically continued to give

$$\langle\sigma_z(t)\sigma_z\rangle_+ = \int_0^\infty \nu(d\lambda)e^{-i\lambda t} \quad . \tag{2.13}$$

How to handle then the Ising model ? First of all we note that the interaction is ferromagnetic. Therefore we expect all kind of inequalities to hold : The continuum is approximated by a lattice with lattice spacing δ. The free measure becomes then the usual nearest neighbor Ising model with a diverging nearest neighbor coupling $\frac{1}{2}\log\delta\varepsilon$. Inequalities carry over to the limit $\delta \to 0$. Except for the case mentioned, it is not wise to break up the free measure. The better strategy is to generalize a known proof to one where the free measure is not necessarily a product measure.

Let me summarize the phase diagram, i.e. uniqueness resp. degeneracy, of the spin-boson Hamiltonian :

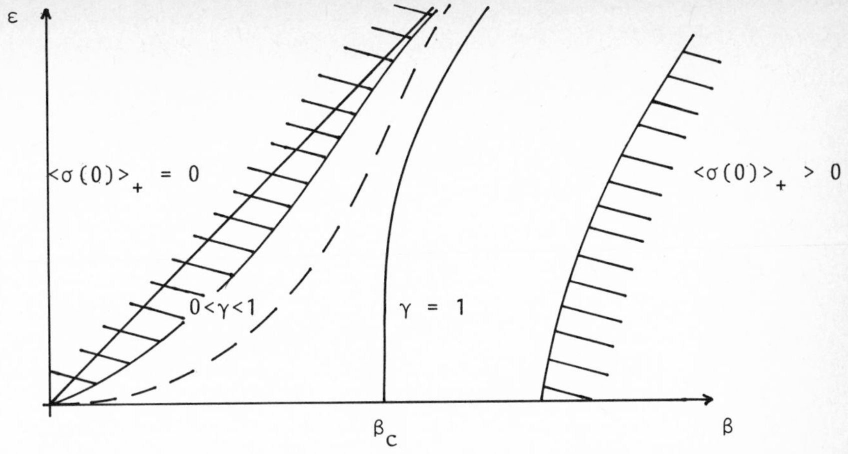

We assume $\varepsilon>0$. For $\varepsilon\to 0$ the model has a transition analoguous to the zero temperature transition of the one-dimensional Ising model. For $\gamma>1$ the effective interaction decays faster than t^{-2} and therefore the model has a unique phase. As for the Ising model with long range interactions the precise boundary for the existence of a phase transition is conjectured to be exactly t^{-2} for large t, but not known rigorously. For $\gamma=1$, i.e. $\lim_{t\to\infty} t^2 W(t)=c>0$, there is a phase transition. For sufficiently strong coupling the ground state is degenerate. We indicate schematically the regions (shaded) in parameter space, where $\langle\sigma(0)\rangle_+=0$ and where spontaneous magnetization, $\langle\sigma(0)\rangle_+>0$, is established. The diagonal $\varepsilon=\beta$ is the mean-field bound. The line (——) of critical points is the one of Anderson and Yuval [17] obtained by a renormalization group argument (for an improved version [38]). For $\gamma=1$ one expects a critical coupling strength β_c below which the ground state is unique, no matter how small ε. For the Ising model with pair interaction $\beta J(i-j)$ M.Aizenman,J.+L.Chayes and C.Newman prove recently [40] that if $J(j)\cong|j|^{-2}$ for large j and if β is sufficiently small, then there is no long range order regardless of the strength of the short range couplings. If their proof extends to the continuum, this would establish the existence of β_c . For $0<\gamma<1$ there is a phase transition and the line of critical point reaches the origin of the (β,ε) plane as $\beta^{1/(1-\gamma)}$. From (2.11) it follows that if the Ising model has long range order, then the total number of bosons in the ground state is infinite. The divergence comes from a slow decay of the spatial boson density, respectively from a large boson density at small momentum (infrared divergence).

The uniqueness of phase follows from an extension of the energy-entropy argument of Simon and Sokal [28]. For $0<\gamma<1$ Dyson's hierarchical model [29,30,31] provides fairly sharp bounds. The most difficult case is the physical $\gamma=1$. For the Ising model on the lattice with $1/j^2$-interaction J.Fröhlich and T.Spencer [32] prove the existence of a spontaneous magnetization at sufficiently low temperatures. I tried hard to extend their proof to the present case - without any success, because their estimates break down at small distances. Therefore one has to average over short distances at the expense of obtaining a lattice Ising model with continuous, but bounded spins. (The steps inbetween are covered by Griffiths inequalities.) As J.L.Lebowitz pointed out to me, the final reduction to Fröhlich/Spencer is achieved through the Wells inequality [33].

Instead of entering into technical details [16], let me point out two, fairly related problems.

Problem 1. Clustering of time correlations.

I would like to show that

$$\lim_{t\to\infty} \langle\sigma_z(t)\sigma_z\rangle_+ = \langle\sigma_z\rangle_+^2 \quad . \tag{2.14}$$

From the spectral representation (2.12), (2.13) we conclude that (2.14) holds for the time average. To prove (2.14) one would have to establish that, except for the Dirac δ at $\lambda=0$, $\nu(d\lambda)$ is absolutely continuous with respect to $d\lambda$. I have no idea of how to extract this information from the Ising model. (2.14) could also be asked at finite temperature ($\langle\sigma_z\rangle=0$ then). There is a corresponding spectral representation but now refering to the finite interval Ising model with periodic boundary conditions. Again, I do not know how to prove absolute continuity of the spectral measure.

Problem 2. Return to equilibrium.

This has been discussed for quantum coherence with dissipation in [34,35]. Suppose we prepare the spin in the state $\binom{1}{0}$. Without coupling to the field ($\beta=0$) the spin tunnels coherently between $\binom{1}{0}$ and $\binom{0}{1}$, i.e. $\langle\sigma_z(t)\rangle=e^{i2\varepsilon t}$. How is this coherent oscillation changed as the coupling is turned on, $\beta>0$? One prescription of the initial state is $\binom{1}{0}$ o (field vacuum). Physically, a more natural choice would be the ground state of (2.1) for $h\to\infty$. From the spectral representation the time average of $\langle\sigma_z(t)\rangle$ tends to $\langle\sigma_z\rangle_+$. For $0<\gamma\leq 1$ and for β somewhat right of the critical line $\langle\sigma_z\rangle_+\cong 1$, essentially, which is interpreted as a (essential) trapping of the initially prepared state. I would like to establish

$$\lim_{t\to\infty} \langle\sigma_z(t)\rangle = \langle\sigma_z\rangle_+ \quad . \tag{2.15}$$

3. A Bounded Particle in Interaction with its Radiation Field

For quantum tunneling and coherence with dissipation Caldeira and Leggett [36] propose the Hamiltonian

$$H = (-\tfrac{1}{2}\Delta+V(q)) \otimes 1 + 1 \otimes \int dk\omega(k)\ a^+(k)\ a(k)$$
$$+ \sqrt{\beta}\, q \otimes \int dk\ \lambda(k)\ (a^+(k)+a(k)) + \beta \int dk\ \lambda(k)^2 \frac{1}{\omega(k)}\, q^2 \otimes 1 \quad . \tag{3.1}$$

Here q is one-dimensional quantum mechanical degree of freedom, which in application to super conducting quantum interference devices (SQID) is either the superconducting current or its phase. $\Delta=\partial^2/\partial q^2$ and V(q) is an external potential. For the tunneling problem V has a metastable minimum, as $V(q)=q^2(1-q)$, and one tries to predict the probability to tunnel out of the minimum both at zero and finite temperatures. For quantum coherence V is a double well potential, e.g. $V(q)=(1-q^2)^2$. The spin-boson Hamiltonian arises then as a two-level approximation of $-\frac{1}{2}\Delta+V$ taking into account only the ground state and the first excited state. 2ε is their energy difference. σ_z plays the role of the position operator. Caldeira and Leggett argue that a linear friction in the classical limit forces $\rho(\omega)\cong\omega$ for small ω.

We consider here only quantum coherence with dissipation. The functional integral for e^{-TH} is

$$\frac{1}{Z}P_{[-T,T]}(dq(\cdot))\ \exp[-\int_{-T}^{T}dt\ V(q(t)) - \beta\,\frac{1}{4}\int_{-T}^{T}dt\int_{-T}^{T}ds\ W(t-s)(q(t)-q(s))^2]. \tag{3.2}$$

Here $t\to q(t)$ is a continuous path and $P_{[-T,T]}$ is the standard Wiener measure on C([-T,T],R), the space of continuous functions on [-T,T]. We pay no attention to boundary conditions. The path measure (3.2) is comparable to (2.6). The Ising configuration $t\to\sigma(t)$ is replaced by the continuous path $t\to q(t)$ and $P_{[-T,T]}(dq(\cdot))\ \exp[-\int_{-T}^{T}dt\ V(q(t))]$ plays the role of the free measure.

If V(q)=V(-q), the properties of (3.2) are analoguous to the spin-boson Hamiltonian. Their proof relies on the fact that under suitable conditions on V all the technical tools used for the spin-boson Hamiltonian extend to (3.2). For $0<\gamma\leq 1$ and sufficiently large coupling β

the particle is self-trapped : If, at zero temperature, the particle is initially localized in the right hand well, say, then it will spend a larger fraction of time in the right hand well than in the left hand well. This fraction of time will tend to one as $\beta\to\infty$. The self-trapping is associated with an infinite number of soft (infrared) bosons.

More generally, we choose as potential $V(q)-hq$ with V double well but not necessarily of perfect reflection symmetry. Then in the (β,ε,h)-space there is a surface of first order transitions terminated by a line of critical points. $V(q)=V(-q)$ simplifies the geometry by forcing this surface to lie in the plane $h=0$.

In (3.1) the coupling of the particle to the Bose field is not translation invariant. If translation invariance is imposed, we arrive at

$$H = \left(-\frac{1}{2}\Delta + V(q)\right) \otimes 1 + 1 \otimes \int dk\ \omega(k)\ a^+(k)\ a(k) + \sqrt{\beta} \int dk\ \lambda(k)\ e^{ikq} \otimes (a(k) + a^+(-k)) \quad . \tag{3.3}$$

Here $\omega(k)=\omega(-k)\geq 0$ and $\lambda(k)=\lambda(-k)^*$. Ultraviolet (short distance) divergencies are avoided by assuming $\int dk\ |\lambda(k)|^2<\infty$. As for the spin-boson Hamiltonian this condition can be relaxed. The infrared (large distance) behavior requires a separate investigation.

Most naturally the Hamiltonian (3.3), generalized in the obvious way to three dimensions, appears in solid state physics as the description of an electron coupled to an ideal crystal. Usually the coupling to a particular mode is dominant. Therefore it suffices to have a Bose field with one component. The best known model is the Fröhlich polaron with $\omega(k)=\omega_o$ and $\lambda(k)=\sqrt{4\pi}/|k|$. Other systems studied are the acoustical polaron with $\omega(k)=|k|$, $\lambda(k)=|k|^{1/2}$, the optical polaron with $\omega(k)=\omega_o$, $\lambda(k)=1$, and the piezoelectric polaron with $\omega(k)=|k|$, $\lambda(k)= |k|^{-1/2}$, all in three dimensions. Also a one-dimensional version of the acoustic polaron is physically realized. For models with $\omega(k)=|k|$ an ultraviolet cut-off, e.g. the inverse lattice constant, is needed.

If we follow the same strategy as for the spin-boson Hamiltonian, the functional integral for the ground state of (3.3) becomes

$$\frac{1}{Z} P_{[-T,T]}(dq(\cdot)) \exp\Big[-\int_{-T}^{T} dt\ V(q(t)) + \beta\,\frac{1}{2}\int_{-T}^{T}dt \int_{-T}^{T} ds\ W(t-s\ ,\ q(t)-q(s))\Big] \tag{3.4}$$

in the limit $T\to\infty$, where

$$W(t,q) \quad = \quad \int dk|\lambda(k)|^2 \, e^{-\omega(k)|t|} \cos kq \qquad , \tag{3.5}$$

[39]. No attention to boundary conditions is paid. Note that if $kq \ll 1$, then $W(t,q)$ becomes

$$\int dk|\lambda(k)|^2 \, e^{-\omega(k)|t|} (1 - \tfrac{1}{2} k^2 q^2) \qquad . \tag{3.6}$$

The 1 cancels by normalization and we are back to (3.2) with $|\lambda(k)|^2 k^2$ replacing $\lambda(k)^2$ in the definition of the effective interaction W . This yields a faster decay of W .

E.Nelson [37] studies the ultraviolet properties of (3.4) in a particular case : $q(t) \in R^3$, the k-integration runs over R^3, $\omega(k) = (k^2+\mu^2)^{1/2}$ with $\mu>0$ and $\lambda(k)^2=\omega(k)^{-1}$. He proves that the exponential of the renormalized action,

$$\exp[\beta \lim_{\kappa\to\infty} \{ \int_{|k|<\kappa} d^3k \, [\tfrac{1}{2} \int_0^T dt \int_0^T ds \, \frac{1}{\omega(k)} \, e^{-\omega(k)|t-s|} \cos(k\cdot(q(t)-q(s)) \; - \; T \, \frac{2}{\omega(k)k^2} \,]\} \,] \qquad , \tag{3.7}$$

is integrable with respect to Brownian motion. I would assume that by now the ultraviolet behavior of (3.4) is more completely understood.

No detailed properties of the path measure (3.4) in the limit $T\to\infty$ have been proved yet. One immediate question is the uniqueness of the infinite volume limit, i.e. the uniqueness of the ground state for the Hamiltonian (3.3). As for unbounded spins on a lattice [41] one possibility is first to define the notion of a regular Gibbs measure essentially by requiring a logarithmically bounded growth of $q(t)^2$ as $|t|\to\infty$ and secondly to prove that the infinite volume Gibbs state obtained from periodic boundary conditions, which are the ones coming from quantum mechanics, is regular. (One also has to take into account that the action for e^{-TH} depends on T, cf. [39] and (4.4) below.) To establish the uniqueness of regular Gibbs measures one picks then two extremal, regular Gibbs states, say μ_1 and μ_2, and restricts them to the interval [-T,T]. If these two measures are absolutely continuous with respect to each other with a density bounded uniformly in T, then $\mu_1=\mu_2$. The bound on the density is easily obtained if the interaction energy between the two half-lines is uniformly bounded. In our particular case this leads to

$$\left| \int_{-\infty}^{0} dt \int_{0}^{\infty} ds\ W(t-s,\ q(t)-q(s)) \right| \leq \int_{-\infty}^{0} dt \int_{0}^{\infty} ds \int dk |\lambda(k)|^2\ e^{-\omega(k)|t-s|}$$
$$= \int dk |\lambda(k)|^2\ \omega(k)^{-2} < \infty \quad . \tag{3.8}$$

Details have not been worked out yet.

To investigate the possibility of a phase transition let V be a double well potential with $V(q)=V(-q)$, sufficiently rapidly increasing as $|q|\to\infty$ and minima at $q=\pm a$. Then the crucial characteristic should be

$$-\ (W(t,2a)\ -\ W(t,0)) = \int dk |\lambda(k)|^2\ (1-\cos 2ak)\ e^{-\omega(k)|t|}$$
$$\cong \quad t^{-(1+\gamma)} \tag{3.9}$$

for large t. If $0<\gamma\leq 1$, then, as before, (3.4) should exhibit a ferromagnetic phase transition for β sufficiently large.

We note that (3.8) and (3.9) do not nearly close up. If $\omega(k)=|k|$, then (3.8) requires $|\lambda(k)|^2<|k|$ whereas (3.9) still allows $|\lambda(k)|^2 <1/|k|$ for small k for the absence of a phase transition. A physical example for such a borderline case is $W(t,q)=1/((|t|+1)^2+q^2)$ which corresponds to $\omega(k)=|k|$ and $|\lambda(k)|^2=1/|k|$, $k,q\in R^3$, with a large k cut-off. In this case the interaction between right and left half-line is no longer uniformly bounded. However, (3.9) tells us that the ground state should be unique.

This leads to

Problem 3. Uniqueness, resp. degeneracy of the ground state of the Hamiltonian (3.3).

V is a symmetric double well potential. If in (3.9) $\gamma>1$, then the Gibbs measure to (3.4) should be unique, whereas for $0<\gamma\leq 1$ it should have long range, ferromagnetic order for β sufficiently large.

As mentioned already, the spin-boson Hamiltonian arises also in the two level approximation of an atom coupled to the radiation field. Let us consider then the full problem, i.e. ground state properties of an electron in a confining external potential coupled to the radiation field. We work in the Coulomb gauge and use the standard non-relativistic quantization of the electromagnetic field. The Hamiltonian is given by

$$H = \frac{1}{2}(-i\nabla-\sqrt{\beta}\sum_{\sigma=1,2}\int d^3k\lambda(k)|k|^{-1/2}e_\sigma(k)[e^{ikq}\otimes a_\sigma(k) + e^{-ikq}\otimes a_\sigma^+(k)])^2 \tag{3.10}$$
$$+ V(q)\otimes 1 + i\otimes\sum_{\sigma=1,2}\int d^3k|k|a_\sigma^+(k)a_\sigma(k) \quad .$$

We have set h=1, c=1 and the mass of the electron m=1. In these units the coupling $\sqrt{\beta}=e$, the charge of the electron. $q\in R^3$ and $\{a_\sigma^+(k),\ a_\sigma(k)|\ \sigma=1,2,\ k\in R^3\}$ is a two component Bose field with commutation relations $[a_\sigma(k),\ a_{\sigma'}^+(k')]=\delta_{\sigma\sigma'}\ \delta(k-k')$. Since the electromagnetic field is transversal, $k\cdot e_\sigma(k)=0$ and $e_\sigma(k)\cdot e_{\sigma'}(k)=\delta_{\sigma\sigma'}$. $\lambda(k)=\lambda(|k|)$ is an ultraviolet cut-off : $\lambda(k)=1$ for small k and $\lambda(k)$ tends to zero for large k with a rate still to be determined.

As noticed already by Feynman, in the functional integral representation for e^{-TH} the vector potential enters linearly in the action. Therefore, as before, the Gaussian integration can be carried out explicitely. Disregarding boundary conditions the result is

$$\frac{1}{Z} P_{[-T,T]}(dq(\cdot)) \exp\Big[-\int_{-T}^{T} dt\ V(q(t)) - \beta\int_{-T}^{T} dq(t)\int_{-T}^{T} dq(s)\ W(t-s,\ q(t)-q(s))\Big] \quad . \tag{3.11}$$

$\int dq(t)\int dq(s)$ is a double stochastic integral. Ito integrals are used throughout. W is the 3x3 matrix given by

$$W_{\alpha\beta}(t,q) = \int d^3k\lambda(k)^2\ |k|^{-1}\ (\delta_{\alpha\beta}-\frac{k_\alpha k_\beta}{k^2})e^{-|k||t|}\ \cos kq\ , \tag{3.12}$$

$\alpha,\beta=1,2,3$. According to [37]

$$\int_{-T}^{T} dq(t)\int_{-T}^{T} dq(s)\ W(t-s,q(t)-q(s)) = 2T\sum_{\alpha=1}^{3} W_{\alpha\alpha}(0,0)$$
$$+ \int_{-T}^{T} dq(t)\Big[\int_{-T}^{t} dq(s)\ W(t-s,\ q(t)-q(s))\Big] + \tag{3.13}$$
$$+ \int_{-T}^{T} dq(s)\Big[\int_{-T}^{s} dq(t)\ W(t-s,\ q(t)-q(s))\Big] \quad ,$$

where the second term are standard iterated integrals. The first term cancels against the normalization. It corresponds to a shift in energy. Therefore (3.11) equals

$$\frac{1}{Z} P_{[-T,T]}(dq(\cdot)) \exp\Big[-\int_{-T}^{T} dt\ V(q(t)) - \beta\underset{\{t\neq s\}}{\int_{-T}^{T} dq(t)\int_{-T}^{T}} dq(s)W(t-s,q(t)-q(s))\Big]\ , \tag{3.14}$$

where $\{t\neq s\}$ indicates the omission of the first term of (3.13).

How do we have to choose the cut-off λ ? V is supposed to be such that large values of $q(\cdot)$ are suppressed. Therefore a necessary condition on λ is the iterated stochastic integral to be well defined. This yields

$$\int d^3k\ \lambda(k)^2\ |k|^{-1}\ e^{-k^2|t|} \tag{3.15}$$

to be integrable for small t. Without cut-off, $\lambda(k)=1$, we would pick up the well-known logarithmic divergence of nonrelativistic QED. Ultraviolet stability requires then a decay of λ as $|k|^{-\delta}$ for some $\delta>0$ and $|k|$ large.

The problem of interest to us here is, even granting a rapid large cut-off, whether the limit $T\to\infty$ of (3.14) is unique. We also would like to estimate the decay of correlations, in particular the $\langle dq(s)\ dq(t)\rangle$ correlation which determines the ground state photon density. I have little to offer. The physical heuristics is that the stochastic integrals $dq(t)dq(s)$ provide two extra power of decay which implies a t^{-4} decay of the effective interaction - too weak to maintain a ferromagnetic ordering.

In a one-dimensional approximation this improved decay can be seen. We assume that the potential V confines very strongly in the 2- and 3-direction. In the 1-direction V is reflection symmetric with a double well, as before. Then in (3.14) the integration over the 2- and 3-component of the Wiener measure can be discarded. The path integral, in this approximation, corresponds to the Hamiltonian

$$\begin{aligned} H = \tfrac{1}{2}\Big(-i\tfrac{\partial}{\partial q} - \sqrt{\beta} \sum_{\sigma=1,2} \int d^3k\ \lambda(k)\ |k|^{-1/2}\ (e_1\cdot e_\sigma(k)) \\ [e^{ik_1 q} \otimes a_\sigma(k) + e^{-ik_1 q} \otimes a_\sigma^+(k)]\Big)^2 + V(q) \otimes 1 \\ + 1 \otimes \sum_{\sigma=1,2} \int d^3k\ |k|\ a_\sigma^+(k)\ a_\sigma(k) \quad . \end{aligned} \tag{3.16}$$

Here $q\in R$ and, for simplicity, we assume a cut-off as $\lambda(k)=e^{-\alpha|k|}$. Finite temperature expectations for (3.16) are computed from the path measure

$$\frac{1}{Z}\ P_T(dq(\cdot))\ \exp[-\int_0^T dt\ V(q(t)) - \beta\int_0^T dq(t) \int_0^T dq(s)\ W_T(|t-s|,q(t)-q(s))]. \tag{3.17}$$

P_T is the Wiener measure on $C([0,T],R)$ conditioned such that $q(0)=q(T)$ and

$$W_T(t,q) = \int d^3k\ \lambda(k)^2\ |k|^{-1}\ (1-\frac{k_1^2}{k^2}) e^{ik_1 q}\ \frac{e^{-|k|t} + e^{-|k|(T-t)}}{1 - e^{-|k|T}}\ , \qquad (3.18)$$

$0 \leq t \leq T$. Note that $W_T(t,q) = W_T(T-t,q)$ and $W_T(t,q) = W_T(t,-q)$.

Since now the stochastic integral in (3.17) is one-dimensional, we can apply twice Ito's lemma, cf.[37]. The boundary terms vanish because $q(0)=q(T)$. Let

$$\frac{\partial^2}{\partial q^2} G_T(t,q) = W_T(t,q) \qquad (3.19)$$

with $G_T(t,q) = G_T(t,-q)$. Then

$$\begin{aligned} &\int_0^T dq(t) \int_0^T dq(s)\ W_T(|t-s|, q(t)-q(s)) \\ &= \int_0^T dt \int_0^T ds\ (-\frac{1}{4}\frac{\partial^4}{\partial q^4} G_T + \frac{\partial^2}{\partial t^2} G_T)\ (|t-s|,\ q(t)-q(s)) \qquad (3.20) \\ &\equiv \int_0^T dt \int_0^T ds\ \tilde{W}_T(|t-s|,\ q(t)-q(s)) \end{aligned}$$

P_T-a.s. . $\tilde{W}_T$ can be computed, essentially explicitely.

Inserting (3.20) in (3.17) the path measure is of the form (3.4). Let $\tilde{W} = \lim_{T\to\infty} \tilde{W}_T$. Then for $q(t)-q(s)=q$ fixed and $|t-s|$ large

$$\tilde{W}(|t-s|,q) = q^2/(t-s)^4 \quad . \qquad (3.21)$$

Since the effective interaction decays as t^{-4}, we expect a unique ground state for the Hamiltonian (3.16). Of course, this is no proof that an atom coupled to the radiation field has a unique ground state. But the evidence tends in this direction.

Problem 4. Uniqueness of the ground state for an electron coupled to the electromagnetic field and confined by an external potential.

Under weak conditions on the cut-off λ and for a general class of external potentials one has to establish the uniqueness of the Gibbs measure corresponding to (3.14).

Problem 5. Decay of correlations.

Of particular interest is the $\langle dq(0)\ dq(t)\rangle$ correlation for (3.14) in the limit $T\to\infty$.

4. The Polaron

We turn to the models briefly summarized under (I) of the Introduction.

Two models are of particular physical interest : (i) The polaron and related models. The Hamiltonian is (3.3) with $V \equiv 0$ including its generalization to more than one dimension, in which case rotation invariance, i.e. $\lambda(k)=\lambda(|k|)$, $\omega(k)=\omega(|k|)$, is imposed. Physically the boson field is interpreted as phonons. We have then the model of a conduction electron moving in an ideal solid. The electron scatters of the phonons which in turn interact with the electron. (ii) An electron coupled to its radiation field as described by the Hamiltonian (3.10) with $V \equiv 0$. I restrict my discussion to the polaronic models because I am more familiar with them.

J.Fröhlich studied these models from a functional analytic point of view in his thesis [42], parts of which have been published [43, 44]. After this monumental work no one seems to have continued this line of research, to my knowledge. Prior work is [45,46,47]. I have nothing to add to this research. In fact, I am not even sure whether these results are connected to the problems I will mention.

I got reinterested in the polaron in an attempt to understand the existence of transport. This leads to a problem rather similar as (1) and (2). But let me list it anyhow as

Problem 6. Finite diffusivity (conductivity) for the polaron.

The quantity of interest is the current-current correlation function in equilibrium defined as

$$\langle p(t)\cdot p(0)\rangle_T = \int_0^T d\lambda \; \mathrm{tr}[e^{-TH} e^{(\lambda-it)H} p\, e^{-(\lambda-it)H} \cdot p] \,/\, \mathrm{tr}\, e^{-TH} \qquad (4.1)$$

with $p=-i\nabla$. As it stands $\mathrm{tr}\, e^{-TH}=\infty$ because the electron is unconfined. Therefore in the Hamiltonian (3.3) with $V=0$ we restrict the electron to the finite box Λ. Then (4.1) makes sense. The definition (4.1) is understood implicitely with the limit $\Lambda \uparrow R^3$. If $\langle p(t)\cdot p(0)\rangle_T$ is absolutely integrable, the static conductivity is defined as

$$\sigma = \int_{-\infty}^{\infty} dt \; \langle p(t) \cdot p(0)\rangle_T \qquad (4.2)$$

in suitable units.

Let $\langle\cdot\rangle_{T,0}$ denote the expectation with respect to the path measure

$$\frac{1}{Z} P_T^0(dq(\cdot)) \exp[\beta \tfrac{1}{2} \int_0^T dt \int_0^T ds \; W_T(|t-s|, q(t)-q(s))] \quad . \qquad (4.3)$$

Here $P_T^0(dq(\cdot))$ is 3-dimensional Brownian motion conditioned that

$q(0)=0=q(T)$ ($\equiv$Brownian bridge) and

$$W_T(t,q) = \int d^3k|\lambda(k)|^2 \, e^{ikq} \frac{e^{-\omega(k)t} + e^{-\omega(k)(T-t)}}{1 - e^{-\omega(k)T}} \quad , \qquad (4.4)$$

$0 \leq t \leq T$, $q \in R^3$. Then with the above limiting convention

$$\begin{aligned} tr[e^{-(t-\lambda)H} \, p \, e^{-\lambda H} \cdot p]/tr \, e^{-TH} &= - <dq(0) \, dq(\lambda)>_{T,0} \\ &= - \lim_{\varepsilon \to} \frac{1}{\varepsilon^2} <(q(\varepsilon)-q(0)) \cdot (q(\lambda+\varepsilon)-q(\lambda))>_{T,0} \quad , \end{aligned} \qquad (4.5)$$

$0<\lambda<T$. Therefore, in principle, σ is obtained by firstly the analytic continuation of $-<dq(0) \cdot dq(\lambda)>_{T,0}$,$<0<\lambda<T$, to the strip $\{\lambda - it | 0 \leq \lambda \leq T$, $t \in R\}$ and secondly integration over the strip. Even for small T I do not know how to handle this analytic continuation.

As a side remark : Physically one would expect that in a dynamical medium diffusion could be more "easily" proved than in a static medium, simply because the motion of the medium destroys correlations. Also high spatial dimension should help because it is more likely for the moving particle to see an environment it has never seen before.

A main feature of the polaron is the spatial shift invariance of the measure in the sense that a path $\{t \to q(t)\}$ and its shift $\{t \to q(t)+a\}$ have the same weight. (In (4.5) the average is over shift invariant observables, which allows then the tie down $q(0)=q(T)=0$.) Some people have speculated, on the basis of variational calculations [48 and references cited in [49]] that at zero temperature this shift symmetry is spontaneously broken. In terms of statistical mechanics this would constitute a roughening transition : We pin $q(-T)=0=q(T)$ and consider the fluctuations at the origin, i.e.$<q(0)^2>_T$. Here the expectation refers to (3.4) with $V \doteq 0$. Then the proposed roughening transition means that for β sufficiently small $<q(0)^2>_T \to \infty$ as $T \to \infty$ whereas for β sufficiently large $<q(0)^2>_T \leq$const. . Such a transition has been proved for the two-dimensional discrete Gaussian model [50]. For the Fröhlich polaron $\omega(k)=\omega_o$ and $\lambda(k)=\sqrt{4\pi}/|k|$ implying that

$$W(t,q) = \frac{e^{-\omega_o|t|}}{|q|} \quad . \qquad (4.6)$$

Because of the exponential decay of the interaction a roughening transition is highly unplausible. In fact [49], if $\int d^3k|\lambda(k)|^2 < \infty$ and $\int d^3k|\lambda(k)|^2/\omega(k)^3 < \infty$, then

$$< q(0)^2 >_T \;\geq\; \text{const.} \;\; T \quad . \qquad (4.7)$$

Strictly speaken this result does not rule out the possibility of a roughening transition in the Fröhlich polaron. However, the missing piece is a short distance (large k) problem which has nothing to do with the claimed transition.

The roughening transition in a related model has been investigated theoretically in great detail [51-54]. Physically the model describes the motion of a quantum particle in an external periodic potential (washboard potential) acted upon by dissipative forces. These are modeled following Caldeira and Leggett. The Hamiltonian is then

$$H = (-\frac{1}{2}\Delta + \cos q) \otimes 1 + 1 \otimes \int dk\ \omega(k)\ a^+(k)\ a(k)$$
$$+ \sqrt{\beta}\ q \otimes \int dk\ \lambda(k)\ (a^+(k)+a(k)) + \beta \int dk\ \lambda(k)^2/\omega(k)\ q^2 \otimes 1. \tag{4.8}$$

Here $q \in R$. Such a Hamiltonian is used for current biased Josephsons junctions, note however the criticism together with proposals for other physical realizations in [54]. (4.8) yields the functional integral

$$\frac{1}{Z} P_{[-T,T]}(dq(\cdot)) \exp\left[- \int_{-T}^{T} dt \cos q(t) - \beta\frac{1}{4} \int_{-T}^{T} dt \int_{-T}^{T} ds\ W(t-s)(q(t)-q(s))^2\right]. \tag{4.9}$$

Following Caldeira and Leggett it is argued that $W(t) \cong t^{-2}$ for large t.

We note that if the cos-term is absent, then (4.9) is a Gaussian theory. Because $W(t) \cong t^{-2}$,

$$< q(0)^2 >_T \cong \log T \tag{4.10}$$

for large T. This is just on the borderline. For an even slower decay of W(t) q(t) would be smooth, i.e. $<q(0)^2>_T$ const. for any $\beta>0$. Therefore the fluctuations are so small that provided the coupling is large enough the washboard potential is able to localize the path q(t) in one of its minima as determined through the boundary conditions. A lattice version of (4.9) is studied in [55] including a Monte Carlo simulation [56]. To my knowledge there is no proof of this roughening transition, yet.

The action (4.9) is too singular as to come from a polaronic model. Since the roughening comes from the large distance interactions, the argument leading to (3.6) suggests that the polaron in one dimension with

$$\omega(k) = |k| \quad , \quad \lambda(k)^2 = \frac{1}{|k|} e^{-\alpha|k|} \quad , \quad \alpha > 0 \quad , \tag{4.11}$$

may have a roughening transition.

Problem 7. Existence of a roughening transition.

The path measure is either (4.9) with $W(t) \sim t^{-2}$ for large t or (3.4) with V=0 and ω, λ as in (4.11).

Another quantity of central interest in polaron theories is the effective mass. There are several definitions in the literature, cf. e.g. the references in [57], and I am not sure whether they actually coincide. I follow here Feynman's original proposal [58]. It is based on the observation that the path measure (3.4), with V=0 and with a sufficiently rapidly decaying interaction, looks on a coarse scale like Brownian motion. The inverse of the diffusion coefficient of this Brownian motion is called the effective mass, $m = m(\beta)$. Clearly $m(0) = 1$. Formally

$$m(\beta)^{-1} = \lim_{t\to\infty} \frac{1}{3t} \{ \lim_{T\to\infty} \langle (q(t)-q(0))^2 \rangle_T \} \quad , \tag{4.12}$$

where the expectation $\langle \cdot \rangle_T$ refers to (3.4) with V=0, $q(-T)=0=q(T)$. More ambitiously one may try to prove the invariance principle : We consider the path measure (3.4) with V=0 conditioned to $q(0)=0$ and take its infinite volume limit $T\to\infty$. $q(\cdot)$ is distributed according to this limiting path measure. Then one has to prove that, in the sense of weak convergence of path measures on $C(\mathbb{R}, \mathbb{R}^3)$,

$$\lim_{\varepsilon\to 0} \varepsilon\, q(\varepsilon^{-2}t) = m(\beta)^{-1/2}\, b(t) \quad , \tag{4.13}$$

where $b(t)$ is two sided standard Brownian motion on $\mathbb{R}^3$.

In an appendix to [49] I give the sketch of a proof that if W is bounded,

$$\int d^3k\, |\lambda(k)|^2 < \infty \quad , \tag{4.14}$$

and if W decays exponentially,

$$\int d^3k\, |\lambda(k)|^2 e^{-\omega(k)|t|} \leq c_1 e^{-c_2|t|} \quad , \tag{4.15}$$

then (4.13) and (4.14) hold. In addition, the effective mass depends analytically on the coupling constant β. The proof is based on a beautiful observation of R.L.Dobrushin. We divide [-T,T] into intervals of unit length, say, T integer. In each half open interval [j,j+1), j=-T,...,T-1, the path q(t) is shifted by the amount q(j). Let $y_j(\cdot)$ be the shifted path in the j-th interval. $t \to y_j(t)$, $0 \leq t \leq 1$, is continuous and $y_j(0)=0$. The $\{t \to y_j(t),\ 0 \leq t \leq 1\}$'s are the "spins". Therefore the single site space is $C([0,1], R^3)$. The path measure (3.4) with V=0 induces a measure for the $y_j(\cdot)$'s . Since Brownian motion has independent increments, the free measure for the $y_j(\cdot)$'s is a product measure with the single site measure given by $P_{[0,1]}$, i.e. Brownian motion over the time interval [0,1]. The integral $\int dt \int ds$ yields a translation invariant, many-body interaction. Because of (4.15) it decays exponentially, however. By a result of Dobrushin [59] this one-dimensional spin system has good mixing properties and its free energy depends analytically on the parameters. Since

$$3m(\beta)^{-1} = \sum_j < y_j(1) \cdot y_o(1) > \tag{4.16}$$

and

$$q(t) = \sum_{j=o}^{[t]-1} y_j(1) + y_{[t]}(t-[t]) , \tag{4.17}$$

[t] the integer part of t, the above claimed properties follow.

The conditions for the validity of the invariance principle are too strong, e.g. the conditions just above (4.7) should suffice, and the proof is not "natural". Of course, Dobrushin provides very detailed information on the path measure. For the central limit theorem more global methods, in particular martingale methods, should be applicable. Therefore I pose

Problem 8. Invariance principle.

A proof of (4.13) under conditions on λ and ω weaker than (4.14), (4.15).

We are back to the Fröhlich polaron with $\omega(k)=\omega_o$ and $\lambda(k) = \sqrt{4\pi}/|k|$. The spatial dimension is three. Let us assume that $m(\beta)$, defined by (4.16), exists. This leads to

Problem 9. The large coupling, $\beta \to \infty$, behavior of $m(\beta)$.

Physical considerations suggest that $m(\beta) \cong \beta^4$ for large β. For the free energy of (3.4) with V=0, i.e. the infinum of the spectrum of the Hamiltonian (3.3) with V=0 a large coupling result is available. Let

$$f_T(\beta) = -\frac{1}{T} \log \int P^o_{[0,T]}\ (dq(\cdot)) \exp\left[\beta\ \frac{1}{2} \int_o^T dt \int_o^T ds\ e^{-\omega_o|t-s|} / |q(t) - q(s)|\right] . \tag{4.18}$$

Then Donsker and Varadhan [60] prove that

$$\lim_{T\to\infty} f_T(\beta) = f(\beta) \tag{4.19}$$

exists and that

$$\lim_{\beta\to\infty} \beta^{-2} f(\beta) = \gamma_p \tag{4.20}$$

with γ_p determined from Pekar's variational problem,

$$\gamma_p = \inf_{\psi, \int d^3x |\psi(x)|^2 = 1} \{ \tfrac{1}{2} \int d^3x |\text{grad } \psi(x)|^2 - \int d^3x d^3y |\psi(x)|^2 \frac{1}{|x-y|} |\psi(y)|^2 \} \quad . \tag{4.21}$$

Properties of this variational problem have been studied by E.Lieb [61]. The crucial observation, cf. also [62], is that $\beta\to\infty$ corresponds to a Kac-type long range, i.e. mean field limit, for (4.18). Consequently the mean field free energy is computable from a variational problem in the "one spin" space. This suggests that the large coupling behavior of the effective mass could also be understood from a mean field type theory. However, (4.16) shows that the large coupling behavior of correlations has to be controled.

Acknowledgements. It is a pleasure to thank M.Aizenman, J.Bricmont, R.L.Dobrushin, R.Dümcke, J.Fröhlich, G.Gallavotti, J.L.Lebowitz, E.Olivieri and E.Presutti for most useful comments.

References

[1] T.Spencer, The Schrödinger equation with a random potential - a mathematical review. In: Percolation and Random Media, Les Houches Summer School 1984 .

[2] L.A.Bunimovich and Ya.G.Sinai, Comm.Math.Phys.78, 479 (1980)

[3] D.W.Jespen, J.Math.Phys.6, 405 (1965)
F.Spitzer, J.Math.and Phys.18, 973 (1969)

[4] Ya.G.Sinai and M.R.Soloveichik, One-dimensional classical massive particle in the ideal gas, preprint.
D.Szász and B.Tóth, Bounds for the limiting variance of the heavy particle in R^1, preprint.

[5] D.Dürr, S.Goldstein and J.L.Lebowitz, Comm.Math.Phys. 78, 507 (1981) and Z.Wahrscheinlichkeitstheorie verw.Gebiete 62, 427 (1983).

[6] H.Kesten and G.C.Papanicalaou, Comm.Math.Phys. 78, 19 (1980) and D.Dürr, S.Goldstein and J.L.Lebowitz, preprint

[7] H.Spohn, J.Stat.Phys. 17, 385 (1977).

[8] H.Spohn, Rev.Mod.Phys. 53, 569 (1980) and J.L.Lebowitz and H.Spohn, J.Stat.Phys. 29, 39 (1982).

[9] E.Presutti, Ya.G.Sinai and M.R.Soloveichik, preprint .

[10] S.Goldstein, J.L.Lebowitz and K.Ravishankar, Comm.Math.Phys. 85, 419 (1982).

[11] O.E.Lanford and D.W.Robinson, Comm.Math.Phys. 24, 193 (1972); H.Narnhofer, Acta Physica Austriaca, Suppl.XI, 527 (1973); D.W.Robinson, Comm.Math.Phys. 31, 171 (1973).

[12] E.B.Davies, Quantum Theory of Open Systems. Academic Press, London (1976).

[13] K.Hepp and E.H.Lieb, Helv.Phys.Acta 46, 573 (1973) and A.Frigerio and V.Gorini, J.Math.Phys. 17, 2123 (1976).

[14] R.Dümcke, Comm.Math.Phys. 97, 331 (1985).

[15] R.Feynman, Rev.Mod.Phys. 20, 367 (1948).

[16] H.Spohn and R.Dümcke, J.Stat.Phys., to appear.

[17] G.Yuval and P.W.Anderson, Phys.Rev. B9, 1522 (1970).

[18] M.Blume, V.J.Emery and A.Luther, Phys.Rev.Lett. 25, 450 (1970).

[19] V.J.Emery and A.Luther, Phys.Rev. B9, 215 (1974).

[20] A.O.Barut, Ed., Foundations of Radiation Theory and Quantum Electrodynamics. Plenum Press New York, 1980.

[21] P.Pfeifer, Chiral Molecules - a Superselection Rule Induced by the Radiation Field, P.h.D.Thesis, ETH Zürich (1980).

[22] L.M.Sander and H.B.Shore, Phys.Rev. B3, 1472 (1979).

[23] R.Beck, W.Götze and P.Prelovsek, Phys.Rev. A20, 1140 (1979).

[24] S.Chakravarty, Phys.Rev.Lett. 49, 681 (1982).

[25] A.J.Bray and M.A.Moore, Phys.Rev.Lett. 49, 1545 (1982).

[26] S.Chakravarty and S.Kivelson, Phys.Rev.Lett. 50, 1811 (1983).

[27] W.Zwerger, Z.Physik B 53, 53 (1983).

[28] B.Simon and A.D.Sokal, J.Stat.Phys. 25, 679 (1981).

[29] F.J.Dyson, Comm.Math.Phys. 12, 91 (1969).

[30] G.Gallavotti and H.Knops, Rev.Nuov.Cim. 5, 341 (1975).

[31] P.Collet and J.P.Eckmann, A Renormalization Group Analysis of the Hierarchical Model in Statistical Mechanics. Lecture Notes in Physics 74, Springer, Berlin, 1978.

[32] J.Fröhlich and T.Spencer, Comm.Math.Phys. 84, 87 (1982).

[33] Appendix in J.Bricmont, J.L.Lebowitz and C.E.Pfister, J.Stat. Phys. 24, 269 (1981).

[34] S.Chakravarty and A.J.Leggett, Phys.Rev.Lett. 52, 5 (1984).

[35] A.J.Leggett and A.Gary, Phys.Rev.Lett. 54, 857 (1985).

[36] A.O.Caldeira and A.J.Leggett, Ann.Phys.(N.Y.) 149, 374 (1983).

[37] E.Nelson, Schrödinger Particles Interacting with a Quantized Scalar Field. In: W.T.Martin and I.Segal, Analysis in Function Space. M.I.T.Press, 1964.

[38] V.Hakim, A.Muramatsu and F.Guinea, Phys.Rev. B30, 464 (1984).

[39] J.Ginibre, in: Statistical Mechanics and Quantum Field Theory. Eds. C.DeWitt and R.Stora, Gordon and Breach, New York (1971).

[40] M.Aizenman, private communication.

[41] J.L.Lebowitz and E.Presutti, Comm.Math.Phys. 50, 195 (1976).

[42] J.Fröhlich, Thesis, ETH Zürich, 1972.

[43] J.Fröhlich, Ann.Inst. Henri Poincaré, A19, 1 (1973).

[44] J.Fröhlich, Fortschritte der Physik 22, 159 (1974).

[45] E.Nelson, J.Math.Phys. 5, 1190 (1964).

[46] J.P.Eckmann, Comm.Math.Phys. 18, 247 (1970).

[47] S.Albeverio, Helv.Phys.Acta 45, 303 (1972).

[48] N.Tokuda, H.Shoji and K.Yoneya, J.Phys. C 14, 4281 (1981).

[49] H.Spohn, J.Phys. A, to appear.

[50] J.Fröhlich and T.Spencer, Comm.Math.Phys. 81, 527 (1981).

[51] A.Schmid, Phys.Rev.Lett. 51, 1506 (1983).

[52] F.Guinea, V.Hakim and A.Muramatsu, Phys.Rev.Lett. 54, 263 (1985).

[53] S.A.Bulgadaev, JETP Lett. 39, 315 (1984).

[54] M.P.A.Fisher and W.Zwerger, Quantum Brownian motion in a periodic potential, preprint.

[55] K.H.Kjaer and H.J.Hilhorst, J.Stat.Phys. 28, 621 (1982).

[56] J.Slurink and H.J.Hilhorst, Physica 120A, 627 (1983).

[57] J.T.Devreese, Path Integrals and Continuum Fröhlich Polarons. In: Path Integrals, ed. by G.J.Papadopoulos and J.R.Devreese. Plenum Press, New York, 1978.

[58] R.P.Feynman, Phys.Rev. 97, 660 (1955).

[59] R.L.Dobrushin, Matth.USSR Sbornik 23, 13 (1974).

[60] M.D.Donsker and S.R.S.Varadhan, Comm.Pure and Appl.Math. 36, 505 (1983).

[61] E.Lieb, Stud.Appl.Math. 57, 93 (1977).

[62] J.Adamowski, B.Gerlach and H.Leschke, Physics Letters 79A, 249 (1980).

WHY DO BOSONS CONDENSE?

J.T. LEWIS
Dublin Institute for Advanced Studies
10 Burlington Road, Dublin 4
Republic of Ireland

1. Introduction

Bose-Einstein condensation was discovered over sixty years ago. Einstein [1] based his prediction on a combination of Bose-Einstein statistics and the classical density of states in phase-space. Three years later, Uhlenbeck [2] objected that the result holds only "when the quantization of translational motion is neglected". The situation remained unclear for ten years until Kramers pointed out the importance of the thermodynamic limit for the sharp manifestation of phase-transitions; whereupon Uhlenbeck [3] withdrew his objection, pointing out that Einstein's expression for the mean particle number density is correct in the thermodynamic limit. In the same year, London [4] introduced the concept of macroscopic occupation of the ground-state and related it to the coherence properties of the condensate. London conjectured that the superfluid phase-transition in He^4 is an example of Bose-Einstein condensation; it remains an open problem: do interacting bosons condense? On p. 39 of his book [5] , London suggested that, as a manifestation of quantum-mechanical complementarity, momentum-space condensation is enhanced by a spatial repulsion among the particles. There is no rigorous proof of this. If we are to make progress with the problem of the existence of boson condensation in interacting systems, we must have a good conceptual understanding of existing results. For this reason it may be useful to review some of these and to ask: why do bosons condense?

First, we review results on the free boson gas. Then we consider two models of an interacting boson gas: the mean field model and the Huang-Yang-Luttinger model of a hard-sphere gas. The mechanism of condensation is radically different in the two models. Finally, we sketch a recent result which shows that, in a scaling limit, the interacting gas behaves like the mean-field model.

2. The Models

The systems we consider are specified by model hamiltonians; since these are diagonal in the occupation number operators of the eigenstates of the single-particle hamiltonian, it is possible to regard the occupation numbers as random variables rather than as operators. We shall do this, paving the way for the use of methods from probability theory. The connections between our model hamiltonians and the full hamiltonian of the interacting system are explained in

The probability space Ω on which our models are defined is the space of terminating sequences of non-negative integers: an element ω of Ω is a sequence $\{\omega(j) \in \mathbb{N} : j = 1, 2, \dots\}$. The basic random variables, the occupation numbers, are the evaluation maps $\sigma_j : \Omega \longrightarrow \mathbb{N}$ given by $\sigma_j(\omega) = \omega(j)$. The total particle number N is the random variable $N(\omega) = \sum_{j \geq 1} \sigma_j(\omega)$. We consider three models:

(1) The Free Gas

We define a sequence $\{H_\ell^0 : \ell = 1, 2, \dots\}$ of free gas hamiltonians

$$H_\ell^0(\omega) = \sum_{j \geq 1} \lambda_\ell(j) \sigma_j(\omega) \tag{2.1}$$

where $\{\lambda_\ell(j) : j = 1, 2, \dots\}$ is an ordered sequence of real numbers: $0 = \lambda_\ell(1) \leq \lambda_\ell(2) \leq \cdots$ associated with a region Λ_ℓ ; the volume of Λ_ℓ is denoted by V_ℓ , and we assume that $V_\ell \longrightarrow \infty$ as $\ell \to \infty$. We must put conditions on the double sequence $\{\lambda_\ell(j) : j = 1, 2, \dots ; \ell = 1, 2, \dots\}$ to ensure the existence of the grand canonical pressure in the thermodynamic limit. We shall assume that the following conditions hold:

(S1): $\varphi(\beta) = \lim_{\ell \to \infty} \frac{1}{V_\ell} \sum_{j \geq 1} e^{-\beta \lambda_\ell(j)}$ exists for all $\beta \in (0, \infty)$.

(S2): $0 < \varphi(\beta) < \infty$ for all $\beta \in (0, \infty)$.

(S3): $\lim_{\beta \to \infty} \varphi(\beta) = 0$.

These conditions hold, for example, when $\lambda_\ell(j) = \varepsilon_\ell(j) - \varepsilon_\ell(1)$ and $\varepsilon_\ell(1) < \varepsilon_\ell(2) \leq \cdots$ are the eigenvalues of the single-particle hamiltonian $-(\hbar^2/2m)\Delta_\ell$, where Δ_ℓ is the Laplacian in a d-dimensional cube Λ_ℓ with periodic boundary conditions ($d \geq 1$) and $V_\ell \to \infty$; they hold also for the subtracted eigenvalues of a wide-class of single-particle hamiltonians associated with a sequence of regions $\{\Lambda_\ell : \ell = 1, 2, \dots\}$ tending to infinity in the sense of van Hove.

(2) The Mean-Field Model

We define a sequence $\{H_\ell^a : \ell = 1, 2, \dots\}$ of hamiltonians

$$H_\ell^a(\omega) = H_\ell^0(\omega) + \frac{a}{2V_\ell} N(\omega)^2 \tag{2.2}$$

with $a > 0$. The term $\frac{a}{2V_\ell} N^2$, which provides a crude caricature of the interaction can be understood classically: it arises in an "index of refraction" approximation in which we imagine each particle to move through the system as if it were moving in a uniform optical medium and so receiving an increment of energy proportional to $\frac{N}{V}$; since a is positive the interaction is repulsive.

(3) The Hard-Sphere Gas Model

We define a sequence $\{\tilde{H}_\ell^a : \ell = 1, 2, \ldots\}$ of hamiltonians

$$\tilde{H}_\ell^a(\omega) = H_\ell^a(\omega) + \frac{a}{2V_\ell}\{N(\omega)^2 - \sum_{j\geq 1}\sigma_j(\omega)^2\} . \tag{2.3}$$

This model was introduced by Huang, Yang and Luttinger [6]. The term $\frac{a}{2V_\ell}\{N(\omega)^2 - \sum_{j\geq 1}\sigma_j(\omega)^2\}$ is purely quantum-mechanical in origin and is a first-order approximation to the effect of Bose-Einstein statistics on the interaction energy of a system of bosons interacting through hard-sphere repulsion; it is least for the configuration in which all particles occupy the same single-particle energy level.

A model is specified by giving a sequence $\{H_\ell : \ell = 1, 2, \ldots\}$ of hamiltonians on Ω and a divergent sequence $\{V_\ell : \ell = 1, 2, \ldots\}$ of positive real numbers. For each hamiltonian H_ℓ we define the grand canonical pressure $p_\ell(\mu)$ by

$$e^{\beta V_\ell p_\ell(\mu)} = \sum_{\omega\in\Omega} \exp[\beta(\mu N(\omega) - H_\ell(\omega))] \tag{2.4}$$

for all values of the chemical potential μ for which the right-hand side of (2.4) is finite. We define the grand canonical measure $\mathbb{P}_\ell^\mu$ by

$$\mathbb{P}_\ell^\mu[A] = \sum_{\omega\in A} \exp[\beta(\mu N(\omega) - H_\ell(\omega)] \tag{2.5}$$

here β is the inverse temperature and is strictly positive.

The first problem for each model is to prove the existence of the pressure $p(\mu) = \lim_{\ell\to\infty} p_\ell(\mu)$ in the thermodynamic limit; the second problem is to compute the total amount of condensate $\Delta(\mu)$ defined by

$$\Delta(\mu) = \lim_{\lambda\downarrow 0} \lim_{\ell\to\infty} \frac{1}{V_\ell} \mathbb{E}_\ell^\mu[N_\ell(\lambda;\cdot)] \tag{2.6}$$

where, for each $\lambda > 0$, $N_\ell(\lambda;\cdot)$ is the random variable defined by

$$N_\ell(\lambda;\omega) = \sum_{\{j:\lambda_\ell(j)<\lambda\}} \sigma_j(\omega) , \tag{2.7}$$

and $\mathbb{E}_\ell^\mu[\cdot]$ denotes the expectation with respect to the measure $\mathbb{P}_\ell^\mu[\cdot]$.

The quantity $\Delta(\mu)$ can be thought of as the density of particles having zero single-particle energy in the thermodynamic limit; it is the order-parameter for the Bose-Einstein phase-transition. It turns out that by solving a slightly generalized version of the first problem, we can solve the second one at the same time (at least in the case of models (2) and (3); the free gas has to be dealt with seperately). Define $H_\ell(\lambda,s;\omega)$ by replacing $\sum_{j\geq 1} \lambda_\ell(j)\sigma_j(\omega)$ by

$$\sum_{\{j:\ \lambda_\ell(j)\leq\lambda\}} \lambda_\ell(j)\sigma_j(\omega) + \sum_{\{j:\ \lambda_\ell(j)>\lambda\}} (\lambda_\ell(j) + s)\sigma_j(\omega)$$

and define $p_\ell(\lambda,s;\mu)$ by

$$\exp\{\beta V_\ell p_\ell(\lambda,s;\mu)\} = \sum_{\omega\in\Omega} \exp[\beta\{(\mu+s)N(\omega) - H_\ell(\lambda,s;\omega)\}] . \tag{2.8}$$

Now

$$\frac{\partial}{\partial s} p_\ell(\lambda,s;\mu)\Big|_{s=0} = \frac{1}{V_\ell}\mathbb{E}_\ell^\mu[N_\ell(\lambda;\cdot)] \tag{2.9}$$

and $s \mapsto p_\ell(\lambda,s;\mu)$ is convex. It follows from the version of Griffiths' Lemma proved by Hepp and Lieb [7] that *if, for* μ *fixed,* λ *fixed and* λ *strictly positive, the limit* $p(\lambda,s;\mu) = \lim_{\ell\to\infty} p_\ell(\lambda,s;\mu)$ *exists for* s *in an open neighbourhood of zero and* $s \mapsto p(\lambda,s;\mu)$ *is differentiable at* $s=0$ *then*

$$\lim_{\ell\to\infty} \frac{1}{V_\ell}\mathbb{E}_\ell^\mu[N_\ell(\lambda;\cdot)] = \left(\frac{\partial p}{\partial s}\right)(\lambda,0;\mu) . \tag{2.10}$$

Hence we have

$$\Delta(\mu) = \lim_{\lambda\downarrow 0}\left(\frac{\partial p}{\partial s}\right)(\lambda,0;\mu) . \tag{2.11}$$

3. Survey of Results on the Free Boson Gas

Using (2.1) with (2.4), we have the following expression for the free-gas pressure $p_\ell^0(\mu)$:

$$p_\ell^0(\mu) = \beta^{-1}\sum_{j\geq 1} \ln(1 - e^{\beta(\mu-\lambda_\ell(j))})^{-1} \tag{3.1}$$

provided that $\mu<0$. It is convenient to re-write it as a Stieltje's integral:

$$p_\ell^0(\mu) = \int_{[0,\infty)} p(\mu|\lambda)\, dF_\ell(\lambda) , \quad \mu<0 , \tag{3.2}$$

where $p(\mu|\lambda)$ is the conditional pressure

$$p(\mu|\lambda) = \beta^{-1}\ln(1 - e^{\beta(\mu-\lambda)})^{-1} \tag{3.3}$$

and $F_\ell(\lambda)$ is defined by

$$V_\ell F_\ell(\lambda) = max\{j : \lambda_\ell(j) \leq \lambda\} \qquad (3.4)$$

Assume that (S1), (S2) and (S3) hold; then for each $\mu<0$ *the limit* $p^0(\mu) = \lim_{\ell\to\infty} p^0_\ell(\mu)$ *exists and is given by*

$$p^0(\mu) = \int_{[0,\infty)} p(\mu|\lambda)\, dF(\lambda) \quad , \qquad (3.5)$$

where $F(\cdot)$ *is the integrated density of states which is determined uniquely by*

$$\varphi(\beta) = \int_{[0,\infty)} e^{-\beta\lambda}\, dF(\lambda) \; . \qquad (3.6)$$

We define $p^0(0)$ by

$$p^0(0) = \lim_{\mu\uparrow 0} p^0(\mu) \quad , \qquad (3.7)$$

and $p^0(\mu)$ for $\mu>0$ by $p^0(\mu) = \infty$; then $\mu \mapsto p^0(\mu)$ *is a closed convex function on* $\mathbb{R}$.

The mean particle number-density is given by

$$\frac{1}{V_\ell}\mathbb{E}^\mu_\ell[N] = (\frac{d}{d\mu}p^0_\ell)(\mu) \; . \qquad (3.8)$$

Now $\mu \mapsto (\frac{d}{d\mu}p^0_\ell)(\mu)$ is an increasing function on the interval $(-\infty,0)$; since $\lambda_\ell(1) = 0$, its range is the whole of $(0,\infty)$. It follows that we can give $\frac{1}{V_\ell}\mathbb{E}^\mu_\ell[N]$ any pre-assigned value ρ in the interval $(0,\infty)$ by choosing μ appropriately in the interval $(-\infty,0)$. These considerations do not apply in general to the limit function p^0 ; in the case where the $\{\lambda_\ell(j)\}$ are the subtracted eigenvalues of $-\frac{1}{2}\Delta_\ell$, where Δ_ℓ is the Laplacian with periodic, Dirichlet or Neumann boundary conditions in a d-dimensional cube of volume V_ℓ , the distribution function F is given by

$$F(\lambda) = \frac{\Gamma(\frac{d}{2}+1)}{(2\pi)^{d/2}}\, \lambda^{d/2} \; . \qquad (3.9)$$

It follows that, for $d>2$, the supremum of $\frac{d}{d\mu}p^0$ on $(-\infty,0)$ is finite and given by

$$\rho_c = \lim_{\mu\uparrow 0}(\frac{d}{d\mu}p^0)(\mu) = \int_{[0,\infty)} (e^{\beta\lambda}-1)^{-1}\, dF(\lambda) \, . \qquad (3.10)$$

(When $(e^{\beta\lambda}-1)^{-1}$ is not integrable with respect to the measure F , we define ρ_c to be $+\infty$). To investigate the behaviour of the free boson gas at mean densities above the critical density ρ_c it is necessary to take the thermodynamic limit at fixed mean density:
let $\mu_\ell(\rho)$ *be the unique real root of*

$$(\frac{d}{d\mu}p^0_\ell)(\mu) = \rho \qquad (3.11)$$

then $\mu(\rho) = \lim_{\ell\to\infty}\mu_\ell(\rho)$ *exists and is equal to the unique real root of*

$$\left(\frac{d}{d\mu}p^{o}\right)(\mu) = \rho \tag{3.12}$$

if $\rho < \rho_c$ *, and is equal to zero otherwise.*

Let $\pi_\ell(\rho) = (p_\ell^o \circ \mu_\ell)(\rho)$ be the pressure at density ρ ; then *the limit* $\pi(\rho) = \lim_{\ell\to\infty} \pi_\ell(\rho)$ *exists and is given by*

$$\pi(\rho) = (p^o \circ \mu)(\rho) . \tag{3.13}$$

The main effort in proving this result goes into showing that, in the thermodynamic limit, particles in the low-lying levels make no contribution to the pressure. The same is not true of the density:

let $\mathbb{E}_\ell^\rho[\cdot]$ *denote the expectation with respect to the measure* $\mathbb{P}_\ell^\mu[\cdot]$ *taken with* $\mu = \mu_\ell(\rho)$ *; then*

$$\Delta^o(\rho) = \lim_{\lambda\downarrow 0}\lim_{\ell\to\infty}\frac{1}{V_\ell}\mathbb{E}_\ell^\rho[N_\ell(\lambda;\cdot)] = (\rho - \rho_c)^+ \tag{3.14}$$

The proof is very short, so we give it:

From the definitions,

$$\frac{1}{V_\ell}\mathbb{E}_\ell^\rho[N(\lambda;\cdot)] = \int_{[0,\infty)} \rho(\mu_\ell(\rho)|\lambda')\,dF_\ell(\lambda') ,$$

where $\rho(\mu|\lambda) = (e^{\beta(\mu-\lambda)} - 1)^{-1}$.

But for $\lambda > 0$ we have

$$\lim_{\ell\to\infty}\int_{[0,\lambda)} \rho(\mu_\ell|\lambda')\,dF_\ell(\lambda') = \rho - \lim_{\ell\to\infty}\int_{[\lambda,\infty)}\rho(\mu_\ell|\lambda')\,dF_\ell(\lambda')$$

$$= \rho - \int_{[\lambda,\infty)}\rho(\mu(\rho)|\lambda')\,dF(\lambda')$$

By the dominated convergence principle,

$$\lim_{\lambda\downarrow 0}\int_{[\lambda,\infty)}\rho(\mu(\rho)|\lambda')\,dF(\lambda') = \int_{[0,\infty)}\rho(\mu(\rho)|\lambda')\,dF(\lambda') = \begin{cases}\rho & , \rho < \rho_c , \\ \rho_c & , \rho \geq \rho_c ,\end{cases}$$

so that (3.14) follows.

The free-energy $f^o(\rho)$ at mean density satisfies $f^o(0) = 0$ and is given by

$$f^o(\rho) = \sup_{\alpha < 0}\{\alpha\rho - p^o(\alpha)\} . \tag{3.15}$$

It is easy to see that $\rho \mapsto f^o(\rho)$ is a decreasing convex function; it is given explicitly by $f^o(\rho) = \mu(\rho)\rho - p^o(\mu(\rho))$. (3.16)

Notice that $f^o(\rho)$ is constant and equal to $-p^o(0)$ for $\rho \geqslant \rho_c$; this linear segment in the graph of f^o signals a first-order phase-transition at $\mu = 0$.

The distribution function $\mathbb{K}^\rho$ for the particle number density is defined by

$$\mathbb{K}^\rho(x) = \lim_{l\to\infty} \mathbb{P}_l^\rho[N/V_l \leqslant x] \quad ; \tag{3.17}$$

it is known as the Kac distribution function. For $\rho < \rho_c$, $\mathbb{K}^\rho$ exists and is the degenerate distribution with mean ρ :

$$\mathbb{K}^\rho(x) = \begin{cases} 0 \ , & x < \rho \ , \\ 1 \ , & x \geqslant \rho \ ; \end{cases} \tag{3.18}$$

in general, for $\rho \geqslant \rho_c$, the sequence does not converge but convergent subsequences always exists and their limits are sensitive to boundary conditions. For $\rho < \rho_c$, the convergence is exponentially fast: define, for each $\mu < 0$ and each Borel subset A of $\mathbb{R}_+$, the measure $\mathbb{K}_l^\mu[\cdot]$ by

$$\mathbb{K}_l^\mu[A] = \mathbb{P}_l^\mu[N/V_l \in A] \quad ; \tag{3.19}$$

then

for each closed subset C *we have*

$$\limsup_{l\to\infty} \frac{1}{\beta V_l} \ln \mathbb{K}_l^\mu[C] \leqslant - \inf_C I^\mu(x) \ , \tag{3.20}$$

and for each open subset G *we have*

$$\liminf_{l\to\infty} \frac{1}{\beta V_l} \ln \mathbb{K}_l^\mu[G] \geqslant - \inf_G I^\mu(x) \ , \tag{3.21}$$

where $I^\mu(\cdot)$ *is given by*

$$I^\mu(\cdot) = \begin{cases} p^o(\mu) + f^o(x) - \mu x \ , & x \geqslant 0 \ , \\ \infty \ , & x < 0 \ . \end{cases} \tag{3.22}$$

This is expressed by saying that $\{\mathbb{K}_l^\mu : l = 1, 2, \ldots\}$ *satisfies the large deviation principle with constants* $\{V_l : l = 1, 2, \ldots\}$ *and rate-function* $I^\mu(\cdot)$.

This will prove to be very important in the sequel. It is a general principle that a rate-function is some kind of relative entropy, so it is of interest to note that we can re-write the expression (3.23) to make it look more like one: for $x \geqslant 0$ we have

$$I^\mu(x) = -(x - \rho_c)^+ \mu - \beta^{-1} \int_{[0,\infty)} s^\mu(x|\lambda)\, dF(\lambda) \ , \tag{3.23}$$

where

$$s^{\mu}(x|\lambda) = \{1+\rho(\mu(x)|\lambda)\} \ln \frac{1+\rho(\mu(x)|\lambda)}{1+\rho(\mu|\lambda)} - \rho(\mu(x)|\lambda) \ln \frac{\rho(\mu(x)|\lambda)}{\rho(\mu|\lambda)} . \tag{3.24}$$

4. The Mean-Field Model

The pressure $p_\ell^a(\mu)$ in the mean-field model is given by

$$e^{\beta V_\ell p_\ell^a(\mu)} = \sum_{\omega\in\Omega} \exp[\beta(\mu N(\omega) - H_\ell^a(\omega)] \tag{4.1}$$

We can re-write this as

$$e^{\beta V_\ell p_\ell^a(\mu)} = e^{\beta V_\ell p_\ell^o(\mu)} \mathbb{E}_\ell^\mu [\exp(-\frac{\beta a}{2V_\ell} N^2)] , \tag{4.2}$$

where $\mathbb{E}_\ell^\mu[\cdot]$ is the expectation with respect to the free-gas probability measure. There is one draw-back with this formula: the series on the right-hand side of (4.1) converges for all real values of μ, so that $p_\ell^a(\mu)$ is defined for all μ in $\mathbb{R}$; on the otherhand, the right-hand side of (4.2) makes sense as it stands for $\mu<0$ only. The way round this is to fix $\alpha<0$ and re-arrange the right-hand side of (4.2) as

$$e^{\beta V_\ell p_\ell^a(\mu)} = e^{\beta V_\ell p_\ell^o(\alpha)} \mathbb{E}_\ell^\alpha [\exp\{\beta(\mu-\alpha)N - \frac{a}{2V_\ell} N^2\}] . \tag{4.3}$$

This can be re-written as

$$e^{\beta V_\ell p_\ell^a(\mu)} = e^{\beta V_\ell p_\ell^o(\alpha)} \int_{[0,\infty)} e^{\beta V_\ell G(x)} \mathbb{K}_\ell^\alpha[dx] , \tag{4.4}$$

where $\mathbb{K}_\ell^\alpha$ is the measure on $\mathbb{R}_+$ defined at (3.19) and $G(\cdot)$ is given by

$$G(x) = (\mu-\alpha)x - \frac{a}{2}x^2 . \tag{4.5}$$

In §3 we noted that, roughly speaking,

$$\mathbb{K}_\ell^\alpha[dx] \sim e^{-\beta V_\ell I^\alpha(x)} dx , \tag{4.6}$$

where $I^\alpha(\cdot)$ is given by (3.22); we would expect, therefore, that for large values of V_ℓ

$$\int_{[0,\infty)} e^{\beta V_\ell G(x)} \mathbb{K}_\ell^\alpha[dx] \sim e^{\beta V_\ell \sup_{x\geq 0}\{G(x) - I^\alpha(x)\}} . \tag{4.7}$$

Looking again at (4.4), we might guess that

$$\lim_{\ell\to\infty} p_\ell^a(\mu) = p^o(\alpha) + \sup_{x\geq 0}\{G(x) - I^\alpha(x)\} = \sup_{x\geq 0}\{\mu x - f^a(x)\} \tag{4.8}$$

where

$$f^a(x) = f^0(x) + \frac{a}{2}x^2 . \tag{4.9}$$

Intuitively, this is obvious: to get the pressure $p^a(\mu)$ in the mean-field model, take the free-energy $f^0(x)$ of the free-gas, add the term $\frac{a}{2}x^2$ to get the free-energy $f^a(x)$ of the mean-field model and then take the Legendre-Fenchel transform (4.7) of $f^a(x)$. It is very satisfying to discover that this argument can be made rigorous using Varadhan's Theorem.

Varadhan's Theorem concerns the asymptotic behaviour of integrals with respect to a sequence of probability measures satisfying the Large Deviation Principle and extends Laplace's method to infinite-dimensional spaces. The application to the mean-field model which we have just sketched uses Varadhan's Theorem applied to measures on a one-dimensional space; in the next section we will use a Banach space version.

Let E be a complete separable metric space and $\{\mathbb{K}_\ell : \ell = 1, 2, \ldots\}$ a sequence of probability measures on the Borel subsets of E; let $\{V_\ell : \ell = 1, 2, \ldots\}$ be a sequence of positive constants such that $V_\ell \to \infty$. We say that

$\{\mathbb{K}_\ell\}$ *obeys the Large deviation principle with rate-function* $I(\cdot)$ *and constants* $\{V_\ell\}$ *if there exists a function* $I : E \to [0,\infty]$ *satisfying:*

(LD1): $I(\cdot)$ *is lower semicontinuous on* E.

(LD2): *For each* $m < \infty$ *the set* $\{x : I(x) \leqslant m\}$ *is compact in* E.

(LD3): *For each closed subset* C *of* E

$$\limsup_{\ell\to\infty} \frac{1}{V_\ell} \ln \mathbb{K}_\ell[C] \leqslant -\inf_C I(x) .$$

(LD4): *For each open subset* G *of* E

$$\liminf_{\ell\to\infty} \frac{1}{V_\ell} \ln \mathbb{K}_\ell[G] \geqslant -\inf_G I(x) .$$

A version of Varadhan's Theorem which is adequate for our needs is the following:

<u>Varadhan's Theorem</u>

Let $\{\mathbb{K}_\ell\}$ *be a sequence of probability measures on* E *satisfying the large deviation principle with constants* $\{V_\ell\}$ *and rate-function* $I(\cdot)$. *Then, for any continuous real-valued function* G *on* E *satisfying*

$$\sup\{G(x) : x \in \bigcup_{\ell\geqslant 1} \operatorname{supp} \mathbb{K}_\ell\} < \infty , \tag{4.10}$$

we have

$$\lim_{\ell\to\infty} \frac{1}{V_\ell} \ln \int_E e^{\beta V_\ell G(x)} \mathbb{K}_\ell[dx] = \sup_E \{G(x) - I(x)\} . \tag{4.11}$$

Returning to the problem of proving the existence of the limit $p^a(\mu) = \lim_{\ell\to\infty} p^a_\ell(\mu)$, our first task is clear: we have to verify that, for fixed $\alpha < 0$, the sequence of measures $\{\mathbb{K}_\ell^\alpha\}$ defined at (3.19) satisfies (LD1) - (LD4) with rate-function $I^\alpha(\cdot)$ given by (3.22); the principle of conservation of difficulty is not violated- this requires some work. The second task, that of verifying that $G(\cdot)$ defined at (4.5) is continuous and satisfies (4.9), is trivial. In this way, we can give a rigorous proof that

$$p^a(\mu) = \sup_{x \geq 0} \{\mu x - (f^0(x) + \tfrac{a}{2}x^2)\} . \tag{4.12}$$

It is easy to check that we can re-cast (4.11) in a form more useful for our next purpose:

$$p^a(\mu) = \inf_{\alpha < 0} \left\{ \frac{(\mu - \alpha)^2}{2a} + p^0(\alpha) \right\} . \tag{4.13}$$

It now follows that

$$p^a(\lambda, s; \mu) = \inf_{\alpha < 0} \left\{ \frac{(\mu + s - \alpha)^2}{2a} + p^0(\lambda, s; \alpha) \right\} , \tag{4.14}$$

where

$$p^a(\lambda, s; \alpha) = \int_{[0,\lambda)} p(\alpha \mid \lambda')\, dF(\lambda') + \int_{[\lambda,\infty)} p(\alpha \mid \lambda' + s)\, dF(\lambda') . \tag{4.15}$$

We verify by direct computation that $s \mapsto p^a(\lambda, s; \mu)$ is differentiable at $s = 0$ so that (2.10) and (2.11) apply; then, using an argument similar to that used in the proof of (3.14), we find that the total amount of condensate is given by

$$\Delta^a(\mu) = (\rho(\mu) - \rho_c)^+ , \tag{4.16}$$

where $\rho(\mu) = (p^a)'(\mu)$. The critical value μ_c at which the phase-transition occurs is now $\mu_c = a\rho_c$. The phase-transition is no longer first-order; this can be seen by looking at the free-energy

$$f^a(x) = f^0(x) + \frac{a}{2} x^2 . \tag{4.17}$$

This has a unique minimum; the linear segment which occurs in the graph of f^0 is absent from the graph of f^a, being swamped by the quadratic term in (4.16). The precise orděr of the phase-transition depends on the behaviour of the integrated density of states near $\lambda = 0$; for hamiltonians for which (3.9) holds, it is third-order when $d = 3$.

5. The Hard-Sphere Gas Model

The pressure $\tilde{p}_\ell^a(\mu)$ in the hard-sphere gas model is given by

$$e^{\beta V_\ell \tilde{p}_\ell^a(\mu)} = e^{\beta V_\ell p_\ell^0(\alpha)} \mathbb{E}_\ell^\alpha [\exp[\beta\{(\mu - \alpha)N - \tfrac{a}{2V_\ell}(2N^2 - \Sigma\, \sigma_j^2)\}] . \tag{5.1}$$

To evaluate $\tilde{p}^a(\mu) = \lim_{\ell \to \infty} \tilde{p}_\ell^a(\mu)$ using Varadhan's Theorem, we introduce random variables $\tilde{X}_\ell(\omega)$ taking values in the positive cone $\ell^1_+ = \{x_j \geq 0 : \sum_{j\geq 0} x_j < \infty\}$ of the real Banach space $\ell^1 = \{x_j \in \mathbb{R} : \sum_{j\geq 0} |x_j| < \infty\}$ and defined by

$$\tilde{X}_\ell(j;\omega) = \begin{cases} V_\ell^{-1} N(\omega) \ , & j = 0 \ , \\ V_\ell^{-1} \sigma_j(\omega) \ , & j \geq 1 \ . \end{cases} \tag{5.2}$$

These random variables induce measures $\tilde{\mathbb{K}}^\alpha_\ell$ on ℓ^1_+ given by

$$\tilde{\mathbb{K}}^\alpha_\ell = \mathbb{P}^\alpha_\ell \circ \tilde{X}^{-1}_\ell \ . \tag{5.3}$$

In terms of the $\tilde{\mathbb{K}}^\alpha_\ell$ we can write (5.1) as

$$e^{\beta V_\ell \tilde{p}^a_\ell(\mu)} = e^{\beta V_\ell p^0_\ell(\alpha)} \int_{\ell^1_+} e^{\beta V_\ell \tilde{G}(x)} \, \tilde{\mathbb{K}}^\alpha_\ell[dx] \ , \tag{5.4}$$

with

$$\tilde{G}(x) = (\mu - \alpha)x_0 - \tfrac{a}{2}(2x_0^2 - \sum_{j\geq 1} x_j^2) \ ; \tag{5.5}$$

if $\{\tilde{\mathbb{K}}^\alpha_\ell\}$ satisfies the large deviation principle with rate-function $\tilde{I}^\alpha[\cdot]$ then we expect to find that $\tilde{p}^a(\mu) = \lim_{\ell\to\infty} \tilde{p}^a_\ell(\mu)$ exists and is given by

$$\tilde{p}^a(\mu) = \sup_{\ell^1_+} \{\tilde{G}(x) - \tilde{I}^\alpha[x]\} \ . \tag{5.6}$$

We can show, in fact, that
if (S1) , (S2) *and* (S3) *hold and, in addition,* $\lim_{\ell\to\infty} \lambda_\ell(j) = 0$ *for each* $j \geq 1$ *then, for each* $\alpha < 0$, *the sequence* $\{\tilde{\mathbb{K}}^\alpha_\ell\}$ *of probability measures on* ℓ^1_+ *satisfies the large deviation principle with constants* $\{V_\ell\}$ *and rate-function* $\tilde{I}^\alpha[\cdot]$ *given by*

$$\tilde{I}^\alpha[x] = \begin{cases} p^0(\alpha) + f^0(x_0 - \sum_{j\geq 1} x_j) - \alpha x_0 \ , & 0 \leq x_0 - \sum_{j\geq 1} x_j \ , \\ \infty \ , & \textit{otherwise.} \end{cases} \tag{5.7}$$

Application of Varadhan's Theorem now yields

$$\tilde{p}^a(\mu) = \sup_{\{x \in \ell^1_+ : x_0 \geq \sum_{j\geq 1} x_j\}} \{\mu x_0 - f^0(x_0 - \sum_{j\geq 1} x_j) - \tfrac{a}{2}(2x_0^2 - \sum_{j\geq 1} x_j^2)\} \tag{5.8}$$

Thus $\tilde{p}^a(\mu)$ satisfies

$$\sup_{\{x_0, x_1 : 0 \leq x_1 \leq x_0\}} \{\mu x_0 - f^0(x_0 - x_1) - \frac{a}{2}(2x_0^2 - x_1^2)\}$$

$$\leq \tilde{p}^a(\mu) \leq \sup_{\{x \in \ell^1_+ : x_0 \geq \sum_{j\geq 1} x_j\}} \{\mu x_0 - f^0(x_0 - \sum_{j\geq 1} x_j) - \frac{a}{2}(2x_0^2 - (\sum_{j\geq 1} x_j)^2\}, \tag{5.9}$$

Since $a > 0$. But both sides of the inequality coincide, so that

$$\tilde{p}^a(\mu) = \sup_{\{x_0, x_1 : 0 \leq x_1 \leq x_0\}} \{\mu x_0 - f^0(x_0 - x_1) - \frac{a}{2}(2x_0^2 - x_1^2)\} . \tag{5.10}$$

In order to compute the total amount of condensate, it is necessary to obtain a more convenient expression for $\tilde{p}^a(\mu)$ than the one given in (5.10). A routine but fairly lengthy analysis of the posibilities yields the following:
the pressure $\tilde{p}^a(\mu)$ *in the hard-sphere boson gas model is given by*

$$\tilde{p}^a(\mu) = \begin{cases} \inf_{\alpha < 0} \{ \frac{(\mu - \alpha)^2}{4a} + p^0(\alpha) \} , & \mu < \mu_t , \\ \sup_{\alpha < 0} \{ \frac{(\mu - \alpha)^2}{4a} - \frac{\alpha^2}{2a} + p^0(\alpha) \} , & \mu \geq \mu_t , \end{cases} \tag{5.11}$$

where

$$\mu_t = \inf_{\alpha < 0} \{ 2\alpha (p^0)'(\alpha) - 1 \} . \tag{5.12}$$

In the mean-field model, there is a phase-transition if and only if there is a phase-transition in the corresponding free gas and this occurs if and only if ρ_c is finite. In the hard-sphere model, there is always a phase-transition; there are two cases to be distinguished:
if $2a(p^0)''(0) \leq 1$ *then* $\mu_t = 2a\rho_c$ *and*

$$(\tilde{p}^a)'_-(\mu) = (\tilde{p}^a)'_+(\mu) , \; (\lim_{\mu \downarrow \mu_t} - \lim_{\mu \uparrow \mu_t})(\tilde{p}^a)''(\mu) = \frac{2(p^0)''(0)}{1 - (2a(p^0)''(0))^2} , \tag{5.13}$$

so that the phase-transition is second-order;
if $2a(p^0)''(0) > 1$ *then* $\mu_t < 2a\rho_c$ *and*

$$(\tilde{p}^a)'_-(\mu_t) = (p^0)'(\alpha_t) - \frac{\alpha_t}{a} < (p^0)'(\gamma_t) = (\tilde{p}^a)'_+(\mu_t) , \tag{5.14}$$

where α_t *is the unique real root of* $2a(p^0)''(\alpha) = 1$ *, while* γ_t *is the unique real root of* $\mu_t = \gamma + 2a(p^0)'(\gamma)$ *, so that the phase-transition is first-order.*

To compute the total amount of condensate, we require $\tilde{p}^a(s, \lambda; \mu)$ defined at (2.8) and given by

$$\tilde{p}^a(s,\lambda;\alpha) = \begin{cases} \inf_{\alpha<0}\left\{\frac{(\mu+s-\alpha)^2}{4a} + p^0(s,\lambda;\alpha)\right\} & , \mu < \mu_t(s,\lambda), \\ \sup_{\alpha<0}\left\{\frac{(\mu+s-\alpha)^2}{4a} - \frac{\alpha^2}{2a} + p^0(s,\lambda;\alpha)\right\} & , \mu \geq \mu_t(s,\lambda). \end{cases} \tag{5.15}$$

For $\mu \neq \mu_t$, $s \mapsto \tilde{p}^a(s,\lambda;\alpha)$ is differentiable at $s=0$ and we find that *the total amount of condensate* $\tilde{\Delta}^a(\mu)$ *is given by*

$$\tilde{\Delta}^a(\mu) = \begin{cases} -\frac{\alpha(\mu)}{a} , & \mu > \mu_t , \\ 0 , & \mu < \mu_t , \end{cases} \tag{5.16}$$

where $\alpha(\mu)$ *is the value of* α *for which the extremum in* (5.15) *is attained.*

If $2a(p^0)''(0) \leq 1$ then $\lim_{\mu\downarrow\mu_t} \tilde{\Delta}^a(\mu) = 0$ but if $2a(p^0)''(0) > 1$ then $\lim_{\mu\downarrow\mu_t} \tilde{\Delta}^a(\mu) = -\frac{\alpha_t}{a} > 0$ and there is a jump in the total amount of condensate at $\mu = \mu_t$.

There is an important difference between the calculation of $\tilde{\Delta}^a(\mu)$ and of $\Delta^a(\mu)$: in the case of Δ^a, the total amount of condensate in the mean-field model, the contribution $(\rho(\mu)-\rho_c)^+$ comes because $\int_{[\lambda,\infty)} \rho(0|\lambda')\, dF(\lambda')$ cannot exceed $\rho_c = \int_{[0,\infty)} \rho(0|\lambda')\, dF(\lambda')$; in the case of $\tilde{\Delta}^a$, the total amount of condensate in the hard-sphere gas model, the contribution $-\frac{\alpha(\mu)}{a}$ comes from the change in the quadratic form in the expression for the pressure at $\mu = \mu_t$.

6. The Van der Waals' Limit

The full quantum mechanical hamiltonian H_ℓ^ϕ for a system of bosons in a cube Λ_ℓ in $\mathbb{R}^d$ interacting through a pair-potential ϕ is an operator on the symmetric Fock space $\mathcal{H}_\ell$ constructed on single-particle hilbert space $L^2(\Lambda_\ell)$ given by

$$H_\ell^\phi = d\Gamma(h_\ell) + U^\phi , \tag{6.1}$$

where $h_\ell = -\frac{1}{2}\Delta_\ell$ is the single-particle kinetic energy, the interaction operator U^ϕ is given by its action on the n-particle subspaces of $\mathcal{H}_\ell$: U^ϕ acts on $\mathcal{H}_\ell^{(n)}$ by multiplication by $\sum_{1\leq i<j\leq n} \phi(y_i - y_j)$, $y_i, y_j \in \Lambda_\ell$. We make the following assumptions about the pair-potential ϕ:

(R1) $\phi : \mathbb{R}^d \longrightarrow \mathbb{R}$ is continuous.

(R2) ϕ is an L^1 function of positive type.

(R3) $\hat{\phi}(0) = \int_{\mathbb{R}^d} \phi(y)\, dy$ is strictly positive.

It follows from (R2) that $\hat{\phi}(0)$ is non-negative, but we shall need more; that is why we have added (R3). It follows also from (R2) that $|\phi(y)| \leq \phi(0)$ for all y, and that $\phi(y) = \phi(-y)$. We will use the following lemma to get an upper bound to the pressure of the interacting system:

Let $\{\Lambda_\ell : \ell = 1,2,\ldots\}$ *be a sequence of nested cubes such that* $V_\ell \to \infty$ *as* $\ell \to \infty$; *let* $\phi : \mathbb{R}^d \to \mathbb{R}$ *be a function satisfying conditions* (R1), (R2) *and* (R3). *Then, given* $\epsilon > 0$, *there exists an integer* $\ell(\epsilon)$ *such that, for each finite set* $\{y_1, \ldots, y_n\}$ *of* n *distinct points of* Λ_ℓ, *the following inequality holds for all* $\ell > \ell(\epsilon)$:

$$\sum_{1 \leq i < j \leq n} \phi(y_i - y_j) \geq \frac{n^2}{2V_\ell} a(1-\epsilon) - \frac{n}{2} b \, , \tag{6.2}$$

where $a = \hat{\phi}(0)$ *and* $b = \phi(0)$.

As we shall see, these constants are best-possible.

The pressure $p_\ell^\phi(\mu)$ of the interacting system in a cube of finite volume V_ℓ is given by

$$e^{\beta V_\ell p_\ell^\phi(\mu)} = \mathrm{trace}[\exp\{-\beta(H_\ell^\phi - \mu N_\ell)\}] \, , \tag{6.3}$$

where N_ℓ is the number operator on $\mathcal{H}_\ell$. But it follows from (6.2) that H_ℓ^ϕ satisfies the following operator inequality for each $\epsilon > 0$:

$$H_\ell^\phi - \mu N_\ell \geq H_\ell^0 + \frac{a}{2V_\ell}(1-\epsilon) N_\ell^2 - (\mu + b) N_\ell \, , \quad \ell > \ell(\epsilon),$$

where H_ℓ^0 is the free-gas hamiltonian.

Since

$$A \geq B \quad \textit{implies} \ \mathrm{trace}[e^{-A}] \leq \mathrm{trace}[e^{-B}] \, , \tag{6.4}$$

we have

$$\begin{aligned} e^{\beta V_\ell p_\ell^\phi(\mu)} &\leq \mathrm{trace}[\exp\{-\beta(H_\ell^0 + \tfrac{a}{2V_\ell}(1-\epsilon)N_\ell^2 - (\mu+b)N_\ell)\}] \\ &= e^{\beta V_\ell p_\ell^{a(1-\epsilon)}(\mu+b)} \, , \end{aligned} \tag{6.5}$$

where $p_\ell^a(\mu)$ is the pressure in the mean-field model. Hence we have

Let ϕ *satisfy* (R1), (R2) *and* (R3); *then*

$$\limsup_{\ell \to \infty} p_\ell^\phi(\mu) \leq p^a(\mu + b) \, , \tag{6.6}$$

where $a = \hat{\phi}(0)$ *and* $b = \phi(0)$.

We sketch a proof of a lower bound to the pressure: we note that

$$\begin{aligned} e^{\beta V_\ell p_\ell^\phi(\mu)} &= \mathrm{trace}[e^{-\beta(H_\ell^\phi - \mu N_\ell)}] \\ &\geq \mathrm{trace}[\sum_{n \geq 0} e^{\beta\mu n} P_\ell^{(n)} e^{-\beta H_\ell^0} e^{-\beta\langle U^\phi\rangle_\ell^{(n)}}] \, , \end{aligned} \tag{6.7}$$

where $P_\ell^{(n)}$ is the orthogonal projection of $\mathcal{H}_\ell$ onto the n-particle subspace $\mathcal{H}_\ell^{(n)}$ and $<\cdot>_\ell^{(n)}$ is the free-gas canonical expectation given by

$$<A>_\ell^{(n)} = \frac{\mathrm{trace}[P_\ell^{(n)} e^{-\beta H_\ell^0} A]}{\mathrm{trace}[P_\ell^{(n)} e^{-\beta H_\ell^0}]} \quad . \tag{6.8}$$

We can re-write (6.7) as

$$e^{\beta V_\ell p_\ell^\phi(\mu)} \geqslant e^{\beta V_\ell p_\ell^0(\alpha)} \int_{[0,\infty)} e^{\beta V_\ell G_\ell(x)} \mathbb{K}_\ell^\alpha[dx] \quad , \tag{6.9}$$

where

$$G_\ell(x) = (\mu - \alpha)x - \frac{1}{V_\ell} < U^\phi >_\ell^{(n)} \quad , \quad \frac{n}{V_\ell} < x \leqslant \frac{n+1}{V_\ell} \quad , \tag{6.10}$$

and $\mathbb{K}_\ell^\alpha$ is the measure on $\mathbb{R}_+$ defined at (3.19). By an extension of Varadhan's Theorem to sequences $\{G_\ell\}$ of functions $G_\ell : E \to \mathbb{R}$ we have

$$\liminf_{\ell\to\infty} p_\ell^\phi(\mu) \geqslant \sup_{[0,\infty)} \{ G(x) - I^\alpha(x) \} \quad , \tag{6.11}$$

where $G(x) = \lim_{\ell\to\infty} G_\ell(x)$. But, for cubes with periodic boundary conditions, it is known that

$$G(x) = (\mu - \alpha)x - u^\phi(x) \quad , \tag{6.12}$$

where

$$u^\phi(\rho) = \frac{1}{2} \int_{\mathbb{R}^d} \phi(y) \{ \rho^2 + (\rho - \rho_c)^+ f_{d/2}(\mu(\rho); y) + (f_{d/2}(\mu(\rho); y))^2 \} dy \,. \tag{6.13}$$

Here

$$f_{d/2}(\mu; y) = \frac{1}{(2\pi)^d} \int_{\mathbb{R}^d} \frac{e^{ik\cdot y}}{e^{\beta(k^2/2 - \mu)} - 1} dk \tag{6.14}$$

and $\mu(\rho)$ is defined at (3.12). Thus we have

$$\liminf_{\ell\to\infty} p_\ell^\phi(\mu) \geqslant \sup_{x \geqslant 0} \{ \mu x - f^0(x) - u^\phi(x) \} \quad . \tag{6.15}$$

Having established the upper and lower bounds for the pressure, we introduce the scaling $\phi^\gamma(y) = \gamma^d \phi(\gamma y)$; as γ decreases to zero, ϕ^γ becomes weaker but its range increases. It is obvious that $\hat{\phi}^\gamma(0)$ is independent of γ while $\lim_{\gamma\downarrow 0} \phi^\gamma(0) = 0$; it follows also from the dominated convergence principle and the Riemann-Lebesgue lemma that $\lim_{\gamma\downarrow 0} u^{\phi^\gamma}(\rho) = \frac{a}{2}\rho^2$ with $a = \hat{\phi}(0)$.

Combining the bounds we have

$$\lim_{\gamma\downarrow 0}\ \liminf_{\ell\to\infty} p_\ell^{\phi^\gamma}(\mu) = \lim_{\gamma\downarrow 0}\ \limsup_{\ell\to\infty} p^{\phi^\gamma}(\mu) = p^a(\mu) , \tag{6.16}$$

where $p^a(\mu)$ is the mean-field pressure. The scaling limit defined above is known as the Van der Waals' limit; we may express the result (6.16) as
the mean-field approximation to the pressure is exact in the Van der Waals' limit.

A similar argument together with some convex analysis yields a result about the total amount of condensate in the Van der Waals' limit. Define the operator $N_\ell(\lambda)$ by

$$N_\ell(\lambda) = \sum_{\{j:\ \lambda_\ell(j) < \lambda\}} n_\ell(j) , \tag{6.17}$$

where $n_\ell(j)$ is the occupation number operator of the jth eigenstate of the single-particle kinetic energy operator h_ℓ ; define the grand canonical expectation $\langle\!\langle \cdot \rangle\!\rangle_\ell^\mu(\phi)$ by

$$\langle\!\langle A \rangle\!\rangle_\ell^\mu(\phi) = \frac{\text{trace}\,[A\, e^{-\beta(H_\ell^\phi - \mu N_\ell)}]}{\text{trace}\,[e^{-\beta(H_\ell^\phi - \mu N_\ell)}]} \tag{6.18}$$

Then it can be shown that

$$\lim_{\lambda\downarrow 0}\ \lim_{\gamma\downarrow 0}\ \liminf_{\ell\to\infty} \frac{1}{V_\ell}\langle\!\langle N_\ell(\lambda)\rangle\!\rangle_\ell^\mu(\phi^\gamma) = \lim_{\lambda\downarrow 0}\ \lim_{\gamma\downarrow 0}\ \limsup_{\ell\to\infty} \frac{1}{V_\ell}\langle\!\langle N_\ell(\lambda)\rangle\!\rangle_\ell^\mu(\phi^\gamma) = \Delta^a(\mu) , \tag{6.19}$$

where $\Delta^a(\mu)$ is the total amount of condensate in the mean-field model, given by (4.16) as $\Delta^a(\mu) = (\rho(\mu) - \rho_c)^+$. In summary:
Condensation persists in the Van der Waals' limit.

7. A Heuristic Derivation of the Models

The full quantum-mechanical hamiltonian (6.1) can be written in terms of annihilation and creation operators $a(k)$, $a^+(k)$ satisfying the canonical commutation relations $[a(k_1), a^+(k_2)] = \delta_{k_1,k_2}$ as

$$H_\ell^\phi = \sum_k \lambda_\ell(k)\, a^+(k) a(k) + \frac{1}{2V_\ell} \sum_{k_1}\sum_{k_2}\sum_q \hat{\phi}(q)\, a^+(k_1) a^+(k_2) a(k_2 - q) a(k_1 + q) , \tag{7.1}$$

where

$$\hat{\phi}(q) = \int_{\Lambda_\ell} \phi(y)\, e^{iq\cdot y}\, dy \quad . \tag{7.2}$$

The operator $n(k) = a^+(k)a(k)$ is interpreted as the occupation number of the eigenstate of the kinetic energy operator with eigenvalue $\lambda_\ell(k)$.

The main approximation made in deriving these models is the neglect of those terms in the hamiltonian which are not diagonal in the $n(k)$. It has the technical advantage of bringing the problem into the realm of probability theory, since we can regard the $n(k)$ as random variables. The diagonal part U_d^ϕ of the interaction operator can be written as

$$U_d^\phi = U_1^\phi + U_2^\phi + U_3^\phi \quad , \tag{7.3}$$

with

$$U_1^\phi = \frac{\hat{\phi}(0)}{2V_\ell}\{N^2 - N\} \, , \tag{7.4}$$

$$U_2^\phi = \frac{\hat{\phi}(0)}{2V_\ell}\{N^2 - \sum_k n(k)^2\} \quad , \tag{7.5}$$

$$U_3^\phi = \frac{1}{2V_\ell}\sum_{k_1}\sum_{k_2}\{\hat{\phi}(k_1 - k_2) - \hat{\phi}(0)\} n(k_1)n(k_2) \quad , \tag{7.6}$$

where $N = \sum n(k)$ is the total number operator. The term $-\frac{\hat{\phi}(0)}{2V_\ell}N$ in (7.4) makes no contribution to the energy density in the thermodynamic limit. If ϕ is very short-range, so that $\hat{\phi}(q)$ is independent of q , the term U_3^ϕ vanishes and we are left with the hard sphere model with interaction operator $U_1^\phi + U_2^\phi$. On the other hand, if we replace ϕ by its spatial average $\bar{\phi}_\ell$, given by

$$\bar{\phi}_\ell(y) = \frac{1}{V_\ell}\int_{\mathbb{R}^d}\phi(y)\, dy = \hat{\phi}(0)/V_\ell \quad ,$$

we have $\hat{\bar{\phi}}_\ell(k_1 - k_2) = \hat{\phi}(0)\, \delta_{k_1,k_2}$ so that U_2^ϕ cancels with U_3^ϕ and we get the mean-field model with interaction $U_1^\phi = \frac{\hat{\phi}(0)}{2V_\ell}N^2$.

8. Notes and Discussion

8.1: This paper is an updated version of the lecture delivered at the Groningen conference in August 1985; it differs from it in two important respects: it uses the theory of large deviations as a unifying theme and it includes new results on the hard-sphere model.

8.2: The general formulation of the free boson gas described here first appeared in van den Berg, Lewis and Pulè [8]; it will be published shortly as [9].

The mean-field model is discussed in Huang's book [10]; it is of interest because the pathological aspects of the free boson gas are removed by the mean-field interaction: the grand canonical partition function converges for all real values of the chemical potential (Ruelle [11]); the grand canonical distribution of the particle number density is asymptotically degenerate for all values of the mean density (Davies [12], Fannes and Verbeure [13]); the fluctuations in the particle number density in the grand canonical ensemble are normal and shape-independent (Buffet and Pulè [14]). It was shown by van den Berg, Lewis and de Smedt [15] that condensation persists in the mean-field model.

The hard-sphere gas model was introduced by Huang, Yang and Luttinger [6] and is discussed in Huang's book [10]; there is an illuminating heuristic discussion in the book of Thouless [16].

The method of computing the total amount of condensate which we describe was introduced in [15]; the concept seems to have been introduced first by Girardeau [17] who called it "generalized condensation". He claimed that, in a model of impenetrable bosons in one-dimension there is generalized condensation but no macroscopic occupation of the ground-state in the sense of London [4]; Schultz [18] showed that this is not the case. Macroscopic occupation of the ground-state is a subtle matter; in the free gas, its magnitude is strongly shape-dependent. The total amount of condensate is much more robust, and it can be computed from a thermodynamic generating function. For a detailed discussion of macroscopic occupation of the ground-state in the free boson gas and further references, see [9].

8.3: The survey of results on the free boson gas follows [9] where full proofs of the assertions are to be found. The first rigorous proof of macroscopic occupation of the ground state in the free boson gas was given in lectures by Kac in 1971; he computed correlation functions and the limiting distribution of the particle number density. The mathematical details were supplied in 1972 in the thesis of Pulè [19] (see also Lewis [20], Canon [21], Lewis and Pulè [22]). Kac's lectures remained unpublished until 1977 when they were incorporated in the review by Ziff, Uhlenbeck and Kac [23].

8.4: The results on the pressure (4.13) and on the total amount of condensate (4.16) were given first in [15]; the reformulation of the proof in terms of the large deviation principle will appear in [24].

Varadhan's Theorem and various generalizations were proved in 1966 by Varadhan

[25]; the method has been used extensively in the asymptotic analysis of function space integrals by Donsker and Varadhan; a brief introduction and survey can be found in the lectures of Varadhan [26]. Bahadur and Zabell [27] attacked the problem of formulating the large deviation principle in the general setting of a topological vector space, basing their approach on Lanford's ideas on entropy [28]. Expositions of these developments are to be found in the Saint-Flour lectures of Azencott [29] and the book of Stroock [30]; an introduction with applications to Statistical mechanics has been provided by Ellis [31].

It is especially clear from the formulation in terms of Varadhan's theorem that when bosons condense, they condense in order to minimize a free-energy. In the free gas, the balance is one of kinetic energy against entropy: if the integrated density of states $F(\lambda)$ goes to zero sufficiently slowly as $\lambda \to 0$ (so that $\int_{[0,\infty)} (e^{\beta\lambda}-1)^{-1} dF(\lambda)$ diverges, to be precise) then this balance can be achieved for all values of the mean density without condensation occuring; if $F(\lambda)$ goes to zero too rapidly as $\lambda \to 0$ (so that $\rho_c = \int_{[0,\infty)} (e^{\beta\lambda}-1)^{-1} dF(\lambda)$ is finite) then once the mean density exceeds ρ_c the balance between kinetic energy and entropy cannot be achieved unless condensation occurs. In the mean-field model the mechanism is the same: the balance is between kinetic energy and entropy; the interaction term changes the critical value of the chemical potential and it changes the order of the phase-transition, but the mechanism is unaltered and so is the total amount of condensate. As was pointed out by Lewis and Pulè [32], the result of Davies [12] on the degeneracy of the Kac distribution in the mean-field model is obvious in the light of the theory of large deviations: the rate-function $I^{\mu}(\cdot)$ for the sequence of distribution functions $\mathbb{K}_\ell^\mu$ of the particle number density is given by $I^{\mu}(x) = p^a(\mu) + f^a(x) - \mu x$; since this has a unique minimum, the sequence $\{\mathbb{K}_\ell^\mu\}$ converges weakly to the degenerate distribution; this holds for all values of μ .

8.5: The expression (5.10) for the pressure $\tilde{p}^a(\mu)$ in the hard-sphere model was obtained by Huang, Yang and Luttinger [6]. The purely quantum-mechanical contribution to the energy from the hard-sphere repulsion is the term $\frac{a}{2V_\ell}\{N^2 - \sum_{j\geq 1}\sigma_j^2\}$; it is smallest when all particles are in the same state so that $N = \sigma_j$ for some j ; the contribution from the kinetic energy term is smallest when all particles are in the lowest $(j=1)$ state; these considerations suggest that the model will display condensation into the ground state. However, this ignores the effect of entropy; a rough calculation, ignoring the kinetic energy (Thouless [16]), suggests that the competition between interaction energy and entropy is so severe that the amount of condensate does not go smoothly to zero as the chemical potential is reduced but has a discontinuous jump. It was argued in [6] that, near the condensation region, the average occupation number of the single-particle ground-state is much greater than those for the excited states and that, consequently, only the first term $-\frac{a}{2V_\ell}\sigma_1^2$ of the sum need be retained. Using this truncated hamiltonian, Huang,

Yang and Luttinger [6] applied the 'method of the largest term'; this is an informal version of Laplace's method and it yields the expression (5.10). Clearly, it is better to work with the untruncated hamiltonian than to introduce the ad hoc assumption that only the first term need be retained. This was carried out by van den Berg, Lewis and Pulè [33]; the proofs of the assertions in § 5 are to be found there. It is clear that in this model it is the quantum-mechanical interaction energy which competes with the entropy; the kinetic energy is relatively unimportant - there is a phase-transition in this model regardless of whether or not ρ_c is finite. The order of the phase-transition is determined by the relative importance of the interaction energy and the kinetic energy, measured by the magnitude of the product $a(p^0)''(0)$. If $a(p^0)''(0) > 1$, the interaction term overshadows the effect of the kinetic energy term and the prediction of Thouless holds: there is a jump in the total amount of condensate at $\mu = \mu_t$; if $a(p^0)''(0) \leq 1$, the total amount of condensate goes smoothly to zero at $\mu = \mu_t$.

8.6: The superstability inequality (6.2) with best-possible constants was proved by Lewis, Pulè and de Smedt [34], improving earlier versions by Ruelle [11] and Lieb [35]. A proof of the trace-inequality (6.4) can be found in [11]. The inequality (6.7) follows from a standard convexity argument. The extension of Varadhan's theorem required here is proved in Varadhan's original paper. The limiting expression (6.13) can be obtained using the results of Canon [21].

The scaling $\phi^\gamma(y) = \gamma^d \phi(\gamma y)$ was introduced by Kac [36] and used by Kac, Uhlenbeck and Hemmer [37] to derive the van der Waals' equation of state for a one-dimensional classical gas. For that reason it has become known as the van der Waals' limit; for classical systems, see also Baker [37], Lebowitz and Penrose [38], Gates and Penrose [39]; for quantum systems, see Lieb [40]. The result on the persistence of condensation in the van der Waals' limit was obtained by Lewis, Pulè and de Smedt [41] where proofs of the main results can be found; here they are reformulated in terms of the theory of large deviations.

8.7: The heuristic derivation of the models was inspired by Thouless [16]. In view of the very different mechanisms which can be seen to be producing condensation in the mean-field model and in the hard-sphere model, it is interesting to note: (1) in the diagonal approximation, one can regard the hard-sphere model as a limiting case as the range of the pair-potential decreases, and the mean-field model can be regarded as a limiting case as the pair-potential becomes infinitely weak but infinitely long-range; (2) the off-diagonal terms in the full hamiltonian make no contribution to the pressure or the condensate in the van der Waals' limit where the mean-field model is exact.

The next step in the investigation of condensation in interacting systems is to investigate a regularized model in which the interaction is taken to be represented by $U_1^{\phi} + U_2^{\phi}$. The existence of the pressure can be proved using the theory of large deviations; however, the solution of the resulting variational problem seems difficult (van den Berg, Lewis and Pulè [42].

References

[1] Einstein, A.: Quantentheorie des einatomigen idealen Gases, Sitzunber. Preus. Akad. Wiss. 1, 3-14 (1925)

[2] Uhlenbeck, G.E.: Thesis, Leyden 1927

[3] Kahn, B., Uhlenbeck, G.E.: On the theory of condensation, Physica 5, 399-415 (1938)

[4] London, F. : On the Bose-Einstein condensation, Phys. Rev. 54, 947-954 (1938)

[5] London, F. : Superfluids, vol. 11. New York: John Wiley and Sons, Inc. 1954

[6] Huang, K., Yang, C.N., Luttinger, J.M. : Imperfect Bose gas with hard-sphere interactions, Phys. Rev. 105, 776-784 (1957)

[7] Hepp, K., Lieb, E. : Equilibrium statistical mechanics of matter interacting with the quantized radiation field, Phys. Rev. A8, 2517-2525 (1973)

[8] van den Berg, M., Lewis, J.T., Pulè, J.V. : A general theory of Bose-Einstein condensation, DIAS-STP-82-35 (1982)

[9] van den Berg, M., Lewis, J.T., Pulè, J.V. : A general theory of Bose-Einstein condensation, Helv. Phys. Acta (in press) (1986)

[10] Huang, K. : Statistical Mechanics. New York: John Wiley and Sons, Inc. 1967

[11] Ruelle, D. : Statistical Mechanics: Rigorous Results. New York: Benjamin 1969

[12] Davies, E.B. : The thermodynamic limit for an imperfect boson gas, Commun. math. Phys. 28, 69-86 (1972)

[13] Fannes, M., Verbeure, A. : The imperfect Boson gas, J. Math. Phys. 21, 1809 (1980)

[14] Buffet, E., Pulè, J.V. : Fluctuation properties of the imperfect Bose gas, J. Math. Phys. 24, 1608-1616 (1983)

[15] van den Berg, M., Lewis, J.T., de Smedt, P. : Condensation in the imperfect boson gas, J. Stat. Phys., 697-707 (1984)

[16] Thouless, D.J. : The Quantum-mechanics of Many-body Systems. New York: Academic Press 1961

[17] Girardeau, M. : Relationship between systems of impenetrable bosons and fermions in one-dimension, J. Math. Phys. 1, 516-523 (1960)

[18] Schultz, T. : Note on the one-dimensional gas of impenetrable point-particle bosons, J. Math. Phys. 4, 666-671 (1963)

[19] Pulè, J.V. : Thesis, Oxford, 1972

[20] Lewis, J.T. : The free boson gas, Proceedings of the LMS Instructional Conference, Bedford College 1971, Mathematical Problems in Mathematical Physics. New York: Academic Press 1972

[21] Cannon, J.T. : Infinite-volume limits of the canonical free Bose gas states on the Weyl algebra, Commun. math. Phys. 29, 89-104 (1973)

[22] Lewis, J.T., Pulè, J.V. : The equilibrium states of the free boson gas, Commun. math. Phys. 36, 1-18 (1974)

[23] Ziff, R.M., Uhlenbeck, G.E., Kac, M. : The ideal Bose gas revisited, Phys. Rep. C32, 169-248 (1977)

[24] van den Berg, M., Lewis, J.T., Pulè, J.V. : The large deviation principle and some models of an interacting boson gas, DIAS-86- (1986)

[25] Varadhan, S.R.S. : Asymptotic probabilities and differential equations, Comm. Pure Appl. Math. 19, 261-286 (1966)

[26] Varadhan, S.R.S. : Large Deviations and Applications. Philadelphia: Society for Industrial and Applied Mathematics 1984

[27] Bahadur, R.R., Zabell, S.L. : Large deviations of the sample mean in general vector spaces, Ann. Prob., 7, 537-621 (1979)

[28] Lanford, O.E. : Entropy and equilibrium states in classical statistical mechanics. Lecture Notes in Physics 20, 1-113. Berlin: Springer 1971

[29] Azencott, R. : Grandes deviations et applications. Ecole d'Eté de Probabilités de Saint-Flour V111-1978, 1-176, Lecture Notes in Mathematics 774 Berlin: Springer 1980

[30] Stroock, D.W. : An Introduction to the Theory of Large Deviations. New York: Springer 1984

[31] Ellis, R. : Entropy, Large Deviations and Statistical Mechanics. New York: Springer 1985

[32] Lewis, J.T., Pulè, J.V. : The equivalence of ensembles in statistical mechanics, Lecture Notes in Mathematics 1095, 25-35 (1984)

[33] van den Berg, M., Lewis, J.T., Pulè, J.V. : The large deviation principle and some models of an interacting boson gas, DIAS-86-12 (1986)

[34] Lewis, J.T., Pulè, J.V., de Smedt, P. : The Superstability of pair-potentials of positive type. J. Stat. Phys. 35, 381-385 (1984)

[35] Lieb, E. : Simplified approach to the ground-state energy of an imperfect Bose gas, Phys. Rev. 130, 2518-2528 (1963)

[36] Kac, M. : On the partition function of a one-dimensional gas, Phys. Fluids 2, 8-12 (1959)

[37] Baker, G.A. : Certain general order-disorder models in the limit of long-range interactions. Phys. Rev. 126, 2072-2078 (1962)

[38] Lebowitz, J.L., Penrose, O. : Rigorous treatment of the van der Waals-Maxwell theory of the liquid-vapour transition J. Math. Phys. 7, 98-113 (1966)

[39] Gates, D.J., Penrose, O. : The van der Waals' limit for classical systems, III Commun. math. Phys. 17, 194-209 (1970)

[40] Lieb, E. : Quantum-mechanical extension of the Lebowitz-Penrose Theorem on the van der Waals' theory J. Math. Phys. 7, 1016-1024 (1966)

[41] Lewis, J.T., Pule, J.V., de Smedt, P. : The persistence of boson condensation in the van der Waals' limit, DIAS-STP-83-48 (1983)

[42] van den Berg, M., Lewis, J.T., Pule, J.V. : The existence of the pressure in a model of an interacting boson gas. DIAS-STP-86-13 (1986)

BLACK HOLES AND QUANTUM MECHANICS

G. 't Hooft
Institute for Theoretical Physics
University of Utrecht
Utrecht, the Netherlands

Certain aspects of computations that produce the spectrum of the well-known Hawking radiation of a black hole are incompatible with the notion that the latter behave as conventional quantum mechanical systems such as unstable particles or bound states. It has been argued before that one might conclude that black holes violate some fundamental principles in quantum mechanics such as a description in terms of pure wave functions. We here take the opposite point of view, name that perhaps corrections to the aforementioned calculations might be necessary due to gravitational interactions between in- and outgoing matter that were previously neglected. Although it is not known whether such corrections can cure the problems, there are strong indications that they are very important. There even is reason to doubt the validity at all of the conventional computation of Hawking's temperature for a finite-size black hole. This is because in- and outgoing matter not only affects the analytic continuation of space-time inside a black hole, but also the topological structure at the horizon.

QUANTUM FIELD THEORY AND GRAVITATION

Rudolf Haag
II. Institut für Theoretische Physik, Universität Hamburg
2000 Hamburg 50

Abstract

The formulation of a generally covariant quantum field theory is described. The scaling limit of physical states and its relation to the metric of space-time is discussed.

One of the long standing challenges concerning the conceptual structure in theoretical physics is the question as to how a synthesis of the ideas of general relativity and quantum theory will look. Between 1945 and 1975 there has been (sociologically and perhaps even ideologically) a rather clean separation between the scientific communities of these two areas. It is a rather recent phenomenon that elementary particle theorists have become actively interested in gravitation theory and conversely, scientists working on differential geometry and general relativity, in elementary particle physics. The success of local gauge theories forced differential geometry on particle physicists. Hawking's suggestion of particle emission from a black hole, the realization that a local formulation of supersymmetry needs a supergravity theory and the projection that the "grand unification" of the weak and strong interactions leads to a mass scale which is not far from the Planck mass (which is defined by the gravitational constant) have contributed to a general feeling that the time is ripe to think about the synthesis of gravitation and quantum field theory.

My talk here will not be a review of the immense amount of work which could be subsumed under the given title. Rather I shall concentrate on some aspect of the problem, in particular the formulation of a generally covariant quantum field theory and the rôle of a scaling limit of states in relation to the space-time geometry.

Let me first recall a few corner stones of the classical theory of gravitation.

i) The effect of a gravitational field on matter.

In Newton's theory there is a gravitational field strength Γ(x,t) due to which a test particle (mass point) experiences an acceleration

$$\frac{d^2\vec{x}}{dt^2} = \vec{\Gamma}(\vec{x},t) \tag{1}$$

The surprising fact is, that the mass of the test particles does not appear in the equation of motion. This universality leads to the principle of equivalence between inertial and gravitational mass.

In Einstein's theory this universality of the orbits of all (uncharged) particles is used to define "straight lines" in the 4-dimensional space time continuum. In a symmetric notation we denote coordinates of a space-time point x by $x^\mu = (t,x)$; $\mu = 0, 1, 2, 3$. To define straight lines in a general manifold we need the concept of "parallel transport", described by the "affine connection coefficients" $\Gamma^\nu_{\mu\rho}(x)$ and Newton's equation (1) is replace by

$$\frac{d^2x^\nu}{ds^2} = - \Gamma^\nu_{\mu\rho}(x) \frac{dx^\mu}{ds} \frac{dx^\rho}{ds} \tag{2}$$

where s is a suitably chosen parameter on the orbit. Γ is interpreted as an intrinsic geometric property of the 4-dimensional space-time continuum. We can choose an arbitrary coordinate system and know how Γ transforms from the one to the other viz:

$$\Gamma'^{\lambda}_{\mu\nu}(x') = \frac{\partial x'^\lambda}{\partial x^\rho} \frac{\partial x^\tau}{\partial x'^\mu} \frac{\partial x^\sigma}{\partial x'^\nu} \Gamma^\rho_{\tau\sigma}(x) + \frac{\partial x'^\lambda}{\partial x^\rho} \frac{\partial^2 x^\rho}{\partial x'^\mu \partial x'^\nu} \tag{3}$$

I have written down this well known formula to remind us of the fact that if Γ vanishes at some point in one coordinate system is does not vanish in another due to the inhomogeneous second term. We will encounter a corresponding transformation formula later. The intrinsic nature of Γ (and related quantities) is usually expressed as the principle of general covariance. The laws do not give any preference to any particular coordinate system. From Γ we can obtain the Riemann curvature tensor and the Ricci Tensor $R_{\mu\nu}$.

ii) The generation of a gravitational field by matter.

In Newton's theory a matter density distribution generates a gravitational field

$$\vec{\Gamma}(\vec{x},t) = - \gamma \int \rho(\vec{x}',t) \frac{\vec{x}-\vec{x}'}{|\vec{x}-\vec{x}'|^3} d^3x' \tag{4}$$

In Einstein's theory it is assumed that space-time is a pseudo-Riemannian manifold equipped with a metric form $g_{\mu\nu}(x)$ which defines both a causal structure and a length scale. This metric form determines the affine connection Γ and allows to define the curvature scalar $R = g^{\mu\nu} R_{\mu\nu}$. The equation (4) of Newton's theory is then replaced by the Einstein equations

$$R_{\mu\nu} - \frac{1}{2} R g_{\mu\nu} = \gamma T_{\mu\nu} \tag{5}$$

where $T_{\mu\nu}$ is the energy momentum tensor of the matter distribution.

I have separated the description of gravitation into the parts i) and ii) because the first, in which the principle of general covariance and the intrinsic-geometric rôle of gravitation appear are more readily incorporated in a quantum theory than the second where the gravitational constant enters. This constant γ adds to $\hbar$ and c one more constant of presumably universal significance. In units in which $\hbar = c = 1$ i.e. where times and lengths are measured in cm and energies and masses in cm^{-1}, γ takes the value 10^{-66} cm². Thus, there is now a natural length unit

$$\lambda_0 = \left(\frac{\hbar\gamma}{c^3}\right)^{\frac{1}{2}} = 10^{-33}\,cm \qquad \text{"Planck length"}$$

and a natural mass unit

$$M_0 = \frac{\hbar}{\lambda_0 c} = (\hbar c \gamma^{-1})^{\frac{1}{2}} = 10^{19} M_{Proton} \qquad \text{"Planck mass"}$$

M_0 is the mass for which the Compton wave length and the Schwarzschild radius become equal. One expects that the physics at lengths below λ_0 will become radically different. In some sence M_0 must act as a natural cut-off, λ_0 as a fundamental length reminiscent of the expectations 40 years ago (when, however, the fundamental length was assumed to be of the order of 10^{-13} cm i.e. larger than λ_0 by a factor 10^{20}). We cannot hope to get any direct experimental evidence about this regime of lengths below λ_0. However there are important implications: the theory should be finite (no renormalization necessary), absolute values of masses should be calculable in terms of M_0. It is tempting to speculate whether such a theory will eventually abandon the space-time continuum in favour of a quantum geometry of elementary events in which irreversibility appears on a fundamental level (rahter than due to subjective ignorance as in statistical mechanics). I shall not pursue this line here but focus attention on the more modest question: if we accept a 4-dimensional space-time continnum and the standard quantum theoretical concepts, how far can we get towards a satisfactory inclusion of gravity; in particular how can we incorporate general covariance?

Let me sketch quickly the frame in which the symbiosis of <u>special</u> relativity and quantum physics may be described. Here the space-time continuum is Minkowski space on which we have a physically distinguished set of coordinate systems (the globally inertial ones). We have the symmetry group of Poincaré transformations P. A group element $(a,\Lambda) \in P$ consists of a translation 4-vector a and a homogeneous Lorentz transformation Λ. In the quantum theory we may take a separable Hilbert space H, assuming that its rays correspond to pure physical states. We represent the action of a Poincaré transformation on the physical states by a unitary operator $U(a,\Lambda)$ acting on H. The U(g) with $g \in P$ form a unitary representation of P (up to a phase factor).

The relation of the quantum theoretic <u>observables</u> to space-time is established in a simple way. To each (open) region $\mathcal{O}$ of space-time we have an algebra $\mathfrak{A}(\mathcal{O})$ of operators acting on H. It is the algebra generated by all observables in $\mathcal{O}$. It is at this point not relevant whether we consider only bounded observables, in which case the algebras $\mathfrak{A}(\mathcal{O})$ may be taken as C^*-algebras or W^*-algebras, or whether we take them as Wightman-Borchers-algebras generated by smeared out field operators whose test functions have support in $\mathcal{O}$. All that is needed is that they are *-algebras over the complex numbers (i.e. addition and multiplication of elements is defined as well as multiplication of an element with a complex number and the adjoint operation $A \to A^*$). For finer analysis, of course, a topology in these algebras has to be used. For many purposes it is preferably to consider the algebras as abstract algebras (not yet represented by operators on a Hilbert space) and P as the abstract Poincaré group. Then a physical state ω is a positive linear form on the algebra $\mathfrak{A} = \cup\,\mathfrak{A}(\mathcal{O})$ (resp. on the covariance algebra generated by $\mathfrak{A}$ and P) and, given one such state, we can reconstruct a Hilbert space and a representation of $\mathfrak{A}$ (resp. $\mathfrak{A}$ and P) by operators acting on H (GNS-construction). Altogether we have a net of local algebras $\mathcal{O} \to \mathfrak{A}(\mathcal{O})$; the Poincaré group acts on $\mathfrak{A} = \cup\,\mathfrak{A}(\mathcal{O})$ by automorphisms in an obvious manner and, in the representations generated from physical states the Poincaré automorphisms are implemented by unitary operators $U(a,\Lambda)$. I shall abstain from writing down the obvious structural relations and mention only three which have relevance to the subsequent discussion:

1) Causality.

a) If $\mathcal{O}_1$, $\mathcal{O}_2$ lie space-like to each other the elements of $\mathfrak{A}(\mathcal{O}_1)$ commute with those of $\mathfrak{A}(\mathcal{O}_2)$.

b) If $\mathcal{O}_2$ lies in the causal shadow of $\mathcal{O}_1$, then $\mathfrak{A}(\mathcal{O}_2) \subset \mathfrak{A}(\mathcal{O}_1)$.

2) Stability.

The infinitesimal generators P_μ of the (represented) translation subgroup U(a) are interpreted as energy-momentum operators of the system. Their simultaneous spectrum is assumed to be contained in the closed forward light cone. Usually one assumes also that there is a distinguished state ω_o, the vacuum, which is Poincaré invariant and hence a ground state for the energy.

3) In a Hilbert space representation we may always pass from the algebras $\mathfrak{A}(\mathcal{O})$ to the associated von Neumann algebras $R(\mathcal{O})$. We usually assume that in the representation generated (via GNS) from the vacuum state, all the $R(\mathcal{O})$ are factors i.e. they have trivial center.

What changes if we go over from Minkowski space to a more general manifold? The first step is the consideration of a pseudo-Riemannian manifold. The metric structure is then fixed and describable by a classical metric field $g_{\mu\nu}$. We want to describe quantum fields living on such a non flat space-time manifold. In that case the concept of the local net of algebras, the causality principles 1) and the assumption 3) can be kept without change. We loose the Poincaré group as a symmetry and thereby the distinguished vacuum state and this means that the stability condition 2) has to be formulated in a different manner. We have made a proposal how this should be done in a recent paper [1]. The basic idea was that by a scaling limit one can define a net of algebras associated with the tangent space at a point of the manifold and that all allowed physical states coalesce to one distinguished state on each tangent space algebra. Under some continuity assumptions this state is invariant under translations in tangent space and therefore the stability requirement can be formulated there. In [1] some of the arguments were somewhat handwaving and some proofs lacked rigour. This has been remedied by K. Fredenhagen [2] and it turned out that the same technique can be applied to a more general situation in which a priori no given Riemannian structure of the space-time manifold is assumed. This is the subject of a joint work by K. Fredenhagen and myself [3] which is still in progress and of which I want to report here some results.

To formulate the theory in a generally covariant way we adopt the point of view which had been advocated for many years by my fondly remembered friend, Hans Ekstein [4]. Ekstein pointed out that one should distinguish between observation procedures and observables. A "procedure" means typically that certain instruments are placed in a certain space-time region. There may be quite different procedures (associated with different space-time regions) which are equivalent in the sense that they yield identical expectation values in all physical states. An "observable" is then an equivalence class of procedures. Two procedures A_1, A_2 are equivalent if $\omega(A_1 - A_2) = 0$ for all physically allowed states ω. This implies that $A_1 - A_2$ generates a proper 2-sided ideal in $\mathfrak{A}$. We have then, for each region $\mathcal{O}$ an ideal $J(\mathcal{O})$ generated by the equivalence relation of procedures. The customary "observable algebras" are the quotients

$$\mathfrak{A}_{obs}(\mathcal{O}) = \mathfrak{A}(\mathcal{O}) / J(\mathcal{O})$$

In the following we shall take $\mathfrak{A}$ as a "flexible algebra" whose elements correspond to procedures. The physical laws may be characterized either by specifying the local ideals $J(\mathcal{O})$ directly (equations of motion, commutation relations ...) or, alternatively, by characterizing the class of all physically relizable states on $\mathfrak{A}$. In the latter case the relevant ideals are formed by the elements of $\mathfrak{A}$ on which all physical states vanish. We shall pursue here this second alternative.

The interpretation of $\mathfrak{A}$ as a flexible algebra of procedures allows us to consider the group of local diffeomorphisms of the manifold as acting by automorphisms on $\mathfrak{A}$. This makes the formulation manifestly covariant under general coordinate transformations. We shall not be concerned here with topological properties of the manifold at large but confine attention to a sufficiently small, contractible region for which a single chart is adequate. The diffeomorphisms mentioned shall be understood to transform this region into itself. We may then interpret such transformations either in the passive sense as changing the description (coordinatization) and leaving the physical situation unchanged or in the active sense, as keeping the reference frame unchanged but going over to a different physical situation (state).

Let us now illustrate this in the simplest case: a scalar quantum field on a 4-dimensional manifold which we do not a priori endow with a (physically distinguished) Riemann structure. We take as the flexible algebra of procedures the free Wightman-Borchers algebra, i.e. the tensor algebra of test functions on the part of the manifold considered. Its elements are formal linear combinations of monomials $f^{(n)}$ where $f^{(n)} = f^{(n)}(P_1, \ldots, P_n)$ is a smooth function on $M \times M \times \ldots \times M$ with support in $\mathcal{O}$ in each argument. In a chart φ which coordinatizes the point $P \in M$ by $x = \varphi(P) \in \mathbb{R}^4$, the monomial is described by the function

$$f_{\varphi}^{(n)}(x_1, \ldots, x_n) = f^{(n)}(\varphi^{-1}(x_1), \ldots, \varphi^{-1}(x_n))$$

Usually, using the field operators $\phi(x)$, the monomial $f^{(n)}$ is written as $\int f_{\varphi}^{(n)}(x_1, \ldots, x_n)\, \phi(x_1) \ldots \phi(x_n)\, d^4 x_1 \ldots d^4 x_n$ but we shall not use this notation because it introduces here unnecessary ambiguities. The algebraic product of $f^{(n)}$ with $g^{(m)}$ is just the tensor product

$$f^{(n)} \cdot g^{(m)} = h^{(n+m)}\ ; \quad h^{(n+m)}(P_1, \ldots P_{n+m}) = f^{(n)}(P_1, \ldots P_n) \cdot g^{(m)}(P_{n+1}, \ldots P_{n+m})$$

A state on the algebra is given by a hierarchy of distributions $\omega^{(n)} \in D'(\mathcal{O}\times\mathcal{O}\ldots\times\mathcal{O})$. $\omega^{(n)}(f^{(n)})$ is the expectation value in ω of the monomial element $f^{(n)}$ of the algebra. In each chart we have, correspondingly, a set of distributions $\omega_\varphi^{(n)}$ on $\mathbb{R}^4\times\ldots\mathbb{R}^4$ and, of course, for two charts

$$\omega_\varphi^{(n)}(f_\varphi^{(n)}) = \omega_\psi^{(n)}(f_\psi^{(n)})$$

$$f_\psi^{(n)}(y_1,\ldots y_n) = f_\varphi^{(n)}(\chi^{-1}y_1,\ldots \chi^{-1}y_n) \tag{6}$$

where $\chi = \psi\varphi^{-1}$ is the transition function between the charts.

To get some handle on the characterization of the physically allowed states we consider their scaling limits at the points of the manifold. In a chart φ (coordinates x) we define the scaling to a point $\bar{x}$ by the 1-parameter group

$$\delta_\lambda x = \bar{x} + \lambda(x-\bar{x}) \tag{7}$$

contracting $\varphi(\mathcal{O})$ to the point $\bar{x}$ for $\lambda \to 0$. This corresponds to a 1-parameter subgroup of diffeomorphisms in M, contracting to $P = \varphi^{-1}(\bar{x})$ in a way which depends on the chosen chart φ . Correspondingly we consider the sequence of test functions

$$(f_\varphi)_{\bar{x},\lambda} = (\lambda^{-4}N(\lambda))^n f_\varphi(\delta_\lambda^{-1}x_1,\ldots \delta_\lambda^{-1}x_n) \tag{8}$$

whose supports shrink to a point for $\lambda \to 0$. $N(\lambda)$ is a suitably chosen scaling factor. We say, that the state ω has a scaling limit of order $N(\lambda)$ if

$$\lim_{\lambda\to 0} \omega_\varphi((f_\varphi)_{\bar{x},\lambda}) \equiv \omega_{\bar{x}}^\varphi(f_\varphi) \tag{9}$$

is finite for all $f \in D'(\mathcal{O}^n)$ and is nonvanishing for some.

One may object that the assumption of the existence of such scaling limits disregards the existence of the Planck length and that therefore part ii) of the gravitation theory cannot be accomplished in an ultimately satisfactory manner once this assumption is made. This is probably true. Nevertheless it is an interesting first step, in which a synthesis of some principles of general relativity with quantum theory is achieved while the other part must be left on a semiclassical level.

One shows easily that the scaling factor $N(\lambda)$ is the same in every chart. It follows also [2] that $N(\lambda)$ is "almost" a power:

$$\lim_{\mu\to 0} \frac{N(\lambda\mu)}{N(\mu)} = \lambda^\alpha \tag{10}$$

We have split off the factor λ^{-4} in (8) so that α is positive if the distributions are singular (compared to the Lebesque measure). For each chart φ and each contraction point $\bar{x}$ the limit distributions (9) define a state on the tensor algebra of smooth functions on $\mathbb{R}^4$. The support restriction of the test functions in the manifold becomes wiped out because δ_λ^{-1} ($\mathcal{O}$) tends to all of $\mathbb{R}^4$ in the limit $\lambda \to 0$. The dependence of $\omega_{\bar{x}}$ on the chart is found to be [2]

$$\omega_{\bar{y}}^{\psi}(F) = \omega_{\bar{x}}^{\varphi}\left(F \circ \mathring{\chi}^{-1}\right) \tag{11}$$

where $\mathring{\chi}$ is the affine (inhomogeneous linear) transformation, obtained from the transition function $\chi = \psi\varphi^{-1}$ between the charts by

$$(\mathring{\chi}x)^\mu = \bar{y} + \frac{\partial \chi^\mu}{\partial x^r}(x^r - \bar{x}^r) \, . \tag{12}$$

We can get, therefore, a chart independent definition of the scaling limits of a state ω at a point $P \in M$ by

$$\omega_P(f) = \omega_{\varphi(P)}^{\varphi}\left(f \circ \mathring{\varphi}^{-1}\right) \tag{13}$$

where f is now in the tensor algebra of smooth functions on the <u>tangent space at P</u> and

$$\mathring{\varphi}(z) = \varphi(P) + d\varphi(z) \tag{14}$$

maps the tangent vector $z \in T_p$ into $\mathbb{R}^4$. [1] If we use in tangent space the frame which corresponds to the coordinatization φ then $\mathring{\varphi}$ is simply

$$\mathring{\varphi}(z) = \bar{x} + z \quad ; \qquad \mathring{\varphi}^{-1}(x) = x - \bar{x} \, . \tag{15}$$

The essential observation which leads to the linear (affine) transformation law (11) is the following. If we expand the transition function χ in the neighborhood of $x = \bar{x}$ in powers of $x - \bar{x}$:

$$(\chi x)^\mu = \bar{y}^\mu + a^\mu{}_r(x^r - \bar{x}^r) + \frac{1}{2} b^\mu{}_{\nu\rho}(x^\nu - \bar{x}^\nu)(x^\rho - \bar{x}^\rho) + \dots \tag{16}$$

1) Notation: φ maps a neighborhood of $P \in M$ into $\mathbb{R}^4$ (chart). $d\varphi$ is the corresponding map of the tangent space at P on the tangent space of $\mathbb{R}^4$ at $\bar{x}$ which is naturally identified with $\mathbb{R}^4$.

and define

$$\chi_\lambda = \delta^{-1}_{\bar{y},\lambda} \circ \chi \circ \delta_{\bar{x},\lambda} \tag{17}$$

where $\delta_{\bar{x},\lambda}$ is the linear contraction (7) to the point $\bar{x}$, $\delta_{\bar{y},\lambda}$ the linear contraction to $\bar{y}$, then

$$(\chi_\lambda x)^\mu = \bar{y}^\mu + a^\mu{}_\nu(x^\nu - \bar{x}^\nu) + \lambda \tfrac{1}{2} b^\mu{}_{\nu\rho}(x^\nu - \bar{x}^\nu)(x^\rho - \bar{x}^\rho) + O(\lambda^2) \tag{18}$$

So χ_λ converges to $\mathring{\chi}$ for $\lambda \to 0$. On the other hand

$$\omega_\psi\left((f)_{\bar{y},\lambda}\right) = \left(\lambda^{-4} N(\lambda)\right)^n \omega_\varphi\left(f \circ \chi_\lambda \circ \delta^{-1}_{\bar{x},\lambda}\right) \to \omega^\varphi_{\bar{x}}(f \circ \mathring{\chi}).$$

There are two properties of the tangent space states which follow generally. The first is the transformation law under dilations in tangent space

$$\omega_P^{(n)}(f^{(n)} \circ \delta_\lambda) = (\lambda^{\alpha-4})^n \, \omega_P^{(n)}(f^{(n)}) \tag{19}$$

$$(f^{(n)} \circ \delta_\lambda)(z_1, \ldots z_n) \equiv f^{(n)}(\lambda z_1, \ldots \lambda z_n). \tag{20}$$

This follows directly from the definition and (10). The second is the invariance under translations in tangent space

$$\omega_P(T_a f) = \omega_P(f) \tag{21}$$

$$(T_a f)(z_1, \ldots z_n) \equiv f(z_1 - a, \ldots z_n - a). \tag{22}$$

This follows if we assume that ω_P depends smoothly on P.

Let us calculate the right hand side of (13) in the chart φ using as a basis in the tangent spaces the one corresponding to the coordinate axes, so $\mathring{\varphi}$ is given by (15). Then

$$\omega^\varphi_{\bar{x}}(f^{(n)} \circ \mathring{\varphi}^{-1}) = \lim \left(\lambda^{-4} N(\lambda)\right)^n \omega \, f\left(\frac{x_1 - \bar{x}}{\lambda}, \ldots \frac{x_n - \bar{x}}{\lambda}\right) \tag{23}$$

If we are allowed to interchange differentiation with respect to $\vec{x}$ and the formation of the limit $\lambda \to 0$ then

$$\frac{\partial}{\partial \bar{x}^{\mu}} \omega_{\bar{x}}^{\varphi}\left(f_{0}^{(n)} \varphi^{-1}\right) = - \lim \left(\lambda^{-4} N(\lambda)\right)^{n} \cdot \lambda^{-1} \omega_{\varphi}\left(P_{\mu} f\big|_{z_i = \frac{x_i - \bar{x}}{\lambda}}\right) \tag{24}$$

where

$$P_{\mu} f = \sum_{i} \frac{\partial f}{\partial z_{i}^{\mu}} . \tag{25}$$

Since the right hand side involves one factor λ more in the denominator this can be finite only if the scaling limit of $\omega\,(P_{\mu}f)$ with the original scaling factor $(N(\lambda)\lambda^{-4})^{n}$ vanishes. This is the infinitesimal form of (21), (22). [1]

From the general properties of ω_P mentioned and the assumption that $\alpha = 1$ (corresponding to the canonical dimension of a scalar field) it follows that the 1-point-distribution $\omega_P^{(1)} = 0$ and the 2-point-distribution $\omega_P^{(2)}$ is given by a (positive) measure on $\mathbb{R}^4$ ("momentum space", intrinsically the cotangent space at P) which is homogeneous of degree-2 with respect to dilations. If, in addition, we believe that there are no quantum observables at a point and that among the physical states there are "primary" ones i.e. states with generate (by the GNS-construction) a Hilbert space representation in which $R(\mathcal{O})$ is a factor, then all the states in one such primary folium (e.g. all vector states of such a representation) give the same scaling limit ω_P because $\bigcap_{\lambda} \mathfrak{A}(\mathcal{O}_{\lambda})$ is the algebra of procedures at the point $\bar{x}$ and this should lie in the center of the algebra. We still have to incorporate the remnants of the causality and stability conditions of the Minkowski space theory. The simplest assumption is the ω_P arising from a primary folium are invariant under a subgroup of linear transformations in tangent space isomorphic to the Lorentz group and that they satisfy the spectrum condition with respect to the translation operators T_a in tangent space. In that case the transformation law of ω_P under dilations with $\alpha = 1$ fixes ω_P uniquely. It must be the vacuum state of a <u>free</u>, <u>massless</u> scalar theory [2] in Minkowski space [1]. The state (or rather the primary folium of states) determines then not only a causal structure but a metric form in the tangent space at each point. The difference of perspective as compared to [1]

1) Actually for (21), (22) the differentiability of ω_P would not be necessary, only some uniform continuity. A precise proof has been given by Fredenhagen [2].

2) This does not mean that the theory itself must be free since $N(\lambda)$ can differ from λ for instance by logarithmic factors as it happens in "asymptotically free" theories.

is that we do not assume a priori that the manifold is equipped with a Riemann structure. Instead each primary folium of states determines a Riemann structure. The physical laws may allow different Riemann structures. They are separated by superselection rules as long as we assume the existence of scaling limits and of physical states which are primary on the algebras of arbitrary small regions.

To finish, let us compute the transformation law of the left hand side of (24) under coordinate change and its relation to an affine connection on M. Denoting this quantity by $(\omega_{\bar{x}}^{\varphi})_{,\mu}$ we have

$$(\omega_{\bar{y}}^{\psi})_{,\mu}(F) = (a^{-1})_{\mu}^{\ \nu}\left[\omega_{\bar{x}}^{\varphi}(F\circ A) - (a^{-1})_{\sigma}^{\ \rho}\, b^{\sigma}_{\ \nu\beta}\, \omega_{\bar{x}}^{\varphi}\left(\sum_j z_j^{\beta}\frac{\partial}{\partial z_j^{\rho}} f\circ A\right)\right]$$

where

$$(Az)^{\mu} = a^{\mu}_{\ \nu} z^{\nu} \;; \qquad a^{\mu}_{\ \nu} = \frac{\partial y^{\mu}}{\partial x^{\nu}}(\bar{x}) \;; \quad b^{\mu}_{\ \nu\rho} = \frac{\partial^2 y^{\mu}}{\partial x^{\nu}\partial x^{\rho}}(\bar{x}).$$

If one assumes that to each point P there exists a coordinate system ψ, depending on P, in which $(\omega^{\psi}_{\psi(P)})_{,\mu} = 0$ (geodesic system at the point) then in another system

$$\nabla_{\mu}\, \omega^{\varphi}_{\varphi(P)} = 0$$

$$\nabla_{\mu} = \frac{\partial}{\partial x^{\mu}} - \Gamma^{\rho}_{\mu\beta} K_{\rho}^{\ \beta}$$

$$K_{\rho}^{\ \beta}\, \omega(f) = \omega\left(\sum_j z_j^{\beta} \frac{\partial}{\partial z_j^{\rho}} f\right)$$

and the Γ have the transformation properties of Christiphel symbols (3). In intrinsic language they define an affine connection which is determined by the primary folium of states considered.

References

1 Haag, R., Narnhofer, H. and Stein, U., "On Quantum Field Theory in Gravitational Background", Commun. Math. Phys. 94, 219 (1984)
2 Fredenhagen, K., Private Communication
3 Fredenhagen, K. and Haag, R., "Generally covariant Quantum Field Theory and scaling limits" (In preparation)
4 Ekstein, H., "Presymmetry", Phys. Rev. 184 (5), 1315 (1969)

A REMARK ON ANTIPARTICLES

H. J. Borchers

Institut für Theoretische Physik

der Universität Göttingen

Bunsenstr. 9

D-3400 Göttingen

I Introduction and Results:

When Dirac found his equation for the spinning electron he run into problems with the interpretation of the negative energy states. This lead him to the postulate that all these states are occupied (Dirac sea). He also observed that a hole in the sea behaves like a particle except that it carries opposite charge. By this observation antiparticles were born. In the general theory of second quantization it has been observed that the condition of positive energy can only be satisfied if one chooses the correct commutation relations for the fields. This observation leads to the theory of spin and commutations (often called spin and statistics).

The question of antiparticles came again up in the theory of superselection sectors which is part of the theory of local observables. This theory was started by Borchers [1], continued by Doplicher, Haag and Roberts [2] and later treated by Buchholz and Fredenhagen [3]. In all these theories one is dealing with positive energy representations of the observable algebra (which will be described in some detail in a moment). One starts from a factor representation, which is not the vacuum representation, and then assumes or proves that there is a vacuum representation closely related to given representation which allows us to construct charged fields connecting the vacuum representation with the representation one started with. Having constructed this charged field its adjoint will then create the antiparticle representation from the vacuum sector. On this way one obtains also other sectors, the multiple particle sectors. This theory is now in a good shape, and a problem which has been open for some time, namely the particles with infinite statistics has been solved by Fredenhagen [4]. This discussion shows us that we are able to construct for many representations the corresponding antiparticle representation.

However, up to now it has not been proved that one obtains by this method to every representation a corresponding antiparticle representation. In the contrary the experience with zero mass field theories shows that one can not expect allways that amount of normality of the two representations which is needed for the construction of the charged fields. This discussion asks for results indicating the existence of antiparticles which do not requires the existence of a vacuumrepresentation with particular properties.

Such a result we want to present here:

Theorem 1:

Let $\{\mathbb{A}(0), \mathbb{A}, \mathbb{R}^d, \alpha\}$ be a theory of local observables and let $\{\pi, U(a), H\}$ be a covariant factor representation fulfilling spectrum condition and assume U(a) is the unique minimal representation fulfilling spectrum condition. Assume the spectrum of U(a) has a lower and an upper mass gap. This means the spectrum of U(a) starts at m_o and it has a mass gap between m_1 and m_2. Then one must have $m_2 \leq 3\, m_o$.

The interpretation of this result is the following: Assume to every factor representation π with spectrum condition can be associated a charge q_π. If p is a point in the spectrum we should have a particle which energy-momentum p and charge q_π. From the antiparticle picture we expect now a particle again with momentum p but with charge $-q_\pi$, and hence there should exist a neutral pair with momentum 2p. Adding this to the original particle we have to expect a state with momentum 3p but the same charge q_π. Hence the spectrum should have the shape described in the theorem. This also shows more, namely in the neutral sector (vacuumsector) the spectrum should be an additive set. This has been proved long ago by A.S. Wightman [5]. The result of the first theorem can be generalized also to the case where π is not a factor representation. In this case one has to introduce the proper notion.

Definition.

Let $\{\mathbb{A}, \mathbb{R}^d, \alpha\}$ be a C^*-dynamical system and let $\{\pi, U(a), H\}$ be a covariant representation such that (α) U(a) is strongly continuous

(β) U(a) commutes with the center of π.

Denote by E(Δ) the spectral projections of U(a) and by F(Δ) their central supports, and by B_r the ball of radius r in $\mathbb{R}^d$. We say

$p_1, \ldots, p_n$ are <u>π-compartible</u> if

$$\prod_1^n F(p_i + B_r) \neq 0 \quad \text{for all } r > 0.$$

With this notations one obtains:

Theorem 2:

Let $\{\mathfrak{A}(0), \mathfrak{A}, \mathbb{R}^d, \alpha\}$ be a theory of local observalbes and $\{\pi, U(a), \mathcal{H}\}$ be a covariant representation fulfilling spectrum condition. Assume U(a) is the unique minimal representation with spectrum condition. If p_i, $i = 1,2,3$ belong to the spectrum of U(a) and if the p_i are π-compatible then

$$p_1 + p_2 + p_3 \in \text{spec } U(a) .$$

II On the definition of energy and momentum in quantum field theory

In the theorems we have mentioned the unique minimal representations of the translation group, which means that we are claiming the existence of the concept of energy and momentum in particle physics although it does neither appear in the axioms of the theory of local observables nor in the definition of representations (states) for particle physics.

We start with the pure cinematical aspect. Let $\{\mathfrak{A}, \mathbb{R}^d, \alpha\}$ be a C^*-dynamical system (no continuity assumption about the group action), K be an open convex proper cone in $\mathbb{R}^d$ and $\hat{K}$ its closed dual cone we look for all representations $\{\pi, U(a), \mathcal{H}\}$ with

(i) U(a) is a continuous unitary representation of the group $\mathbb{R}^d$

(ii) $U(a)\ \pi(x)\ U^*(a) = \pi(\alpha_a x) \quad , \quad x \in \mathfrak{A}$

(iii) Spectrum $U(a) \subset \hat{K}$.

These representation can be characterized. The answer is given in:

Theorem 3: [6]

Let $\{\pi, U(a), \mathcal{H}\}$ be a representation of $\{\mathfrak{A}, \mathbb{R}^d, \alpha\}$ with the above requirement then:

(a) U(a) can be chosen to belong to $\pi(\mathfrak{A})''$

(b) Denote by $E_o(K) = \{\omega \in E(\mathfrak{A})$ with

(i) $\omega(x\ \alpha_a(y))$ is continuous in a for all $x, y \in \mathfrak{A}$.

(ii) $\omega(x\ \alpha_a(a))$ is boundary value of an analytic function $F_{x,\phi}(z)$ holomorphic in $T(K) = \{z \in \mathbb{C}^d ; \ \mathrm{Im}\, z \in K\}$.

(iii) Exist $M > 0$ with $|F_{x,\phi}(z)| > \|x\|\ \|y\|\ e^{M\,|\mathrm{Im}\, z|}$ }

Denote by E(K) the norm closure of $E_o(K)$.

Then E(K) is a folium.

(c) Given a representation $\{\pi, H\}$ then one can find U(a) acting on H such that $\{\pi, U(a), H\}$ is a representation with an above properties if and only if the vector states of $\{\pi, H\}$ belong to E(K).

By this theorem the representation U(a) is not unique defined. In the one dimensional case Olesen and Pedersen [7] have constructed a unique minimal representation of the group. This does not extend to the higher dimensional case except either the cone K has a specia shape or there is a close relation of the structure of the algebra and the cone K. This is the case when the algebra $\mathbb{A}$ describes some dynamics. This situation is fulfilled for the theory of local observables defined by the following axioms:

(A) To every bounded open region $0 \subset \mathbb{R}^d$ exist a C^*-algebra $\mathbb{A}(0)$ such that $0_1 \subset 0_2$ implies $\mathbb{A}(0_1) \subset \mathbb{A}(0_2)$. The C^*-indeuctive limit will denoted by $\mathbb{A}$.

(B) The vector group $\mathbb{R}^d$ acts on $\mathbb{A}$ such that

$$\alpha_a \mathbb{A}(0) = \mathbb{A}(0 + a) \quad \text{holds.}$$

(C) If 0_1 and 0_2 are spacelike to each other then $\mathbb{A}(0_1)$ and $\mathbb{A}(0_2)$ commute pointwise with each other. A system fulfilling these axioms will be denoted by $\{\mathbb{A}(0), \mathbb{A}, \mathbb{R}^d, \alpha\}$. That this represents a special theory provided the cone K coincides with the light-cone V is given in the following result proved by Borchers and Buchholz [8].

Theorem 4:

Let $\{\mathcal{A}(\mathcal{O}), \mathcal{A}, \mathbb{R}^d, \alpha\}$ be a theory of local observables and let V be the forward lightcone ($\hat{V} = \bar{V}$). Let $\{\pi, U(a), \mathcal{H}\}$ be a covariant representation fulfilling spectrum condition with respect to the cone $\bar{V}$. Then there exist a unique and minimal representation $U^o(a)$ on $\mathcal{H}$ implementing α_a and which fulfills the spectrum condition. The minimality means:

let $$U(a) = e^{i(P,a)}$$

$$U^o(a) = e^{i(P^o, a)}$$

then one has

$$(P^o,t) \leq (P,t) \qquad \text{for every} \quad t \in V$$

in the order of operators and U(a) any group representation fulfilling spectrum condition such that $\{\pi, U(a), \mathcal{H}\}$ is a covariant representation.

By this result we can take P^o as energy and momentum operators. At least we have a unique definition. That this describes exactly the energy and momentum operators needed for physics is supported by the following result again due to Borchers and Buchholz [8], [9].

Theorem 5:

The common spectrum of the operators P^o defined in Theorem 4 is a Loerentz invariant set.

III Indication of the proof of the additivity property of the spectrum

We do not want to give all details of the proof, but, we want to make the proof understandable. This we can achieve by making some simplifications. These are:

(a) As in Theorem 1 we assume that we are dealing with a factor representation

(b) We assume that the spectrum consist of an isolated hyperboloid of mass m_o and a continuum starting at mass $m_1 > m_o$.

The statement of Theorem 1 then reads $m_1 \leq 3\, m_o$.

For proving this let U(a) be the unique minimal representation of the translation group, let $x \in \mathfrak{A}(0)$ and $\phi \in H_\pi$. Define

$$F^+_{x,\phi}(a) = (\phi, \pi(x^*)\, U(a)\, \pi(x)\phi)$$

$$F^-_{x,\phi}(a) = (\phi, U(a)\, \pi(x)\, U^{-1}(a)\, \pi(x^*)\phi)$$

and

$$F_{x,\phi}(a) = F^+_{x,\phi}(a) - F^-_{x,\phi}(a) .$$

Locality implies $F_{x,\phi}(a) = 0$ if 0 and $0 + a$ are spacelike with respect to each other. In order to have a simple description we take for 0 an order interval $D_t = (-t + V) \cap (t - V)$ with $t \in V$. Then $F_{x,\phi}(a) = 0$ for $a \in D'_{2t}$. This implies we can write

$$F_{x,\phi}(a) = G^{+}_{x,\phi}(a) - G^{-}_{x,\phi}(a)$$

with

$$\operatorname{supp} G^{+}_{x,\phi}(a) \subset -2t + V$$

$$\operatorname{supp} G^{+}_{x,\phi}(a) \subset 2t - V .$$

Taking now the Fourier transform of G^{+} and G^{-} we see:
$\tilde{G}^{+}_{x,\phi}(p)$ is boundary value of an analytic function $\hat{G}^{+}_{x,\phi}(z)$ holomorphic in $T^{+} = \{z \in \mathbb{C}^d \; ; \; \operatorname{Im} z \in V\}$;
$\tilde{G}^{-}_{x,\phi}(p)$ is boundary value of $\hat{G}^{-}_{x,\phi}(z)$ holomorphic in $T^{-} = -T^{+}$, and

$$\tilde{G}^{+}_{x,\phi}(p) = \tilde{G}^{-}_{x,\phi}(p) \quad \text{for} \quad p \in \Gamma \quad \text{with}$$

$$\Gamma = \mathbb{R}^d \setminus \operatorname{supp} \tilde{F}_{x,\phi}(p) \quad .$$

This implies we are dealing with an edge of the wedge problem with coincidence domain Γ. The edge of the wedge theorem [10] tells us that $\hat{G}^{+}_{x,\phi}(z)$ and $\hat{G}^{-}_{x,\phi}(z)$ are analytic continuations from each other provided Γ is not empty.

In order to continue the discussion we have to make a choice for ϕ. Let $E(\Delta)$ be the spectral projections of $U(a)$ then we say $\operatorname{supp} \Psi \subset \Delta$ if $E(\Delta)\ \phi = \phi$. Let $t_o \in V$ with $t_o^2 = 1$ and take $\Delta = D_{t_o(m_o-\varepsilon),\, t_o(m_o+\varepsilon)}$ then we get the following support for $\tilde{F}_{x,\phi}(p)$;

$$\operatorname{supp} \tilde{F}_{x,\phi}(p) \subset \operatorname{supp} \tilde{F}^{+}_{x,\phi}(p) \cup \operatorname{supp} \tilde{F}^{-}_{x,\phi}(p)$$

$$\operatorname{supp} F^{+}_{x,\phi}(p) \subset \operatorname{spec} U(a) = \{p \,;\, p^2 = m_o^2 \,,\; p > 0\} \cup \{p ;\, p^2 \geq m_1^2 \,,\; p > 0\}$$

$$\operatorname{supp} \tilde{F}^{-}_{x,}(p) \subset \{p \,;\, (p - 2\,t_o(m_o + \varepsilon))^2 \geq m_o^2 \,,$$

$$p < 2\,t\,(m_o + \varepsilon)\} \quad .$$

If we replace $\tilde{G}^{\pm}(p)$ by $(p^2 - m_o^2)\,\tilde{G}^{\pm}(p)$ then the isolated hyperboloid drops and one obtains a larger coincidence domain $\hat{\Gamma}$ where $\hat{\Gamma}$ is bounded by:

(a) the upper branch of the hyperboloid
$$p^2 = m_1^2 \quad , \quad p > 0$$

(b) the lower branch of the hyperboloid

$$(p - 2\,t_o(m_o + \varepsilon))^2 = m_o^2 \quad , \quad p < 2\,t_o(m_o + \varepsilon)$$

which tells us that $\hat{\Gamma}$ is of the Jost-Lehmann-Dyson type. Therefore according to Bros, Messiah, and Stora [11] the envelope of holomorphy of the edge of the wedge problem $T^{+} \cup T^{-} \cup \hat{\Gamma}$ is given by the Jost-Lehmann-Dyson formula [12]. From this we learn that the coincidence domain $\hat{\Gamma}$ can be enlarged to $\hat{\Gamma}_1$ if $m_o < m_1 - 2(m_o + \varepsilon)$. The boundary of $\hat{\Gamma}_1$ is given by:

(a) the upper branch of the hyperboloid

$$p^2 = m_1^2 \quad , \quad p_o > 0$$

(b) the lower branch of the hyperboloid
$$(p - 2\,t_o(m_o + \varepsilon))^2 = (m_1 - 2(m_o + \varepsilon))^2 \;.$$

Or with other words there is an enlargement of $\hat{\Gamma}$ if

$m_1 - 2(m_o + \varepsilon) > m_o$ or

$m_1 - 3\,m_o > 2\,\varepsilon$.

Remark now that $\{\pi(x)\Psi\ ;\ x \in \mathfrak{A}(0)$ for some 0 and $\text{supp}\ \Psi \in \Delta\}$ is total in H. From this we conclude

$$2\,\Delta - \text{spec}\ U(a) \subset \text{supp}\ (p^2 - m_o)^2\ \tilde{F}^-_{x,\phi}(p)$$

which implies that

$$\text{spec}\ U(a) \subset \{p \in V\ ;\ p^2 > (m_1 - 1\,m_o - 2\,\varepsilon)^2\}\ .$$

But this implies the equation

$$m_1 - 3\,m_o \leq 0$$

References

[1] H.J. Borchers: Local Rings and the Connection of Spin with Statistics
C.M.P. 1, 281 (1965)

[2] S. Doplicher, R. Haag, and J. Roberts: Local Observables and Particle Statistics
C.M.P. 23, 199-230 (1971)
C.M.P. 35, 49 (1974)

[3] D. Buchholz and K. Fredenhagen: Locality and the Structure of Particle States
C.M.P. 84, 1 (1982)

[4] K. Fredenhagen: On the Existence of Antiparticles
C.M.P. 79, 141 (1981)

[5] A.S. Wightman: Recent Achievements of Axiomatic Field Theory
Trieste Lectures 1962, Theorem 12

[6] H.J. Borchers: Translation Group and Spectrum Condition.
C.M.P. 96, 1-13 (1984)

[7] D. Olesen and G.K. Pedersen: Derivations of C^*-Algebras have Semi-Continuous Generators.
Pacific J. Math. 53, 563-572 (1974)

[8] H.J. Borchers and D. Buchholz: The Energy-Momentum Spectrum of Local Field Theories with Broken Lorentz-Symmetry
C.M.P. 97, 169-185 (1985)

[9] H.J. Borchers: Locality and Covariance of the Spectrum
FIZIKA: V.J. Glaser memorial volume, in print

[10] H.J. Brehmermann, R. Oehme, and J.G. Taylor: Proof of Dispersion Relation in Quantized Field Theories.
Phys. Rev. 109, 2178-2190 (1958)

[11] J. Bros, A. Messiah, and R. Stora: A Problem of Analytic Completion related to the Jost-Lehmann-Dyson Formula
J. Math. Phys. 2, 639 (1961)

[12] F.J. Dyson: Integral Representation of Causal Commutators
Phys. Rev. 110, 1460 (1958).

NOTES ON THE CANONICAL ANTICOMMUTATION RELATIONS

Richard V. Kadison

1. Introduction. Since the late 1920s and early 1930s, when Dirac and Von Neumann phrased the basic assumptions of quantum mechanics in the formalism of Hilbert spaces, it has been accepted procedure to identify the observables of a physical system with self-adjoint operators on a Hilbert space. Computation with the mathematical model requires that we consider functions and, in particular, polynomials in these observables. These functions occur, for example, when we try to describe the Hamiltonian of the system. The family of self-adjoint operators representing our observables must possess some algebraic structure. One may simply assume that each self-adjoint operator represents some observable - this works reasonably well if the system being studied has finitely many degrees of freedom and we have no need to consider the mathematical model of it in other than irreducible representations. That assumption is not adequate for systems with infinitely many degrees of freedom - the study of quantized fields and quantum statistical mechanics (after passing to the thermodynamical limit) requires other models.

In this article, some of the models that have come to be useful for studying systems with infinitely many degrees of freedom will be described along with some of the powerful techniques and results that have been developed during the more than fifty years that this subject has been studied. The canonical commutation and anticommutation relations, associated with infinite systems of particles satisfying Bose-Einstein and Fermi-Dirac statistics, respectively, and their representations by operators on Hilbert spaces is a recurring theme in the investigation of such systems. A particular operator algebra is remarkably suited to the analysis of the representations of the canonical anticommutation relations. We shall describe this connection. The results we discuss will be illustrated by applying them to this algebra and producing information about representations of the canonical anticommutation relations. All of these results are to be found in the text and exercises of [6] (specific reference to results in [6] will be made where appropriate).

2. Notation and preliminaries. Our Hilbert space $\mathcal{H}$ has scalar field $\mathbb{C}$, the complex numbers. The inner product of two vectors x and y in $\mathcal{H}$ is denoted by $\langle x,y \rangle$. It is linear in x and conjugate linear in y. The length or norm of a vector x is $\langle x,x \rangle^{\frac{1}{2}}$ and is denoted by $\|x\|$. The bound or norm of a (continuous) linear transformation T of $\mathcal{H}$ into itself (bounded operator) is $(\sup\{ \|Tx\| : \|x\| \le 1, x \in \mathcal{H} \})$ denoted by $\|T\|$. The family of all bounded operators on $\mathcal{H}$ is denoted by $\mathfrak{B}(\mathcal{H})$. The adjoint of an operator T in $\mathfrak{B}(\mathcal{H})$ is denoted by T^* (and characterized by the equality, $\langle Tx,y \rangle = \langle x,T^*y \rangle$ for all x and y in $\mathcal{H}$).

A family $\mathfrak{F}$ of operators in $\mathfrak{B}(\mathcal{H})$ is said to be self-adjoint when $\mathfrak{F} = \mathfrak{F}^*$ $(= \{ T^* ; T \in \mathfrak{F}\})$. A subset $\mathfrak{A}$ of $\mathfrak{B}(\mathcal{H})$ that contains each linear combination $aT+S$ and product TS of operators T and S in $\mathfrak{A}$ and is self-adjoint ($T^* \in \mathfrak{A}$ if $T \in \mathfrak{A}$) is said to be a (self-adjoint) operator algebra. If $\mathfrak{A}$ is a self-adjoint operator algebra such that $T \in \mathfrak{A}$ when $\|T-T_n\| \to 0$ and each $T_n \in \mathfrak{A}$, we say that $\mathfrak{A}$ is a C*-algebra. We assume that our operator algebras contain the identity element I (for each x in $\mathcal{H}$, $Ix = x$) If $\mathfrak{A}$ is a self-adjoint operator algebra such that $T \in \mathfrak{A}$ when $\|(T-T_n)x\| \to 0$ for each x in $\mathcal{H}$ and each $T_n \in \mathfrak{A}$, we say that $\mathfrak{A}$ is a von Neumann algebra. A factor is a von Neumann algebra $\mathfrak{M}$ whose center $(\{ T \in \mathfrak{M} : TS = ST \text{ for all } S \text{ in } \mathfrak{M}\})$ consists of scalar multiples of I.

3. Matricial operator algebras. The algebra $\mathfrak{B}(\mathcal{H})$ is a factor (hence, a von Neumann algebra and a C*-algebra). If $\mathcal{H}$ is n-dimensional with n finite, then $\mathfrak{B}(\mathcal{H})$ is isomorphic to $M_n(\mathbb{C})$, the algebra of $n \times n$ complex matrices.

A class of C*-algebras that has come to be useful for models in quantum physics was introduced by Glimm [2]. These are the matricial C*-algebras. Given a sequence of positive integers $r(1), r(2), \cdots$, and an infinite dimensional Hilbert space $\mathcal{H}$, one can construct a family $\{\mathfrak{A}_j\}$ of C*-subalgebras $\mathfrak{A}_j$ of $\mathfrak{B}(\mathcal{H})$ such that each $\mathfrak{A}_j$ contains I, $A_jA_k = A_kA_j$ when $A_j \in \mathfrak{A}_j, A_k \in \mathfrak{A}_k$ and $j \ne k$ (we say that $\mathfrak{A}_j$ and $\mathfrak{A}_k$ commute, in this case), and $\mathfrak{A}_j$ is isomorphic to $M_{r(j)}(\mathbb{C})$ $(j = 1,2,\cdots)$. The norm closure of the algebra generated by $\mathfrak{A}_1, \mathfrak{A}_2, \cdots$, is a matricial C*-algebra. Glimm shows (see [6:Section 12.1]):

THEOREM Two matricial C*-algebras $\mathfrak{A}$ and $\mathfrak{B}$, generated by $\mathfrak{A}_1, \mathfrak{A}_2, \cdots$ and $\mathfrak{B}_1, \mathfrak{B}_2, \cdots$, with orders $r(1), r(2), \cdots$ and $s(1), s(2), \cdots$, respectively, are isomorphic if and only if each prime power p^m that divides some product $r(1)\cdots r(j)$ also divides some product $s(1)\cdots s(k)$.

If each $r(j)$ is 2 and each $s(j)$ is 3, then $\mathfrak{A}$ and $\mathfrak{B}$ are not isomorphic. If each $r(j)$ is 2 and $s(1) = 2$, $s(2) = 4$, $s(3) = 8$, $\cdots$, then $\mathfrak{A}$ and $\mathfrak{B}$ are isomorphic. The case where each $r(j)$ is 2 gives rise to the <u>CAR algebra</u>, which is of special interest in quantum physics. The representations of the CAR algebra and those of canonical anticommutation relations are very closely related.

<u>4. The canonical anticommutation relations.</u> The system of relations

$$C_jC_k + C_kC_j = 0 \qquad (j,k = 1,2,\cdots)$$

$$C_jC_k^* + C_k^*C_j = 0 \qquad (j \neq k)$$

$$C_jC_j^* + C_j^*C_j = I \qquad (j = 1,2,\cdots)$$

in the infinite set of variables $C_1, C_2, \cdots$, is called the <u>canonical anticommutation relations</u>. A set of operators $C_1^o, C_2^o, \cdots$, on a Hilbert space, that satisfies the canonical anticommutation relations is said to be a <u>representation</u> of the canonical anticommutation relations.

A representation of a C*-algebra $\mathfrak{A}$ is a homomorphism of $\mathfrak{A}$ into $\mathfrak{B}(\mathcal{H})$ that preserves adjoints for some Hilbert space $\mathcal{H}$. The connection between the representations of the canonical anticommutation relations and representations of the CAR algebra is established with the aid of <u>Pauli spin matrices</u>. We write σ_x, σ_y, σ_z, for the matrices

$$\begin{bmatrix} 1 & 0 \\ 0 & -1 \end{bmatrix}, \quad \begin{bmatrix} 0 & i \\ -i & 0 \end{bmatrix}, \quad \begin{bmatrix} 0 & 1 \\ 1 & 0 \end{bmatrix},$$

respectively. With $\mathfrak{A}$ the CAR algebra generated by commuting subalgebras $\mathfrak{A}_1, \mathfrak{A}_2, \cdots$, each isomorphic to $M_2(\mathbb{C})$, we identify $\mathfrak{A}_n$ with $M_2(\mathbb{C})$ and write $\sigma_x^{(n)}$, $\sigma_y^{(n)}$, $\sigma_z^{(n)}$, for the Pauli matrices in $\mathfrak{A}_n$. Let A_n be $\sigma_z^{(1)} \cdots \sigma_z^{(n-1)}(\sigma_x^{(n)} - \sigma_y^{(n)})/2$. In this notation, the following theorem describes the way the representations of the canonical anticommutation relations are tied to those of the CAR algebra.

THEOREM The elements $A_1, A_2, \cdots$, in $\mathfrak{A}$ satisfy the canonical anticommutation relations and generate $\mathfrak{A}$ as a C*-algebra. If φ is a representation of $\mathfrak{A}$ on the Hilbert space $\mathcal{H}$, then $\varphi(A_1), \varphi(A_2), \cdots$, is a representation of the canonical anticommutation relations (on $\mathcal{H}$). If $C_1, C_2, \cdots$, is a representation of the canonical anticommutation relations on a Hilbert space $\mathcal{H}$, then there is a unique representation φ of the CAR algebra $\mathfrak{A}$ on $\mathcal{H}$ such that $\varphi(A_1) = C_1$, $\varphi(A_2) = C_2$, $\cdots$.

5. Some irreducible representations. There are two basic structures needed for the characterization of representations of C*-algebras [4]. If $\mathfrak{A}$ is a C*-algebra and φ is a representation of $\mathfrak{A}$ on a Hilbert space $\mathcal{H}$, then $\varphi(\mathfrak{A})$ is a self-adjoint operator algebra on $\mathcal{H}$. (It is, in fact, a C*-algebra - that is, closed with respect to taking limits relative to the operator norm [6: 4.1.9].) If we adjoin to $\varphi(\mathfrak{A})$ all the operators in $\mathcal{B}(\mathcal{H})$ that are limits of operators in $\varphi(\mathfrak{A})$ on vectors in $\mathcal{H}$ (as described in Section 2), the resulting family $\varphi(\mathfrak{A})^-$ is a von Neumann algebra. The combination of the "type decomposition" for $\varphi(\mathfrak{A})^-$ and that for $\varphi(\mathfrak{A})'$, the von Neumann algebra consisting of those elements of $\mathcal{B}(\mathcal{H})$ that commute with every element of $\varphi(\mathfrak{A})$ (and, hence, of $\varphi(\mathfrak{A})^-$), is one of the basic structures involved. It need not concern us, for the present, since we consider irreducible representations in this section. When φ is irreducible, $\varphi(\mathfrak{A})^- = \mathcal{B}(\mathcal{H})$ (equivalently, $\varphi(\mathfrak{A})'$ consists of scalar multiples of I). (We may take either of these conditions as our definition of irreducibility.) If $\mathfrak{A}$ is the CAR algebra, then $\mathcal{H}$ is necessarily separable when φ is irreducible. (This is true, more generally, when $\mathfrak{A}$ has a countable number of generators as a C*-algebra.) In this way, the considerations of the type decompositions of $\varphi(\mathfrak{A})^-$ and $\varphi(\mathfrak{A})'$ disappear.

The second basic structure needed is measure-theoretic in nature. While it has a general description [4], it usually appears in more convenient special forms in special cases. We shall take advantage of such special structure for our construction of inequivalent irreducible representations of the CAR algebra.

The CAR algebra $\mathfrak{A}$ is generated by an infinite commuting family of self-adjoint subalgebras $\{\mathfrak{A}_n\}_{n=1,2,\ldots}$ with each $\mathfrak{A}_n$ containing the identity element I of $\mathfrak{A}$ and * isomorphic to $M_2(\mathbb{C})$. Choose matrix units $\{E^n_{jk}\}_{j,k=1,2}$ for each $\mathfrak{A}_n$ such that

$(E^n_{jk})^* = E^n_{kj}$. (For example, we may choose for E^n_{jk} the element of $\mathfrak{A}_n$ corresponding to the matrix in $M_2(\mathbb{C})$ with 1 in row j and column k and 0 at all other entries.) The subalgebra $\mathfrak{B}_n$ of generated by $\mathfrak{A}_1, \cdots, \mathfrak{A}_n$ is * isomorphic to $M_{2^n}(\mathbb{C})$. The set of all products $E^1_{j(1)k(1)} \cdots E^n_{j(n)k(n)}$ is a (self-adjoint) system $\{F^n_{jk} : j,k = 1,\cdots, 2^n\}$ of $(2^n \times 2^n)$ matrix units for $\mathfrak{B}_n$. As n varies these systems of matrix units fulfill certain compatibility conditions:

$$F^n_{jk} = \sum_{h=1}^{r} F^m_{(j-1)r+h,(k-1)r+h} \quad (j,k = 1,\cdots,2^n)$$

where $n \le m$ and $r = 2^{m-n}$. It is reasonably clear that such a compatible family of (self-adjoint, $2^n \times 2^n$, $n = 1,2,\cdots$) matrix unit systems acting on a Hilbert space $\mathcal{H}$ generate the CAR algebra $\mathfrak{A}$ and, thereby, give rise to a representation of $\mathfrak{A}$ and of the canonical anticommutation relations.

We construct our irreducible representations of the CAR by describing such systems of matrix units on the Hilbert space $\mathcal{H}$ $L_2(S, \mathcal{S}, m)$ $(= L_2)$, where S is the half-open interval $[0,1)$, $\mathcal{S}$ is the σ-algebra of Borel subsets of S and m is a (σ-finite) positive measure on $\mathcal{S}$. Let D be the subset of $[0,1)$ consisting of rationals with denominator some power of 2 (the "dyadic rationals"). Provided with addition modulo 1, S is a group and D is a subgroup. For each d in D, let g_d be translation by d (modulo 1) on S so that $\{g_d : d \in D\}$ is a group G of transformations of S. Assume that our measure m has been chosen invariant under each element of G. Define F^n_{jk} $(j,k = 1,\cdots,2^n)$ acting on L_2 by

$$(F^n_{jk}f)(s) = \begin{cases} f(s+2^{-n}(k-j)) & (2^{-n}(j-1) \le s < 2^{-n}j) \\ 0 & \text{elsewhere on } S \end{cases}$$

where $f \in L_2$. Computations show that $\{F^n_{jk} : j,k = 1,\cdots,2^n\}$ is a self-adjoint system of $2^n \times 2^n$ matrix units and that these systems form a compatible family as n varies. With h a bounded measurable function on S, let M_h denote the " multiplication" operator on L_2 that transforms f (in L_2) to the product hf. Further computation shows that F^n_{jj} is M_h where h is the characteristic function of $[2^{-n}(j-1), 2^{-n}j)$. It follows that the only operators commuting with all F^n_{jk} are those M_h such that h is (almost everywhere) invariant under each element of G. The condition that such invariant functions are constant (almost everywhere) is equivalent to the condition that G act ergodically on S (relative to m - see [6 : 8.6.6]). Thus the representation π_m of the CAR on

$L_2(S, \mathcal{S}, m)$ is irreducible if and only if G acts ergodically on S.

We say that a point s of S is an "atom" for m when $m(\{s\}) > 0$. A point s in S is an atom for m if and only if the multiplication operator corresponding to the characteristic function of the one-point set $\{s\}$ is a non-zero projection $P(m)$ on L_2. For each n, there is a unique integer j in $1, \cdots, 2^n$ such that $2^{-n}(j-1) \leq s < 2^{-n}j$.

The multiplication operator corresponding to the characteristic function of $[2^{-n}(j-1), 2^{-n}j)$ is a projection in $\pi_m(\mathfrak{A})$ (the algebra on L_2 generated by all F^n_{jk}) and corresponds to a projection P_n in $\mathfrak{A}$ (in fact, to the matrix unit E^n_{jj}). Now $\wedge_n \pi_m(P_n) = P(m)$. Thus s is an atom for m if and only if $\wedge_n \pi_m(P_n) \neq 0$. If there is a unitary operator U such that $U^* \pi_m(A) U = \pi_{m'}(A)$ for each A in $\mathfrak{A}$ (that is, if π_m and $\pi_{m'}$ are unitarily equivalent for our two G-invariant measures m and m'), then

$$U^*P(m)U = U^* \wedge_n \pi_m(P_n)U = \wedge_n U^* \pi_m(P_n)U = \wedge_n \pi_{m'}(P_n) = P(m') .$$

Of course, then, $P(m) \neq 0$ if and only if $P(m') \neq 0$. Thus m and m' have the same set of atoms when π_m and $\pi_{m'}$ are unitarily equivalent.

For specific examples, choose a point s in S and let G(s) be its "orbit" under G. Let m_s be the measure that assigns to each Borel set, as measure, the number of points of G(s) it contains. Since G(s) is an orbit, m_s is G-invariant and the atoms of m_s are precisely the points of G(s). Now G is a countable group and S has the cardinality of the continuum. Two orbits either coincide or are disjoint. Thus there are a continuum of disjoint orbits and so a continuum of measures m_s with mutually disjoint sets of atoms. Each of these continuum measures is ergodic under G and therefore gives rise to an irreducible representation of the canonical anticommutation relations. No two of these representations are unitarily equivalent. (See [6 : pp. 759-766] for details.)

6. States and representations. The most effective method for producing representations of C*-algebras involves the GNS construction for a state of that algebra. A linear functional ρ on a C*-algebra $\mathfrak{A}$ is said to be a state of $\mathfrak{A}$ when $\rho(A) \geq 0$ for each positive operator A in $\mathfrak{A}$ and $\rho(I) = 1$. If $\mathfrak{A}$ acts on the Hilbert space $\mathcal{H}$ and x is a unit vector in $\mathcal{H}$, then the mapping $A \to \langle Ax, x \rangle$ provides us with an example of a state.

We denote this state by ω_x when $\mathfrak{A}$ is $\mathcal{B}(\mathcal{H})$ and $\omega_x|\mathfrak{A}$ otherwise. We refer to it as a vector state of $\mathfrak{A}$. The state ρ is said to be faithful when $\rho(A) > 0$ if $A > 0$. With ρ a faithful state of $\mathfrak{A}$, the mapping $(A,B) \to \rho(B^*A)$ defines a (positive definite) inner product $\langle\ ,\ \rangle_\rho$ on $\mathfrak{A}$ relative to which it has a Hilbert space completion $\mathcal{H}_\rho$. With $\pi_\rho(A)(B)$ defined as AB, $\pi_\rho(A)$ is a linear operator on $\mathfrak{A}$ and

$$\langle\pi_\rho(A)(B),\ \pi_\rho(A)(B)\rangle_\rho = \langle AB, AB\rangle_\rho = \rho(B^*A^*AB).$$

Now $A^*A \le \|A\|^2 I$ so that $B^*A^*AB \le \|A\|^2 B^*B$, and

$$\rho(B^*A^*AB) \le \|A\|^2\rho(B^*B) = \|A\|^2\langle B,B\rangle_\rho .$$

Thus $\|\pi_\rho(A)B\|_\rho^2 \le \|A\|^2\|B\|_\rho^2$ and $\|\pi_\rho(A)\| \le \|A\|$. It follows that $\pi_\rho(A)$ extends (uniquely) to a bounded linear operator, we denote again by $\pi_\rho(A)$, on $\mathcal{H}_\rho$. It is easy to check that π_ρ is a representation of $\mathfrak{A}$ on $\mathcal{H}_\rho$. We call π_ρ the GNS (Gelfand-Neumark-Segal) representation for ρ. This construction has an extension to general states (not necessarily faithful) that need not concern us. Note that $\rho(A) = \langle\pi_\rho(A)(I), I\rangle_\rho$ for each A in $\mathfrak{A}$, so that ρ becomes a vector state when "transported" to $\pi_\rho(\mathfrak{A})$. Note too that $\{\pi_\rho(A)(I) : A \in \mathfrak{A}\} = \mathfrak{A}$ and $\mathfrak{A}$ is dense in $\mathcal{H}_\rho$. The "vector" I in $\mathfrak{A}$ is said to be cyclic under $\pi_\rho(\mathfrak{A})$ and π_ρ is said to be a cyclic representation of $\mathfrak{A}$. With ρ a faithful state of $\mathfrak{A}$, π_ρ is a faithful representation of $\mathfrak{A}$ - that is, $\pi_\rho(A) = 0$ only if $A = 0$ (for $\pi_\rho(A)(I) = A$).

The class of states of the CAR algebra $\mathfrak{A}$ known as product states is of special interest in quantum statistical mechanics. Suppose that $\mathfrak{A}$ is generated by the commuting family $\{\mathfrak{A}_r\}$ with each $\mathfrak{A}_r$ * isomorphic to $M_2(\mathbb{C})$. Choose a self-adjoint system of 2x2 matrix units $\{E_{jk}\}$ in $\mathfrak{A}_r$. Then each A in $\mathfrak{A}_r$ has the form $\Sigma\, c_{jk}E_{jk}$ with c_{jk} a complex number. Define $\rho_r(A)$ to be $a_r c_{11} + (1-a_r)c_{22}$, where $0 < a_r \le \frac{1}{2}$. Then ρ_r is a state of $\mathfrak{A}_r$. With A_j in $\mathfrak{A}_j$, define $\rho(A_1\cdots A_n)$ to be $\rho_1(A_1)\cdots\rho_n(A_n)$ It is not difficult to show that ρ extends to a state ρ of $\mathfrak{A}$, the product state $\otimes\rho_r$. While far from apparent, it the case that the GNS representation π_ρ for ρ has the following properties: $\pi_\rho(\mathfrak{A})^-$ is a factor on $\mathcal{H}_\rho$, and with x_ρ the unit vector in $\mathcal{H}_\rho$ corresponding to I (so that $\rho(A) = \langle\pi_\rho(A)x_\rho, x_\rho\rangle_\rho$ for each A in $\mathfrak{A}$ and $\pi_\rho(\mathfrak{A})x_\rho$ is dense in $\mathcal{H}_\rho$), $\omega_{x_\rho}|\pi_\rho(\mathfrak{A})^-$ is faithful. With $\pi_\rho(\mathfrak{A})^-$ a factor, we say that ρ is a factor state of $\mathfrak{A}$. Our program in the final sections is to study these factor representations of the canonical anticommutation relations arising from the product states as described. For this purpose, we develop some of the essential theory of factors.

7. Types of factors. An initial separation of factors into types that are algebraically distinct (non-isomorphic) can be effected by studying their lattices of projections. If $\mathfrak{M}$ is a factor and E is a minimal projection in $\mathfrak{M}$ (that is, $E \neq 0$ and if $0 \leq F \leq E$ with F a projection in $\mathfrak{M}$, then $F = 0$ or $F = E$), then $\mathfrak{M}$ is isomorphic to $\mathfrak{B}(\mathcal{H})$ for some Hilbert space $\mathcal{H}$. In particular, I in $\mathfrak{M}$ is the sum of an orthogonal family of minimal projections in $\mathfrak{M}$. If n is the (possibly infinite) cardinal number of that family of minimal projections, then $\mathcal{H}$ has dimension n. In this case, we say that $\mathfrak{M}$ is of type I_n.

Factors need not have minimal projections. Let G be a (discrete) group all of whose conjugacy classes (other than that of the group identity) are infinite. Let $\mathcal{H}$ be $l_2(G)$ (square-summable, complex-valued functions on G with the inner product $\langle f,h\rangle = \sum_{g \in G} f(g)\overline{h(g)}$). For each g in G, define $(L_g f)(g')$ to be $f(g^{-1}g')$ for each g' in G, where $f \in \mathcal{H}$. Then each L_g is a unitary operator on $\mathcal{H}$ and $\{L_g : g \in G\}$ generates a factor $\mathfrak{M}$. If x_0 is the element of $\mathcal{H}$ that takes the value 1 at the group identity and 0 at all other elements of G, then $\omega_{x_0}|\mathfrak{M}$ is a state τ of $\mathfrak{M}$ with very special properties. Most importantly, $\tau(AB) = \tau(BA)$ (an easy computation). We say that τ is a <u>tracial state</u> on $\mathfrak{M}$ (although we may happen on it in many different ways). It is also the case that $\mathfrak{M}$ has no minimal projections. If it did, $\mathfrak{M}$ would be isomorphic to $\mathfrak{B}(\mathcal{H})$ and I would be the sum of n minimal projections. The value of τ at all minimal projections is the same positive number b (easily seen) and $\tau(I)(= 1)$ would be nb - from which, n is finite. Thus $\mathcal{H}$, $\mathfrak{B}(\mathcal{H})$, and $\mathfrak{M}$, would all have finite linear dimension. But $\{L_g\}$ is an infinite linearly independent family in $\mathfrak{M}$ (not difficult). We call factors with no minimal projections and a tracial state, <u>factors of type</u> II_1. Specific examples are obtained by choosing for G the free group on n (> 1) generators or the group of those permutations of the integers that move at most a finite number of integers. The factors of type II_1 obtained from these two groups can be shown to be non-isomorphic. Factors of type II_1 need not be isomorphic to one another.

A factor $\mathfrak{M}$ may have no minimal projections and may have no tracial state but possess a family of projections $\{E_a\}$ with the following properties: $\Sigma E_a = I$, $\{A : A \in \mathfrak{M}, E_a A E_a = A\}$ $(= \mathfrak{M}_a)$ has a tracial state τ_a ($\mathfrak{M}_a$ is necessarily a factor). In this case, with H a positive element of $\mathfrak{M}$, $\Sigma\, \tau_a(H)$ $(= \tau(H))$ is in $[0,\infty]$. The mapping τ of $\mathfrak{M}^+$, the positive elements in $\mathfrak{M}$, into $[0,\infty]$ is a <u>tracial</u> <u>weight</u> on $\mathfrak{M}$ - that is, $\tau(H+K) = \tau(H) + \tau(K)$, $\tau(aH) = a\,\tau(H)$ when $a > 0$, and $\tau(A^*A) = \tau(AA^*)$ for each A in $\mathfrak{M}$. In addition, τ is <u>semi</u>-<u>finite</u> - that is, each T in $\mathfrak{M}$ is the limit (on vectors) of linear combinations of elements in $\mathfrak{M}^+$ at which τ takes finite values. Finally, τ is normal - that is there is a family of vectors $\{x_b\}$ such that $\tau(H) = \Sigma\langle Hx_b, x_b\rangle$ for each H in $\mathfrak{M}^+$. We say that $\mathfrak{M}$ is a factor of type II_∞ when it has no minimal projections, has no tracial state but has a non-zero normal semi-finite tracial weight.

Specific examples of factors of type II_∞ arise from specific examples of factors of type II_1. If $\mathfrak{M}$ is a factor of type II_1 acting on a Hilbert space $\mathcal{H}$ and $\mathcal{K}$ is the countable (Hilbert-space) direct sum of $\mathcal{H}$ with itself, then each element of $\mathfrak{B}(\mathcal{K})$ corresponds to an infinite matrix all of whose entries lie in $\mathfrak{B}(\mathcal{H})$. The elements in $\mathfrak{B}(\mathcal{K})$ whose matrix representations have all entries in $\mathfrak{M}$ form a factor of type II_∞. Moreover each factor of type II_∞ has this form (namely, infinite matrices with entries in a factor of type II_1).

Finally, there are the factors that possess no non-zero normal semi-finite tracial weights. These are the factors of type III. They play the dominant role in the operator algebra formulation of quantum field theory and quantum statistical mechanics. Specific examples were first obtained from an ergodic-theoretic construction [7]. We won't describe these, since we shall be constructing examples by other means.

The type classification of factors provides us with a means of distinguishing among factor representations of the canonical anticommutation relations. We say that a factor representation π of a C*-algebra $\mathfrak{A}$ is of type I_n, II_1, II_∞, or III, when the factor $\pi(\mathfrak{A})^-$ has the corresponding type. With π_ρ the GNS representation for the state ρ of $\mathfrak{A}$ we say that ρ is of type I_n, II_1, II_∞, or III, when π_ρ has the corresponding type.

8. Modular theory and the T-invariant. A deep result of Tomita's (see [6 : 9.29]) associates with each faithful state ω of a von Neumann algebra $\mathcal{R}$ some structure (the modular structure of $\mathcal{R}$ and ω) that will be critical in helping us distinguish factor representations of the canonical anticommutation relations. It is easiest to describe this modular structure after using the GNS representation π_ω of $\mathcal{R}$ for ω. Through this representation, we may assume that $\mathcal{R}$ acts on a Hilbert space $\mathcal{H}$ and that $\omega(A) = \langle Au,u\rangle$ for each A in $\mathcal{R}$ where u is a unit vector in $\mathcal{H}$ such that $\mathcal{R}u$ is dense in $\mathcal{H}$ (we say that u is generating for $\mathcal{H}$) and $T = 0$ when $Tu = 0$ and $T \in \mathcal{R}$ (we say u is separating for $\mathcal{R}$). The mapping S_0 that assigns A^*u to Au for each A in $\mathcal{R}$ is a conjugate-linear operator on the dense domain $\mathcal{R}u$. Its adjoint contains the conjugate-linear operator F_0 that assigns A'^*u to $A'u$, where A' is an operator in the commutant $\mathcal{R}'$ of $\mathcal{R}$ (those bounded operators commuting with all operators in $\mathcal{R}$). Now $\mathcal{R}'u$ is dense in $\mathcal{H}$ (this follows from the fact that u is separating for $\mathcal{R}$), so that S_0 has a closure S. Let Δ be S^*S. Then S has a polar decomposition $J\Delta^{\frac{1}{2}}$, where J is a conjugate-linear isometry of $\mathcal{H}$ onto itself. From the fact that $S_0 = S_0^{-1}$, it follows that $J = J^{-1} = J^*$. Tomita's main result states that $J\mathcal{R}J = \mathcal{R}'$ and $\Delta^{it}\mathcal{R}\Delta^{-it} = \mathcal{R}$ for each real t. In particular, with $\sigma_t(A)$ defined to be $\Delta^{it}A\Delta^{-it}$ for each A in $\mathcal{R}$, we have that σ_t is a * automorphism of $\mathcal{R}$ for each real t. Moreover, $\sigma_{t+t'} = \sigma_t\sigma_{t'}$ for each pair of real numbers t and t'. We refer to the one-parameter group of * automorphisms $t \to \sigma_t$ of $\mathcal{R}$ as the modular automorphism group for ω (or u).

The state ω and the one-parameter automorphism group $\{\sigma_t\}$ are interrelated by a condition introduced into the infinite-system formulation of quantum statistical mechanics to describe equilibrium states by Haag, Hugenholtz, Winnink [3,7]. It results from the construction of σ_t that for each pair of elements A and B in $\mathcal{R}$ there is a complex-valued function f defined, continuous, and bounded on the strip $\{z \in \mathbb{C} : 0 \le \text{Im } z \le 1\}$ $(= \Omega)$, and analytic on the interior of that strip, such that

$$f(t) = \omega(\sigma_t(A)B), \qquad f(t+i) = \omega(B\sigma_t(A)) \qquad (t \in \mathbb{R}).$$

When a one-parameter group of automorphisms $\{\sigma_t\}$ and a state ω of a von Neumann algebra $\mathcal{R}$ fulfill this condition, we say that the group satisfies the modular condition relative to ω. It follows from this condition (together with some complex function theory) that $\omega(\sigma_t(A)) = \omega(A)$ for each A in $\mathfrak{R}$ and all real t. (We say that ω is invariant under σ_t.) Using the assumption that ω is faithful, the modular condition yields that $\sigma_t(H) = H$ for some H in $\mathfrak{R}$ and all real t, if and only if $\omega(AH) = \omega(HA)$ for all A in $\mathfrak{R}$. An element such as H is said to lie in the <u>centralizer</u> of ω. (See [6 : 9.2.13 and 9.2.14].) In addition, there is a <u>unique</u> (continuous) one-parameter group of * automorphisms of $\mathfrak{R}$ (the <u>modular group</u>) that satisfies the modular condition relative to a given faithful normal state ω of $\mathfrak{R}$ [6 :9.2.16]. A useful extension of this last result [6 : 9.2.17] asserts that it suffices to find our complex-valued function f satisfying the continuity, boundedness, and analyticity condition just for A and B in a self-adjoint subalgebra of $\mathfrak{R}$ provided each element of $\mathfrak{R}$ is a limit (on vectors) of elements of this subalgebra.

The relation between the modular automorphisms corresponding to two faithful normal states of $\mathfrak{R}$ is established in [1]. If $\{\alpha_t\}$, $\{\beta_t\}$ are the corresponding modular automorphism groups, there is, for each t, a unitary operator U_t <u>in $\mathfrak{R}$</u> such that, for each A in $\mathfrak{R}$, $\alpha_t(A) = U_t\beta_t(A)U_t^*$, $U_{s+t} = U_s\beta_s(U_t)$, and $U_tx \to x$ for each x in $\mathcal{H}$ as $t \to 0$. (The mapping $t \to U_t$ is the <u>Connes cocycle</u> relating α_t and β_t.) In particular, α_t is an inner * automorphism of $\mathfrak{R}$ (that is, there is a unitary operator V in $\mathfrak{R}$ such that $\alpha_t(A) = VAV^*$ for each A in $\mathfrak{R}$) if and only if β_t is inner. Thus the set $T(\mathfrak{R})$ of real numbers t such that σ_t is inner for the modular automorphism group $\{\sigma_t\}$ of $\mathfrak{R}$ relative to a faithful normal state of $\mathfrak{R}$ is the same for all faithful normal states of $\mathfrak{R}$. It follows that this set (which is trivially seen to be a subgroup of $\mathbb{R}$) is an (isomorphism) invariant for $\mathfrak{R}$. It is the <u>T-invariant</u> [1], and will help us to distinguish certain factor representations of the canonical anti-commutation relations. (See [6 : 13.1.9].)

<u>9. Some general properties of the T-invariant.</u> If $\mathfrak{M}$ is a factor not of type III, a sequence of Radon-Nikodym results (in the non-commutative setting of operator algebras), involving the normal tracial weight, leads to the conclusion that $T(\mathfrak{M}) = \mathbb{R}$, that is, each modular automorphism group of $\mathfrak{M}$ consists entirely of inner automorphisms. The T-invariant is not at all sensitive to factors of types other than III. (See [6 : 9.2.21].) In the case of a type III factor $\mathfrak{M}$ acting on a separable Hilbert space, almost the reverse situation prevails. A measure-theoretic,

cohomological argument shows that all the automorphisms of a one-parameter group can be inner only if the group is implemented by a one-parameter group of unitary operators in $\mathfrak{M}$ ([5] - see also [6 : 14.4.3-14.4.10]). In this event, the factor will have a non-zero, normal, tracial weight and cannot be of type III [6 : 9.2.21]. Thus, for a factor $\mathfrak{M}$ of type III acting on a separable Hilbert space, $T(\mathfrak{M})$ must be different from $\mathbb{R}$. Indeed, complicated measure- and group-theoretic considerations show that $T(\mathfrak{M})$ must have Lebesgue measure 0.

As to what subgroups of $\mathbb{R}$ appear as $T(\mathfrak{M})$ for some factor $\mathfrak{M}$ acting on a separable Hilbert space, it is known that each countable subgroup of $\mathbb{R}$ arises in this way. Suppose G is an arbitrary subgroup of $\mathbb{R}$ (perhaps $\mathbb{R}$ itself) considered as a discrete group. With ω a faithful state of a factor $\mathfrak{M}$ acting on a Hilbert space $\mathcal{H}$ and $\{\sigma_t\}$ the corresponding modular automorphism group, a special construction allows us to realize G as $T(\mathfrak{N})$ for some factor $\mathfrak{N}$. To effect this construction, we introduce the Hilbert space direct sum $\Sigma \oplus \mathcal{H}_t$ (= $\mathcal{K}$) of copies $\mathcal{H}_t$ of $\mathcal{H}$ (one for each real t). The elements of $\mathcal{K}$ are functions x on $\mathbb{R}$ such that $x(t) \in \mathcal{H}_t$ for each real t (and $\Sigma \|x_t\|^2 < \infty$). We define operators $\Phi(A)$ on $\mathcal{K}$ for each A in $\mathfrak{M}$ by $(\Phi(A)x)(t) = Ax(t)$ and V(t) by $V(t)x(s) = \Delta^{it}x(s-t)$, where $\Delta^{it}A\,\Delta^{-it} = \sigma_t(A)$ for each A in $\mathfrak{M}$. Then the von Neumann algebra $\mathfrak{R}$ generated by $\{\Phi(A), V(t) : A \in \mathfrak{M}, t \in \mathbb{R}\}$ is called the <u>crossed-product</u> of $\mathfrak{M}$ by the automorphism group $\{\sigma_t\}$. Let $\mathfrak{N}$ be the von Neumann subalgebra of $\mathfrak{R}$ generated by $\Phi(\mathfrak{M})$ and $\{V(t) : t \in G\}$. (Then $\mathfrak{N}$ is isomorphic to the crossed-product of $\mathfrak{M}$ by the automorphism group $\{\sigma_t : t \in G\}$.) If σ_t is an outer automorphism of $\mathfrak{M}$ for t in G different from the identity, then $\mathfrak{N}$ is a factor [6:13.1.5] Relative to appropriately defined (normal) states of $\mathfrak{R}$ and $\mathfrak{N}$, $\{V(t) : t \in \mathbb{R}\}$ implements the modular automorphism groups of $\mathfrak{R}$ and $\mathfrak{N}$. (See [6 : 14.4.19].)

In the next section, we construct certain factor representations of the canonical anticommutation relations and compute the the T-invariant for the factors arising. Among these is a factor $\mathfrak{M}$ (of type III) for which $T(\mathfrak{M}) = \{0\}$ (no σ_t is inner other than the identity automorphism). If we perform the foregoing construction with this factor as $\mathfrak{M}$, then $\mathfrak{R}$ becomes a factor of type III and $T(\mathfrak{R}) = \mathbb{R}$ - but $\mathcal{K}$ is <u>not separable</u>. In this same case, $T(\mathfrak{N}) = G$, the restriction of $\mathfrak{N}$ to $\Sigma_{t \in G} \oplus \mathcal{H}_t$ is a * isomorphism, and $\Sigma_{t \in G} \oplus \mathcal{K}_t$ is separable.

10. Some factor representations of the CAR. With $\mathfrak{A}$ the CAR algebra and ρ the product state of $\mathfrak{A}$ determined by the sequence of numbers $\{a_r\}$ in $(0,\frac{1}{2}]$, as described in Section 6, we noted that π_ρ is a factor representation $\mathfrak{A}$. We introduce b_r as $\log(a_r^{-1}(1-a_r))$. Then:

(1) $$T(\pi_\rho(\mathfrak{A})^-) = \{t \in \mathbb{R} : \sum_{r=1}^{\infty} [1 - |a_r^{1+it} + (1-a_r)^{1+it}|] < \infty\}$$
$$\{t \in \mathbb{R} : \sum_{r=1}^{\infty} e^{-b_r} \sin^2(\tfrac{1}{2}b_r t) < \infty\}$$

(2) If $a_r \to \frac{1}{2}$ and $b_r^2 \to \infty$, then $\pi_\rho(\mathfrak{A})^-$ is a factor of type II_1 and $T(\pi_\rho(\mathfrak{A})^-) = \mathbb{R}$.

(3) If $\Sigma a_r < \infty$ (equivalently, $\Sigma e^{-b_r} < \infty$), then $\pi_\rho(\mathfrak{A})^-$ is a factor of type I_∞.

(4) If $a_r \to 0$ and $\Sigma a_r = \infty$, then $\pi_\rho(\mathfrak{A})^-$ is a factor of type III.

(5) If $a_r = (r+1)^{-1}$, then $T(\pi_\rho(\mathfrak{A})^-) = \{0\}$. In this case, $\pi_\rho(\mathfrak{A})^-$ serves as the factor $\mathfrak{m}$ we needed in Section 9.

With a_r as in (5), b_r is $\log r$. If we alter the a_r (equivalently, the b_r) slightly, the T-invariant can change significantly. Let $[x]$ denote the largest integer not exceeding x.

(6) If $b_r = [\log r]$ $(r = 3,4,\cdots)$, then
$$T(\pi_\rho(\mathfrak{A})^-) = \{0, \pm 2\pi, \pm 4\pi, \pm 6\pi, \cdots\}.$$

(7) If $b_r = n!$ when $[e^{n!}] < r \leq [e^{(n+1)!}]$, then $T(\pi_\rho(\mathfrak{A})^-)$ contains each rational multiple of 2π (but is not $\mathbb{R}$).

(See [6 : 13.1.15,13.4.9-13.4.15] for details.)

REFERENCES

1. A. Connes, Une classification des facteurs de type III, Ann. Sci. École Norm. Sup. Paris 6 (1973), 133-252.
2. J. G. Glimm, On a certain class of operator algebras, Trans. Amer. Math. Soc. 95 (1960), 318-340.
3. R. Haag, N. M. Hugenholtz, M. Winnink, On the equilibrium states in quantum statistical mechanics, Commun. math. Phys. 5 (1967), 215-236.
4. R. V. Kadison, Unitary invariants for representations of operator algebras, Ann. of Math. 66(1957), pp. 304-379.

5. R. V. Kadison, Transformations of states in operator theory and dynamics, Topology, 3, Suppl. 2(1965), 177-198.
6. R. V. Kadison and J. R. Ringrose, Fundamentals of the Theory of Operator Algebras, Vols. I,II, Academic Press, New York 1983, 1986.
7. D. Kastler, J. C. T. Pool, E. Thue Poulsen, Quasi-unitary algebras attached to temperature states in statistical mechanics. A comment on the work of Haag, Hugenholtz and Winnink, Commun. math. Phys. 12(1969), 175-192.
8. J. von Neumann, On rings of operators. III, Ann. of Math. 41(1940), 94-161.

Department of Mathematics
University of Pennsylvania
Philadelphia, Pennsylvania 19104-6395

*With partial support of the National Science Foundation (USA).

A DIFFERENTIAL - GEOMETRIC SETTING FOR BRS TRANSFORMATIONS AND ANOMALIES

by Daniel KASTLER [1)]

This is a report of a common work with Raymond Stora (part II of reference [1] with the same title). BRS transformations and anomalies occured in the study of perturbative gauge theories - the anomalies being cohomological obstructions to (gauge invariant) renormalization (and also, positively, devices for writing effective Lagrangian in phenomenological field theories) Although originally quantum items, both BRS transformations and anomalies however ultimately appear as purely classical (-differential-geometric) objects, which we may consider as items in the theory of smooth principal vector bundles. This is the point of view which is adopted in this lecture.

These notes are organized as follows : section 1 recalls facts about principal vector bundles, fixes notation, and defines the gauge group **G** and its Lie algebra **L** in an intrinsic way. Section 2 describes the M_v-valued de Rham complex $\Lambda^*(P,M_v)$ of the total space P of the principal bundle at hand (where M_v is the set of $v \times v$ matrices accomodating the structure group $\mathcal{G}$ (assumed linear) and its Lie algebra $\mathcal{L}$). $\Lambda^*(P,M_v)$ is at the same time a complex with differential d and a representation space of the gauge group **G** (and thus of its Lie Algebra **L**)[1)]. Section 3 then constructs the Chevalley cohomology of **L** with values in the representation space $\Lambda^*(P,M_v)$, the frame for calculation of anomalies (the genral concept of Chevalley cohomology is described in Appendix 2). This construction is by means of a double complex [2)] endowed

----

1) in addition, it is a differential graded Lie Algebra under the "Schouten product $[\wedge]$" defined below.

2) equal as a set to $\Lambda^*(P,M_v) \otimes \Lambda^*(L,C^\infty(M))$; M the base of our principal bundle.

with two commuting derivations δ and d, δ being the Chevalley cohomology operator, whilst d stems from the exterior derivative of $\Lambda^*(P,M_v)$. The corresponding total complex leads to the variant s of δ (anticommuting with d) and the corresponding "total derivation" Δ = s+d. This setting now allows (in section 4) a display of (the classical aspect of) the BRS relations involving a connection one-form a and the Maurer-Cartan form ω of $\mathcal{G}$, both elements of the bicomplex : these relations in fact consist in expressing sa and sω in terms of a, ω and dω (using a product $[\wedge]$ defined on the bicomplex in terms of the wedge product of $\Lambda^*(L,C^\infty(M)$ and the Schouten product of $\Lambda^*(P,M_v))$. Section 5 then describes succinctly the homotopy formula allowing an elegant calculation of the anomalies.

The description of this setting is completed with Section 4, but still fails to reveal the relationship of s with the exterior derivative of **G** , which is then described in Section 6.

1. The principal bundle $\mathbb{P}$ = (P→ M,G)

We recall that a <u>smooth principal bundle</u> consists in a "total space" P smoothly mapped onto the "base" M by the "projection" π (P and M are smooth manifolds), moreover in such a way that the anti-images under π of the points of the base are the orbits in P of the smooth right action R of the "<u>structural</u> (Lie) -<u>group</u>" $\mathcal{G}$.

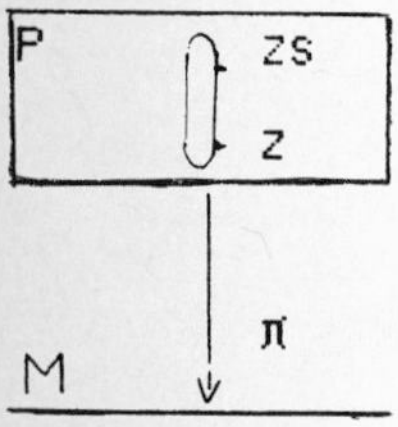

Notation : $R_s z = zs$, $z \in P$, $s \in \mathcal{G}$

The action R is proper in the sense :

$zs = zs' \Rightarrow s = s'$

The gauge group $\mathcal{G}$ is defined as the group of automorphisms of the bundle $\mathbb{P}$ (i.e. diffemorphisms $\Psi : P \to P$ such that $\pi(\Psi(z) = \pi(z)$ and $\Psi(zs) = \Psi(z)s$ for all $z \in P$ and $s \in \mathcal{G}$). The first property entails that one must have

(1,1) $\Psi(z) = z\, g(z)$

where $z \to g(z)$ is a smooth map : $P \to \mathcal{G}$ Ad-equivariant in the sense

(1,2) $g(zs) = s^{-1} g(z) s,$

as results from the second property. We shall identify **G** with the set of these maps - the multiplication of **G** then consists in the point-wise multiplication

(1,3) $(gg')(z) = g(z) g'(z)$, $g, g' \in \mathcal{G}$, $z \in P$

Accordingly, the Lie algebra **L** of **G** is identified with the smooth maps Ω : $P \to \mathcal{L}$, $\mathcal{L}$ the Lie algebra of G, Ad-equivariant in the sense :

(1,4) $\Omega(zs) = s^{-1} \Omega(z) s$, $z \in P$, $s \in \mathcal{G}$.

Here $s^{-1}.s$ is the tangent map to the map $s^{-1}.s$ r.h.s. of (1.2). However, if, as we assume, $\mathcal{G}$ is a linear group i.e. a subgroup of the group of unvertibles of M_ν, M_ν the set of complex $\nu \times \nu$ matrices, we have $\mathcal{G}$, $\mathcal{L} \subset M_\nu$ with the product of $\mathcal{G}$ the product of matrices (and the Lie bracket of $\mathcal{L}$ the commutator of matrices) and all products r.h.s. of (1,2), (1,3) , (1,4) can simply be considered as matrix products in M_ν.

2. The M_ν-valued de Rham complex $\Lambda^*(P,M_\nu)$ of P

With $\Lambda^p(P,M_\nu)$ the set of differential p-forms on P with values in the set M_ν of $\nu \times \nu$ complex matrices, we set :

(2.1) $\Lambda^*(P,M_\nu) = \bigoplus_p \Lambda^p(P,M_\nu) = M_\nu \otimes \Lambda^*(P,\mathbb{C})$

where $\Lambda^*(P,\mathbb{C})$ is the usual (complex valued) de Rham complex. The identification

(2.2) $\Lambda^p(P,M_\nu) = M_\nu \otimes \Lambda^p(P,\mathbb{C})$

is obtained from

(2,3) $(m \otimes \alpha)(\xi_1, \ldots, \xi_p) = \alpha(\xi_1, \ldots, \xi_p)\, m$, $\xi_i \in \mathcal{X}(P)$, $i=1, \ldots p$

with $\mathcal{X}(P)$ the Lie algebra of smooth vector fields on P . $\Lambda^*(P,M_\nu)$ is equipped with the exterior derivation

(2,4) $d = id_{M_\nu} \otimes d$,

the two products $\dot{\wedge}$ and $[\wedge]$ respectively specified as follows [3]: for $m, m' \in M_\nu$, $\alpha, \alpha' \in \Lambda^*(P,\mathbb{C})$:

(2,5) $(m \otimes \alpha)\ \dot{\wedge}\ (m' \wedge \alpha') = mm' \otimes \alpha \wedge \alpha'$

(2,6) $[\dot{u} \otimes \alpha \wedge m' \otimes \alpha'] = [m,m'] \otimes (\alpha \wedge \alpha')$,

and the following representations r, ϱ of G, resp. $\mathbf{G}$:

(2,7) $r(s) = \mathrm{Ad}s \otimes R^*s \quad , s \in G$

(2,8) $\varrho(g) = id_{M_\nu} \otimes (g^{-1}) \quad , \ g \in G$

Under these premises, we have that (cf. Appendix A):

| $(\Lambda^*(P,M_\nu)\ ,\ d,\ \dot{\wedge})$ is an associative GDA
| $(\Lambda^*(P,M_\nu)\ , d, [\wedge]$ is D G L .
| $r(s)$, $s \in G$, and $\varrho(g)$, $g \in \mathbf{G}$, are commuting GDA – and DG L –automorphisms.

Denoting by $A^*(P,M_\nu)$ the fixpoints of $\Lambda^*(P,M_\nu)$ under r:

(2,9) $A^p(P^*,M_\nu) = \{\lambda \in \Lambda^*(P,M_\nu)\ ;\ r(s)\lambda = \lambda \text{ for all } s \in \mathcal{G}\}$,

i.e.

(2,9a) $\lambda \in \Lambda^p(P,M_\nu) \iff$

$$\lambda(zs, Z_1 s, \dots, Z_p s) = s^{-1}\lambda(z, Z_1, \dots, Z_p)\, s$$
$$\text{for all } s \in \mathcal{G},\ z \in P,\ Z_1, \dots, Z_p \in T^P_z$$

we then have that

| $(\Lambda^*(P,M_\nu)\ , d, \dot{\wedge})$ is a ϱ–invariant sub GDA of $(\Lambda^*(P,M_\nu), d, \dot{\wedge})$
| $(\Lambda^*(P,M_\nu)\ , d\,[\wedge\,])$ is a ϱ–invariant sub D G L of $(\Lambda^*(P,M_\nu), d, [\wedge\,])$
| One has $\mathbf{G}$, $\mathbf{L} \in \Lambda^\circ(P,M_\nu)$, namely
| (2.10) $\mathbf{G} = A^\circ(P, \mathcal{G})$
| (2.11) $\mathbf{L} = A^\circ(P, \mathcal{L})$

----

3) $\wedge$ denotes the wedge product of differential forms , whilst $[m,m'] = mm' - m'm$.

We call $[\ \wedge\ ]$ the <u>Schouten product</u> (familiar for a convenient writing of covariant derivations, the Bianchi identity, etc).

(here $\mathrm{Ad}s : s^{-1}.s$, R^*_s and $(g^{-1})^*$ denoting the pull backs of differential forms by R_s, resp. g^{-1}) .

3. Cohomology of L with values in $(\Lambda^*(P,M_\nu), \rho)$

With ρ denoting also the associated representation of **L**

(3,1) $\rho(\Omega) = d/dt\big|_{t=0} \rho(e^{t\Omega})$, $\Omega \in \mathbf{L}$,

$\Lambda^*(P,M_\nu)$ is a representation space of the Lie algebra **L**.

We now apply the standard construction of Appendix C where we make, $\mathbf{L} \to \mathbf{L}$, $(V,\rho) \to (\Lambda^*(P,M_\nu), \rho)$

(or $A^*(P,M_\nu), \rho)$. Thus, we work within

(3,2) $$\wedge^*(\mathbf{L}, \Lambda^*(P,M_\nu) = \bigoplus_{p,\alpha} \Lambda^{p\alpha}$$

with

(3.3) $$\Lambda^{p\alpha} = \wedge^\alpha(\mathbf{L}, \Lambda^p(P,M_\nu)) = \Lambda^p(P,M_\nu) \otimes \Phi^\alpha$$

where the tensor product is over $\mathbb{C}^\infty(M)$

(3,4) $$\Phi^\alpha = \wedge^\alpha(\mathbf{L}, \mathbb{C}^\infty(M))$$

identification with tensors products being specified by

(3,5) $$\begin{cases} (\lambda \otimes \varphi)(\Omega 1, \dots, \Omega p) = \varphi(\Omega 1, \dots, \Omega p)\lambda \\ \lambda \in \Lambda^p(P,M_\nu), \varphi \in \Phi^\alpha, \Omega_1, \dots, \Omega_p \in \mathbf{L}. \end{cases}$$

We now consider the operators δ (<u>Chevalley coboundary</u>) and d (<u>exterior derivative</u>) : for $U \in \Lambda^{p\alpha}$ and $\Omega_0, \dots, \Omega_\alpha \in \mathbf{L}$:

(3,6) $$(\delta U)(\Omega_o, \dots, \Omega_\alpha) = \alpha \bigoplus_{i=0}^{\alpha} (-1)^i \sum{}_\rho(\Omega_i) U(\Omega_o, \dots, \hat{\Omega}_i, \dots, \Omega_\alpha)$$

$$+ \sum_{o\leq i<j\leq\alpha} (-1)^{i+j} / o\leq i<j\leq\alpha \;\; U([\Omega_i,\Omega_j], \Omega_o, \dots, \hat{\Omega}_i, \dots, \hat{\Omega}_j, \dots, \Omega_\alpha)$$

(3,7) $$(dU)(\Omega_o, \dots, \Omega_\alpha) = d\{U(\Omega_o, \dots, \Omega_\alpha\}$$

1.e.

(3,7a) $$d = d \otimes id_{\Phi*} + id_{\wedge^p(P,M\nu)} \otimes d$$

the last operator d acting on values.

We now have

$(\Lambda^{**} = \sum_{p,\alpha} \Lambda^{p\alpha}, d, \delta)$ is a double complex, i.e. one has

(3,8) $d^2 = \delta^2 = d\delta - \delta d = 0$

p is called the <u>degree of form</u>, α the <u>ghost number.</u>

As customary, we consider the associated total complex (Λ^*, Δ) where

(3,9) $^n\Lambda = \bigoplus_{p+\alpha=n} \Lambda^{p\alpha}$ (n is the <u>total grading</u>)

(3,10) $\Delta = d + s$

with

(3,11) $sU = (-1)^p \delta U \quad , U \in \Lambda^{p*}$

the definition implying

(3,12) $\Delta^2 = ds + sd = 0$

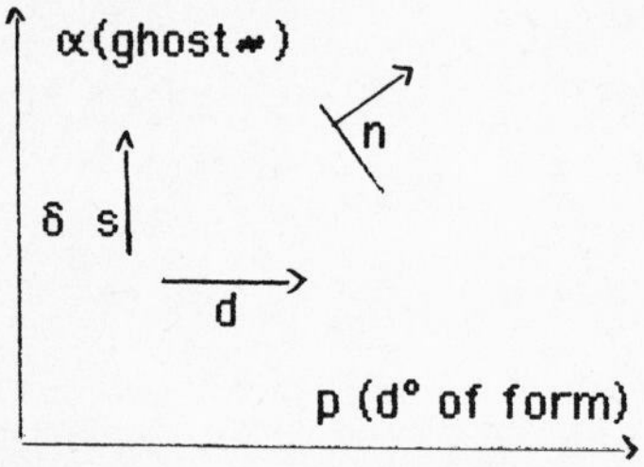

We furthermore equip. $^*\Lambda = \Lambda^*(\mathbf{L}, \Lambda^*(P,M_\nu))$ with the two products $\dot{\wedge}$, $[\wedge]$ induced by the products $\dot{\wedge}$, resp. $[\wedge]$ of $\Lambda^p(P,M_\nu)$ and the wedge product $\wedge$ of Φ^* : for $(\lambda , \mu \in \Lambda^p(P,M_\nu)$, q, $\psi \in \Phi^\alpha$

(3,13) $(\lambda \otimes \varphi) \dot{\wedge} (\mu \otimes \psi) = (-1)^{\alpha q} (\lambda \dot{\wedge} \mu) \otimes (\varphi \wedge \psi)$

(3,14) $[\lambda \otimes \wedge \mu \otimes \psi] = (-1)^{\alpha q} [\lambda \wedge \mu] \otimes (\varphi \wedge \psi)$

We now have that [4)]

$(^*\Lambda, \Delta, \dot{\wedge})$ is an associative GDA with sub GDA $(^*A, \Delta, \dot{\wedge})$

$(^*\Lambda, \Delta, [\wedge])$ is a DGL with sub DGL $(A^*, \Delta, [\wedge])$

4) also the replacing Δ bu its components s and d.

4. B.R.S. relations

Let a be a potential (connection one-form) a is an $\mathcal{L}$-valued, Ad-equivariant one-form on P : thus one has

(4,1) $a \in A^1(P,M_\nu) \sim A^{10}$

On the other hand we consider the Maurer-Cartan form of $\mathcal{G}$ under its "tautological aspect", i.e. as the identity map : $\mathcal{L} \to \mathcal{L}$. In this way

(4,2) ω ; $\underline{L \text{ id } L}$ $\subset A°(P,M\nu)$ appears as belonging to A^{01}

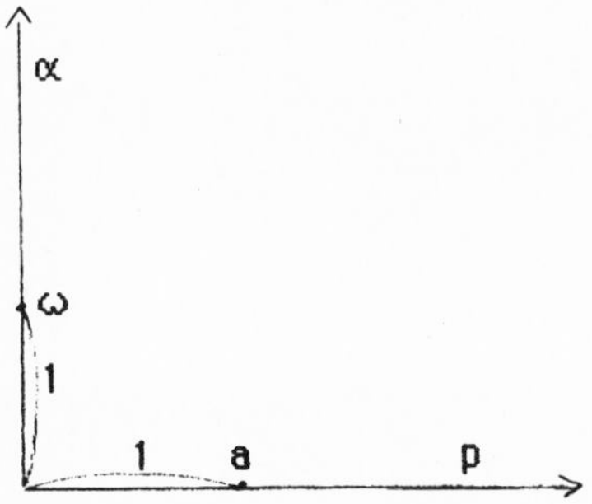

Having accomodated both a and ω in the double complex A^{**}, the following <u>BRS relations</u> make sense (and hold in reality) :

$$(4,3)\quad \begin{cases} sa = -d\omega - [a \wedge \omega] \\ s\omega = -1/2\ [\omega \wedge \omega] \end{cases}$$

5. The Homotopy Formula

We now give a brief indication on how one calculates anomalies. Setting

$$(5,1)\quad \begin{cases} A = a + \omega \\ \mathcal{F}^A = \Delta A + 1/2\ [A \wedge A] \end{cases}$$

it follows immediately from the BRS relations that one has

(5,2) $\mathcal{F}^A = F^a = da + 1/2\ [a \wedge a]$

If, with $B \in {}^1A$, we define the corresponding covariant derivation by

(5,3) $\mathcal{D}^B \lambda = \Delta\lambda + [B \wedge \lambda]$

we then have, for $t \in \mathbb{R}$, the two following relations :

(5,4) $\mathcal{D}^{tA}\ \mathcal{F}^{tA} = 0$ (Bianchi identity)

and

(5,5) $\mathcal{D}^{tA} A = d/dt\ \mathcal{F}^{tA}$

whence

(5,6) $\mathcal{D}^{tA}(A \curlywedge \mathcal{F}^{tA} \curlywedge \ldots F^{tA})$

$$= d/dt\ (\mathcal{F}^{tA} \curlywedge \mathcal{F}^{tA} \curlywedge \ldots \curlywedge \mathcal{F}^{tA})$$

(we work here within the double complex $\Lambda^{**}(L\ ,\Lambda^*(P\ S(M_\nu)))) =$ $S(M_\nu) \otimes \Lambda^*(P, \mathbb{C}) \otimes \Phi^*$ where $S(M_\nu)$ denotes the symmetric algebra over M_ν, and the product $\curlywedge$ is obtained by combining the symmectric product V of $S(M_\nu)$ with the wedge product of $\Lambda^*(P,\mathbb{C})$ and that of Φ^*, in a way extending (2,6). For P a polynomial over $\mathcal{L}$ (i.e. an element of the dual of $S(\mathcal{L})$, assumed Ad-invariant of order k, we then obtain

(5,7) $P\ \mathcal{D}^{tA}(A\ \mathcal{F}^{tA} \ldots \mathcal{F}^{tA})$

$$= d/dt\ P\{\mathcal{F}^{tA} \ldots \mathcal{F}^{tA}\}$$

$$= 1/k\ d/dt\ P\{\mathcal{F}^{tA} \ldots \mathcal{F}^{tA}\}$$

$$= \Delta P(A\ \mathcal{F}^{tA} \ldots \mathcal{F}^{tA})$$

Integration w.r.t from 0 to 1 then yields the <u>homotopy formula</u>

(5,8) $\Delta^{2k-1} Q = P(F^a \ldots F^a)$

where

(5,9) ${}^{2k-1} Q = P\{A\ \mathcal{F}^{tA} \ldots \mathcal{F}^{tA}\}$

(we took account of (5,2)

Decomposition of this in the double complex now yields

$s\ Q^{0,2k-1} = 0$

$d\ Q^{0,2k-r} + s\ Q^{1,2k-3} = 0$

$d\ Q^{2k-3,2} + s\ Q^{2k-2,1} = 0$

. . . . .

etc.

Assuming the principal bundle $\mathbb{P}$ trivial with compact base M (euclidean case) and pulling back by a section σ^*, yields after integration over M (by Stokes theorem) the relation

$$(4,4) \qquad s \int_M \sigma^* Q^{2k-2,1} = 0$$

yielding a solution of the Wess-Zumino compatibility condition.

6. Description in terms of ϱ-equivariant smooth forms on G

We now consider another double complex which will turn to be equivalent to that of Section 3 above : this new double complex is obtained by considering the set of $\Lambda^*(P,M_\nu)$ - valued smooth forms on the gauge group **G** :

$$(6,1) \quad \Lambda^*(\mathbf{G},\Lambda^*(P,M_\nu)) = \bigoplus_{p\alpha} \Lambda^\alpha(G,\Lambda^p(P,M_\nu))$$

with

$$(6,2) \quad {}^n\Lambda(\mathbf{G},\Lambda^*(P,M_\nu)) = \bigoplus_{p+\alpha=n} \Lambda^\alpha(G,\Lambda^p(P,M_\nu))$$

and the operators δ,d,s specified as follows

(6,3) δ = exterior derivative of **G**

$$(6,4) \quad \begin{cases} (dS)(g,z_1, \dots, z_\alpha) = dS(g,z_1, \dots, z_\alpha) \\ S \in \Lambda^*(\mathbf{G}, \Lambda^*(P,M_\nu)),\ g \in \mathbf{G},\ z_1, \dots, z_\alpha \in T^{\mathbf{G}}_g \end{cases}$$

$$(6,5) \quad sS = (-1)p\ \delta S \ ,\ S \in \Lambda^*(\mathbf{G},\Lambda^p(P,M_\nu))$$

giving rise to

(6,6) $\delta^2 = d^2 = d\delta - \delta d = ds + sd$

so that $\Delta = d+s$ again fufills $\Delta^2 = 0$.

We furthermore equip $\Lambda^*(\mathbf{G},\Lambda^*(P,M_\nu))$ with the products $\wedge$ and $[\ \wedge\ ]$ defnined as follows for $S \in \Lambda^\alpha(\mathbf{G},\Lambda^p(P,M_\nu)$, $T \in \Lambda^\beta(\mathbf{G},\Lambda^q(P,M_\nu))$ $g \in \mathbf{G}$ and $Z_1, \dots, Z_{\alpha+\beta} \in T^{\mathbf{G}}_g$ we set

$$(6,7)\ (S \dot{\wedge} T)(g\,;Z_1, \dots, Z_{\alpha+\beta}) = (-1)^{\alpha q}\ \frac{1}{\alpha!}\ \frac{1}{\beta!} \sum_{\sigma\in\Sigma_{\alpha+\beta}} \mathcal{X}(\sigma)$$

$$S(g\,;Z_{\sigma 1}, \dots, Z_{\sigma\alpha})\ \dot{\wedge}\ T(g\,;Z_{\sigma(\alpha+1)}, \dots, Z_{\sigma(\alpha+\beta)})$$

$$(6,8)\ [S \wedge T](g\,;Z_1, \dots, Z_{\alpha+\beta}) = (-1)^{\alpha q}\ \frac{1}{\alpha!}\ \frac{1}{\beta!} \sum_{\sigma\in\Sigma_{\alpha+\beta}} \mathcal{X}(\sigma)$$

$$[S(g,Z_{\sigma 1}, \dots, Z_{\sigma\alpha}) \wedge T(g\,;Z_{\sigma(\alpha+1)}, \dots, Z_{\sigma(\alpha+\beta)})]$$

Finally we define $r(s)$, $s\in G$, and $\rho(g)$, $g\in\mathbf{G}$, acting on $S \in \Lambda^*(\mathbf{G},\Lambda^*(P,M_\nu))$ by

(6,9) $(r(s)S)(h\,;z_1, \dots, z_\alpha) = r(s)\ S(h;z_1, \dots, z_\alpha)$

(6,10) $(\rho(g)S)(h;z_1, \dots, z_\alpha) = \rho(g)\ S(g^{-1}h\,;g^{-1}z_1, \dots, g^{-1} z_\alpha)$

where $h \in \mathbf{G}$, $z_1, \dots, z_\alpha \in T^{\mathbf{G}}_h$.

We then have that

| $({}^*\Lambda(\mathbf{G},\Lambda^*(P,M_\nu)), \Delta, \wedge)$ is an associative GDA
| $({}^*\Lambda(\mathbf{G},\Lambda^*(P,M_\nu)), \Delta, [\ \wedge\ ])$ is a DGL
| $\rho(g)$, $g \in G$, and $r(s)$, $s\in G$, are commuting GDA – and DGL – automorphisms.

The ρ–<u>equivariant</u> elements of $\Lambda^*(\mathbf{G}, \Lambda^*(P,M_\nu))$ are the fixpoints for ρ : hence S is ρ– equivariant whenever one has, for g, $h\in \mathbf{G}$, $z_1, \dots z_\alpha \in T^{\mathbf{G}}_h$:

(6,11) $S(gh;gZ_1, \dots, gZ_\alpha) = \rho(g)\ S(h\,;Z_1, \dots, Z_\alpha)$

Each equivariant S is determined by its value at the origin e of G : indeed with h = e yields

(6,12) $S(g\,;\,g\Omega_1, \dots , g\Omega_\alpha) = \rho(g)\, S(e\,;\,\Omega_1, \dots , \Omega_\alpha)$

for $\Omega_1, \dots , \Omega\alpha \in T\mathbf{G}_e = \mathbf{L}$.

This suggest, to define, for $U \in \Lambda^{p\alpha}$

(6,13) $U^1(g;g\Omega_1, \dots, g\Omega_\alpha) = \rho(g)U(\Omega_1, \dots, \Omega_\alpha)$

Then $U^1 \in \Lambda^\alpha(\mathbf{G},\Lambda^p(P,M_\nu))$ is ρ-equivariant.

The foregoing entails that

The map $U \to U^1$ commutes with d and ρ (hence with s and Δ) ; and with the products $\wedge$, $[\wedge]$.

ρ-equivariant, $\Lambda^*(P,M_\nu)$-valued smooth forms on $\mathbf{G}$ yield in this way an equivalent version of the Chevalley cohomology of section 3, where δ is now interpreted as the exterior derivative of $\mathbf{G}$.

We conclude in describing the ρ-equivariant version of the BRS relations. For a and ω as in Section 4 we have, from (6,13)

(6,14) $a1 = \mathbf{g} \wedge a \wedge \mathbf{g}^{-1} - dg \wedge g^{-1}$

and

(6,15) $\omega^1 = dg \wedge g^{-1}$

where $\mathbf{g}$ denotes the identity map from $\mathbf{G}$ to $\mathbf{G} \in A^\circ(P,M\nu)$. The BRS relations now read

$$
(6,16)\quad \begin{cases} s a^1 = -d\omega^1 - [a^1 \wedge \omega^1] \\ s\omega^1 = -1/2\,[\omega^1 \wedge \omega^1] \end{cases}
$$

and can be derived as such within the ρ-equivariant formalism of this section (cf. [2]).

<u>Remark</u> : Defining instead the representation of $\mathbf{G}$ on $\Lambda^*(\mathbf{G},\Lambda^*(P,M_\nu))$ by

(6,17) $\rho(g) = \rho(g)\, R^*g$

(instead of $\rho(g)\, L^*g_{-1}$) (L and R the action of $\mathbf{G}$ in itself by left, resp. right translations) consideration of fixpoints yields the opposite kind of ρ-equivariant forms where (6,12) is replaced by

(6,12') $S(hg,Z,g,\dots,Z_\alpha g) = \rho(g^{-1})\, S\,(h\,;Z_1,\dots,Z_\alpha)$

One has then the following replacements

(6,13') $U^r(g,Z_1,\dots,Z_\alpha) = \rho(g^{-1})\, U\,(Z_1 g^{-1},\dots,Z_\alpha g^{-1})$

(6,14') $a^r = \mathbf{g}^{-1} \wedge a \wedge \mathbf{g} + \mathbf{g}^{-1} \wedge d\mathbf{g}$

(6,15') $\omega^r = \mathbf{g}^{-1} \wedge \delta \mathbf{g}$

leading to the following form of the BRS-relations

$$sa^r = d\omega^r + [a^r \wedge \omega^r]$$

(6,16')

$$s\omega^r = 1/2\,[\omega^r \wedge \omega^r]$$

this beingh tied up with the fact that one has now

(6,17) $(\delta U)^r = -\delta U^r$

Appendix A : Graded differential algebras

A GDA (graded differential algebra) is a real (or complex) vector space **A** such that :

(i)
$$\mathbf{A} = \bigoplus_{n\in\mathbb{N}} \mathbf{A}^n$$

(where $\mathbf{A}^n$ is the linear subspace of elements of grade n)

(ii) there is a bilinear product : $a,b\in\mathbf{A} \rightarrow ab\in\mathbf{A}$ such that [1)]

$$\mathbf{A}^p\,.\,\mathbf{A}^q \subset \mathbf{A}^{p+q}$$

(iii) there is a linear operator $d : \mathbf{A} \rightarrow \mathbf{A}$, the differential of grade 1 :

$$d(\mathbf{A}^n) \subset \mathbf{A}^{n+1}$$

which is a graded derivation in the sense that :

$$d(ab) = da\,.\,b + (-1)^p\, a.db \;,\quad a \in \mathbf{A}^p \,,\, b \in \mathbf{A}$$

of square zero

$$d^2 = 0$$

We shall need particularly the two following types of GDAs.

1) the product is not necessarily associative – see below.

II/ the associative GDAs, with associative product

$$a(bc) = (ab)c \qquad , a,c \in \mathbf{A}$$

with the subclass of GCDAs (graded commutative differential algebras) for which the associative product (usually denoted by a wedge-like symbol $\wedge$) fulfills

$$b \wedge a = (-1)^{pq}\, a \wedge b \qquad , a \in \mathbf{A}^p \ , \ b \in \mathbf{A}^q$$

III/ the DGLs (differential graded Lie algebras) whose product (usually denoted by a bracket-like symbol [,] fulfills

$$[b,a] = -(-1)^{pq}\,[a,b]$$

$$(-1)^{pr}\,[a,[b,c]] + (-1)^{qp}\,[b,[c,a]] + (-1)^{rq}\,[c,[a,b]]$$

for any $a \in \mathbf{A}^p$, $b \in \mathbf{A}^q$, $c \in \mathbf{A}^r$

A is trivially graded whenever one has $A = A^\circ$, i.e. all $A_n = 0$ for $n \neq 0$. In that case a GCDA is abelian, and a DGL is a Lie algebra.

Appendix B. : The exterior derivative.

Let M be a smooth manifold, with $\mathbb{C}^\infty(M)$ the abelian algebra of smooth functions on M, and $\mathcal{X}(M)$ the Lie algebra of smooth functions on M , and $\mathcal{X}(M)$ the Lie-algebra of smooth vector fields on M. The set $\Lambda^p(M,\mathbb{C})$ of complex p-forms on M can be considered as that of alternate $\mathbb{C}^\infty(M)$-p-linear forms on $\mathcal{X}(M)$ with values in $\mathbb{C}^\infty(M)$:

$$\Lambda^p(M,\mathbb{C}) = \wedge^p{}_{C\infty(M)}\ (\mathcal{X}(M)\ ,\ \mathbb{C}^\infty(M))$$

The exterior derivative is then the linear operator d on

$$\Lambda^*(M,\mathbb{C}) = \bigoplus_p \Lambda^p(M,\mathbb{C})$$

defined as follows : for $\lambda \in \Lambda^p(M,\mathbb{C})$ and $\xi_i \in \mathcal{X}(M)$, $i = 0, 1, \dots, p$, one has

$$(d\lambda)\,(\xi_o,\xi_1, \dots, \xi_p) = \sum_{i=}^{p} (-1)^i /_{i=0}\ \xi^i \{\lambda(\xi_o, \dots, \xi_i, \dots \xi_p)$$

$$+ \sum_{0 \leq i < j \leq \alpha} (-1)^{i+j}\ \ \lambda([\xi_i,\xi_j], \xi_o, \dots, \xi_i, \dots, \xi_j, \dots \xi_p)$$

where arguments provided with a caret $\circ$ have to be omitted.

With $\wedge$ the wedge product of differential forms $(\Lambda^*(M,\mathbb{C}), d, \wedge)$ is then a GCDA called the <u>de Rham algebra</u> (or <u>de Rham complex</u>) of M.

<u>Appendix C.</u> : Chevalley cohomology of a Lie algebra **L** with values in a representation space (V,ρ).

Let A be a complex abelian algebra with unit, and let **L** be a Lie algebra over A(i.e. **L** is a left A-model and the Lie bracket [,] is A-bilinear). Let V be a left A-module carrying a representation ρ of **L** by module endomorphisms of A (i.e.

$$\rho(a\xi + b\eta) = a\rho(\xi) + b\rho(\eta)\ ,\ a,b \in A,\ \xi,\ \eta \in \mathbf{L}$$

$$\rho([\xi,\eta]) = \rho(\xi)\,\rho\,(\eta) - \rho(\eta)\,\rho(\xi)$$

The <u>Chevalley coboundary operator</u> δ is the linear operator on

$$\wedge^*_A(\mathbf{L},V) = \bigoplus_{\alpha} \wedge^{\alpha}_A(\mathbf{L},V)$$

(where $\wedge^{\alpha}_A(\mathbf{L},A)$ is the set of alternate V-valued A-α-linear forms on **L**) defined as follows : for $\varphi \in \wedge^{\alpha}_A(\mathbf{L},V)$, and $\Omega_o, \Omega_1, \dots, \Omega_\alpha \in \mathbf{L}$, we have

$$(\delta\varphi)\,(\Omega_o, \dots, \Omega_\alpha) = \sum_{i=0}^{\alpha} (-1)^i\ \ \rho(\Omega_i)\ \{\ \varphi(\Omega_o, \dots, \Omega_i, \dots, \Omega_\alpha)$$

$$+ \sum_{0 \leq i < j \leq \alpha} (-1)^{i+j}\ \ \varphi([\Omega_i, \Omega_j], \Omega_o, \dots, \Omega_i, \dots \Omega_j, \dots \Omega_\alpha)$$

One has in particular, for $\varphi \in \Lambda^{\circ}{}_{A}(\mathbf{L},V) = V$

$(\delta\varphi)\,(\Omega_0) = \varrho(\Omega_0)\,\varphi\,,$

and for $\varphi \in \Lambda^{1}{}_{A}(\mathbf{L},V)$

$(\delta\varphi)\,(\Omega_0,\Omega_1) = \varrho(\Omega_0)\,\varphi(\Omega_1) - \varrho(\Omega_1)\,\varphi\,(\Omega_0) - \varrho([\Omega_0,\,\Omega_1])$

δ is actually a coboundary operator : one shows that

$$\delta^2 = 0$$

The 1-cocycles of the corresponding cohomology are the $\varphi_1 \in \Lambda^{1}{}_{A}(\mathbf{L},V)$ fulfilling

$$\varrho(\Omega_0)\,\varphi_1(\Omega_1) - \varrho(\Omega_1)\,\varphi_1(\Omega_0) - \varphi_1\,([\Omega_0,\Omega_1]) = 0\,,$$

the 1-coboundaries being the $b \in \Lambda'_{A}(\mathbf{L}\,,\,V)$ of the form

$$b(\Omega_0) = \varrho(\Omega_0)\,v \qquad , v \in V$$

BIBLIOGRAPHIE

****

[1] D. Kastler and R. Stora. A differential geometric setting for BRS relations and Anomalies I and II.

POSTER ABSTRACTS

RENORMALIZATION OF A HIERARCHICAL FERMION MODEL IN TWO DIMENSIONS

by T.C. Dorlas, Institute for Theoretical Physics, University of Groningen Netherlands

ABSTRACT

Fermion fields on a lattice are easier to renormalize then Boson fields because the former do not have an infinite range of possible values. This becomes particularly transparent in a hierarchical model since the interaction stays strictly local in renormalizing such a model. The model considered here is obtained by dividing a square lattice into blocks, blocks of blocks and so forth at each level or scale. The blocks at the ℓ-th level will contain $2^{2\ell}$ lattice sites. The 'kinetic part' of the action is then given by $<\bar{\psi},C^{-1}\psi>$, where C is the hierarchical covariance,

$$C(x,y) = 2^{2-\rho(x,y)} \qquad (x,y \in \mathbb{Z}^2)$$

and $\rho(x,y)$ is the logarithmic distance between x and y, i.e. the scale of the smallest block containing x and y.

A local interaction between two fields ψ_1 and ψ_2 can be written as

$$V(\psi) = \sum_{x \in \mathbb{Z}^2} \nu(\psi(x))$$

with

$$\nu(\psi) = r(\bar{\psi}_1\psi_1 + \bar{\psi}_2\psi_2) + g\,\bar{\psi}_1\psi_1\bar{\psi}_2\psi_2 \,.$$

After renormalization one obtains an interaction of the same kind with

$$r' = r'(r,g)$$

$$g' = g'(r,g)\,.$$

We have here an exact transformation in a 2-dimensional parameter space.

From this transformation it can be deduced that

1. the thermodynamic limit exists;
2. there are 3 fixed points, one of which is attractive and two of which are hyperbolic;
3. in the neighbourhood of the fixed point (0,0) the model is asymptotically free.

ELABORATION

The subdivision of the lattice into blocks is illustrated in Fig. 1. In the renormalization transformation the blocks at the first level are the sites of a new lattice at which the transformed field ψ' lives. The block to which a point x of the original lattice belongs, will be denoted by $\dot{x}$.

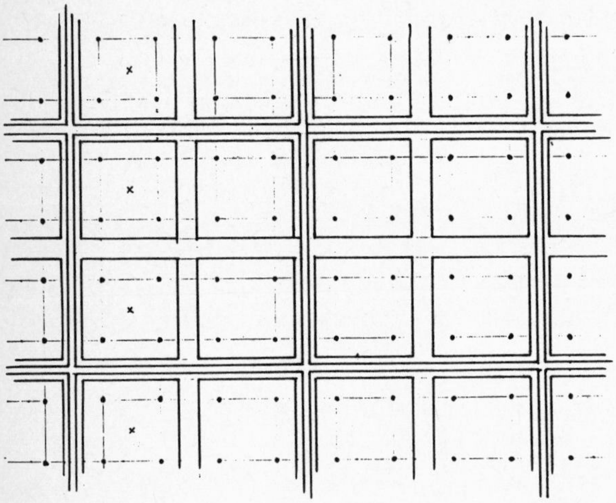

Fig.1

The covariance C can be split into a block-spin part and a fluctuation part as follows

$$C(x,y) = \frac{1}{2} C(\dot{x},\dot{y}) + \Gamma(x,y) ,$$

where

$$\Gamma(x,y) = \begin{cases} 1 & \text{if } \dot{x} = \dot{y} \\ 0 & \text{otherwise}. \end{cases}$$

This corresponds with the following splitting of the fields

$$\begin{cases} \psi(x) = \frac{1}{\sqrt{2}} \psi'(\dot{x}) + \zeta(x) \\ \bar{\psi}(x) = \frac{1}{\sqrt{2}} \bar{\psi}'(\dot{x}) + \bar{\zeta}(x). \end{cases}$$

The covariance C defines a Gaussian state on a Grassmann-algebra according to the Berezin-formula

$$\frac{\int \bar{\psi}(x)\psi(y) \exp[<\bar{\psi}, C^{-1}\psi>] \prod_z d\psi(z) d\bar{\psi}(z)}{\int \exp[<\bar{\psi}, C^{-1}\psi>] \prod_z d\psi(z)\, d\bar{\psi}(z)} =$$

$$= C(x,y) \equiv <\bar{\psi}(x)\, \psi(y)>_C \equiv \omega_C(\bar{\psi}(x)\, \psi(y)).$$

More general n-point functions are given by Wick's formula, e.g.

$$<\bar{\psi}(x_1)\ \bar{\psi}(x_2)\, \psi(y_2)\, \psi(y_1)>_C = C(x_1,y_1)\, C(x_2,y_2) - C(x_1,y_2)\, C(x_2,y_1).$$

Making use of the above splitting one can perform the functional integral in stages. One writes for a functional $F(\psi)$ of the fields ψ and $\bar{\psi}$,

$$\omega_C(F(\psi)) = \left(\omega_C' \otimes \omega_\Gamma\right)\left(F\left(\frac{1}{\sqrt{2}}\psi' + \zeta\right)\right).$$

Performing the expectation ω_Γ w.r.t. the variable ζ one obtains a new functional of ψ', and this transformation can be repeated.

The Pauli-principle forbids the local interaction of a fermion field with itself: $\psi^2 = 0$.

We therefore consider two fields ψ_1 and ψ_2 in interaction with a local potential $V(\psi) = \sum_x \nu(\psi(x))$. Here ψ is a shorthand for $(\psi_1, \bar{\psi}_1, \psi_2, \bar{\psi}_2)$.

Executing the first stage of integration over ζ one obtains the renormalized potential $V'(\psi')$

$$\exp\left[-V'(\psi')\right] = \left(\omega_\Gamma^{(\zeta_1)} \otimes \omega_\Gamma^{(\zeta_2)}\right)\left(\exp\left[-V\left(\frac{1}{\sqrt{2}}\psi' + \zeta\right)\right]\right) =$$

$$= \prod_{\dot{x}}\left(\omega_{\Gamma_0}^{(1)} \otimes \omega_{\Gamma_0}^{(2)}\right)\left(\exp\left[-\sum_{x\in\dot{x}} \nu\left(\frac{1}{\sqrt{2}}\psi'(\dot{x}) + \zeta(x)\right)\right]\right).$$

Here we have used the fact that Γ is zero between the blocks, an artefact of hierarchical models. This implies that ω_Γ is a product state $\omega_\Gamma = \prod_{\dot{x}} \omega_{\Gamma_0}$. We conclude that $V'(\psi') = \sum_{\dot{x}} \nu'(\psi'(\dot{x}))$ is also local, with ν' given by

$$\exp\left[-\nu'(\psi')\right] = \left\langle \exp\left[-\sum_{x\in\dot{x}} \nu\left(\frac{1}{\sqrt{2}}\psi' + \zeta(x)\right)\right]\right\rangle_{\Gamma_0 \oplus \Gamma_0}.$$

The expectation is simple to do because in Γ_0 all points within a block are equivalent. This means that we can replace the exponential of the sum over block-sites by the exponential per site in the power 4, the number of sites per block:

$$\left\langle \exp\left[-\sum_{x\in\dot{x}} \nu\left(\frac{1}{\sqrt{2}}\psi' + \zeta(x)\right)\right]\right\rangle_{\Gamma_0 \oplus \Gamma_0} =$$

$$= \left\langle \left(\exp\left[-\nu\left(\frac{1}{\sqrt{2}}\psi' + \zeta\right)\right]\right)^4 \right\rangle_{1\oplus 1}$$

$$= \omega_1^{(\zeta_1)} \otimes \omega_1^{(\zeta_2)} \left(\exp\left[-4r\left\{\left(\frac{1}{\sqrt{2}}\bar{\psi}_1' + \bar{\zeta}_1\right)\left(\frac{1}{\sqrt{2}}\psi_1' + \zeta_1\right)\right.\right.\right.$$

$$\left. + \left(\frac{1}{\sqrt{2}}\bar{\psi}_2' + \bar{\zeta}_2\right)\left(\frac{1}{\sqrt{2}}\psi_2' + \zeta_2\right)\right\}$$

$$\left.\left. -4g\left(\frac{1}{\sqrt{2}}\bar{\psi}_1' + \bar{\zeta}_1\right)\left(\frac{1}{\sqrt{2}}\psi_1' + \zeta_1\right)\left(\frac{1}{\sqrt{2}}\bar{\psi}_2' + \bar{\zeta}_2\right)\left(\frac{1}{\sqrt{2}}\psi_2' + \zeta_2\right)\right]\right)$$

$$= (1-4r)^2 - 4g - (2r + 2g - 8r^2)(\bar{\psi}_1'\psi_1' + \bar{\psi}_2'\psi_2') - (g - 4r^2)\bar{\psi}_1'\psi_1'\bar{\psi}_2'\psi_2'$$

$$= \left[(1-4r)^2 - 4g\right] \exp\left[-r'(\bar{\psi}_1'\psi_1' + \bar{\psi}_2'\psi_2') - g\bar{\psi}_1'\psi_1'\bar{\psi}_2'\psi_2'\right]$$

with

$$r' = \frac{2(r+g) - 8r^2}{(1-4r)^2 - 4g}$$

$$g' = \frac{g - 4r^2}{(1-4r)^2 - 4g} + \left\{ \frac{2(r+g) - 8r^2}{(1-4r)^2 - 4g} \right\}^2 .$$

This is an exact transformation in a 2-dimensional space of coupling parameters r and g.

In Fig. 2 the iteration of this mapping around the fixed point (0,0) is shown. There is another fixed point at $(-\frac{1}{4},0)$. All points in the neighbourhood of (0,0) are attracted to $(-\frac{1}{4},0)$ except for the critical line starting out from (0,0) in the direction $\begin{pmatrix}-2\\1\end{pmatrix}$. This can be seen from the linearized transformation $\begin{cases} r' = 2(r+g) \\ g' = g \end{cases}$. To first order g is a marginal parameter; taking account of all orders it is irrelevant, the model is asymptotically free.

There is a third fixed point: (1,5/2) which, like (0,0), has one attractive and one strongly repulsive direction. This means there is a possibility of obtaining a non-Gaussian theory as well.

The linearized equations around (1,5/2) are:

$$r' - 1 = 38(r-1) - 6\left(g - \frac{5}{2}\right)$$

$$g' - \frac{5}{2} = 120\ (r-1) - 19\left(g - \frac{5}{2}\right)$$

Eigenvalues $\lambda_{\pm} = \dfrac{19 \pm 3\sqrt{41}}{2}$ $\quad \left(\lambda_- = -0.1 \ , \ \lambda_+ = 38.1\right)$

Eigendirections $\begin{pmatrix}1\\6.35\end{pmatrix}$, $\begin{pmatrix}1\\3.15\end{pmatrix}$.

In Fig. 3 the direction $\begin{pmatrix}1\\3.15\end{pmatrix}$ is clearly visible. This line is the set of non-Gaussian renormalized theories.

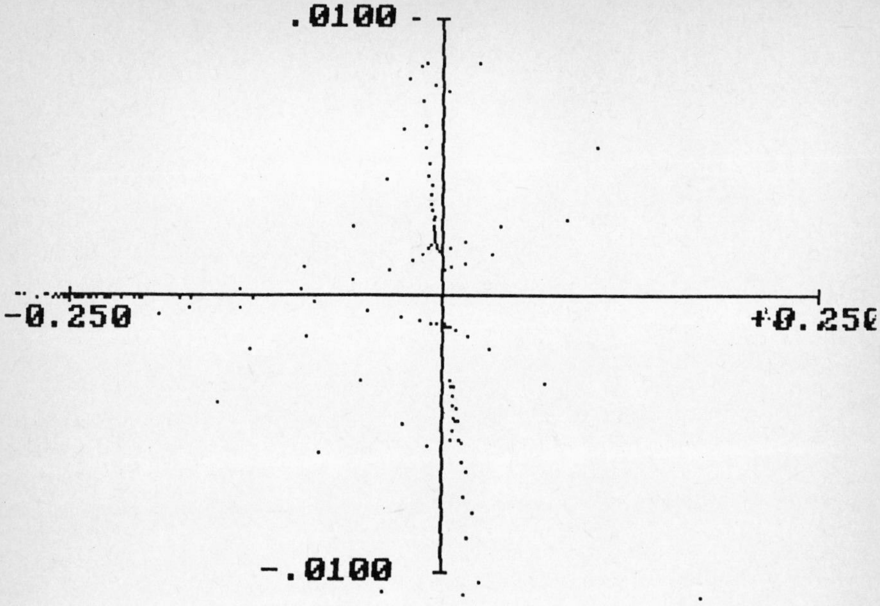

Fig.2

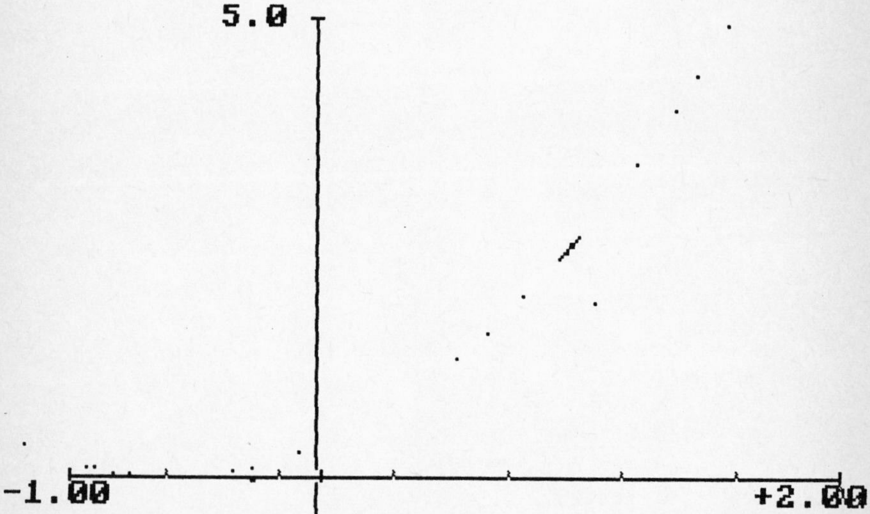

Fig.3

GRIBOV COPIES AND ABSENCE OF SPONTANEOUS SYMMETRY BREAKING IN COMPACT U(1) LATTICE HIGGS MODELS

Christian Borgs and Florian Nill
Max-Planck Institut für Physik und Astrophysik
München
B.R.D.

We consider abelian Higgs models with compact gauge group U(1) and gauge fixing term $\frac{1}{\alpha}\sum_x \cos\partial_\mu A_\mu(x)$. We note that the gauge fixed action still is invariant under the following subgroup Grib of gauge transformations

$$\text{Grib}: = \{g = e^{i\phi} | e^{i\Delta\phi} = 1\!\!1\}$$

which in particular contains the global gauge transformations Glob. If Glob is not broken spontaneously, the expectation value of the Higgs field is zero, i.e. $\langle\Phi\rangle = 0$. If also Grib is unbroken, then in addition the two point function is zero at noncoinciding points, $\langle\bar{\Phi}_x\Phi_y\rangle = 0$ for $x \neq y$.
We show that Grib resp. Glob are unbroken in the full Higgs model, whenever they are unbroken in a modified plane rotator model with action $S = \frac{1}{\alpha}\sum_x \cos(\Delta\phi)_x$, which in a way consists of the longitudinal degrees of freedom of the gauge field. We then prove by a cluster expansion that for α large Grib is unbroken (and hence $\langle\bar{\Phi}_x\Phi_y\rangle = 0$, for $x \neq y$) in any dimension $d \geq 1$. Moreover, in $d = 4$ we prove by a spin wave argument analogous to McBryan and Spencer's version of the Mermin-Wagner theorem that Glob is unbroken (and hence $\langle\Phi\rangle = 0$) for all $\alpha > 0$.

Reference:

C. Borgs, F. Nill, preprint MPI-PAI/PTh 59/85, submitted to Nucl.Phys. B.

RENORMALIZATION GROUP APPROACH FOR THE ISING MODEL AS AN APPROXIMATE SOLUTION OF THE DIAPHANTIAN SYSTEM OF EQUATIONS

S. Rabinovich
School of Physics and Astronomy
Tel Aviv University
Ramat Aviv 69978
Israel

The decimated expression for the partition function for 2D square Ising lattice (1) has been considered:

$$Z_N(K,h)=(e^h+e^{-h})^{N/2} \sum_{\substack{k_{-4},k_{-2}\\ k_2,k_4}} P_{N/2}(k_{-4},k_{-2},k_2,k_4)\left(e^{-h}\,\frac{e^{-4K+h}+e^{4K-h}}{e^h+e^{-h}}\right)^{k_{-4}}$$

$$\left(e^{-h}\,\frac{e^{-2\;+h}+e^{2\;-h}}{e^h+e^{-h}}\right)^{k_{-2}}\left(e^{h}\,\frac{e^{2\;+h}+e^{2\;-h}}{e^h+e^{-h}}\right)^{k_2}\left(e^{h}\,\frac{e^{4\;+h}+e^{-4\;-h}}{e^h+e^{-h}}\right)^{k_4}$$

The combinatorial coefficients $P_{n/2}(k_{-4},k_{-2},k_2,k_4)$ might be determined as the number of integer solutions of the system of equations:

$$\sum_{j,\ell}^{N/2}(\sigma_{j,\ell}+\sigma_{j+1,\ell}+\sigma_{j,\ell+1}+\sigma_{j+1,\ell+1})^s=4^s k_4+2^s k_2+(-2)^s k_{-2}+(-4)^s k_{-4},\quad s=1,2,3,4$$

and $\sigma_{j,\ell}=\pm 1$.

Calculating these powers, neglecting the products of four σ's and assuming that the contribution of the next nearest neighbours is the same as the nearest neighbours one (2), we obtain the relation:

$$Z_N(K,h)\simeq(e^h+e^{-h})^{N/2}\left[\frac{(e^{2K+h}+e^{-2K-h})(e^{2K-h}+e^{-2K+h})}{(e^h+e^{-h})^2}\right]^{N/8}$$

$$\left[\frac{(e^{4K+h}+e^{-4K-h})(e^{4K-h}+e^{-4K+h})}{(e^h+e^{-h})^2}\right]^{N/32} Z_{N/2}(K',h'),$$

where $K'=\frac{3}{16}\ln\left[\frac{(e^{4K+h}+e^{-4K-h})(e^{4K-h}+e^{-4K+h})}{(e^h+e^{-h})^2}\right]$

and $h'=h+\frac{1}{2}\ln\left(\frac{e^{4K+h}+e^{-4K-h}}{e^{4K-h}+e^{-4K+h}}\right)$

The critical point will be found as a root of the equation:

$K=\frac{3}{8}\ln h\cos\ 4K$ (for h=0) and $K_c=0.506981$

References:

1. S. Rabinovich, in this volume

2. A. Casher and M. Schartz (1978), Phys.Rev., B18, 7, 3440.

TRANSFORMATION OF THE TRANSFER MATRIX FOR THE ISING MODEL UNDER DECIMATION

S. Rabinovich
School of Physics and Astronomy
Tel Aviv University
Ramat Aviv 69978
Israel

Transformation of the transfer matrix for the Ising model, induced by the decimation procedure, has been considered. The consideration is based on the decimation for the number of states $W_N^{n,s}$ (1), defined as a number of integer solutions of the following system of equations:

$$\begin{cases} \sum_{<j,\ell>} \sigma_j \sigma_\ell = s, \\ \sum_j \sigma_j = n, \end{cases}$$

where $\sigma_j = \pm 1$.

Then, for example, for a 2D square lattice, the partition function equals $Z_N(\mathrm{Tr}(T^N))$, where $T_{\alpha,\beta} = (e^h)^{n'(\alpha)} (e^K)^{s'(\alpha,\beta)}$. $n'(\alpha)$ and $s'(\alpha,\beta)$ are some combinatorial functions of two spin chains α and β. The decimation leads to the expression $Z_N = \mathrm{Tr}(R^{N/2})$, where

$$R_{\alpha,\beta} = (e^h + e^{-h}) \left(e^{-h} \frac{e^{-4K+h} + e^{4K-h}}{e^h + e^{-h}} \right)^{k'_{-4}(\alpha,\beta)} \left(e^{-h} \frac{e^{-2K+h} + e^{2K-h}}{e^h + e^{-h}} \right)^{k'_{-2}(\alpha,\beta)}$$

$$\left(e^{h} \frac{e^{2K+h} + e^{-2K-h}}{e^h + e^{-h}} \right)^{k'_{2}(\alpha,\beta)} \left(e^{h} \frac{e^{4K+h} + e^{-4K-h}}{e^h + e^{-h}} \right)^{k'_{4}(\alpha,\beta)}$$

$k'_{-4}(\alpha,\beta)$, $k'_{-2}(\alpha,\beta)$, $k'_{2}(\alpha,\beta)$ and $k'_{4}(\alpha,\beta)$ are some other combinatorial functions of spin chains α and β.

For the 1d Ising chain the traces are easily calculated.

For the 2d honeycomb lattice the decimation leads to the connection with the partition function for the 2D triangle lattice

$$Z_N^{hon}(K) = (2^4 \cosh 3K \cosh^3 K)^{N/8}\, Z_{N/2}^{tr}\left(\tfrac{1}{4} \ln \frac{\cosh 3K}{\cosh K}\right),$$

has been obtained early (2) by the star-triangle transformation.

References:

1. G. Freiman (1983), private communication.
2. M.E. Fisher (1959), Phys. Rev., 113, 969.

SOME RESULTS CONCERNING THE LOCALIZATION PROBLEM IN ONE-DIMENSIONAL QUASIPERIODIC SYSTEMS

Dimitri Petritis
Institut de Physique Théorique
Université de Lausanne
1015 Lausanne
Switzerland

One dimensional quasiperiodic tiling of the real line can be obtained by using the general method of projection introduced by Duneau and Katz [1].
The obtained tiling sequence depends on the slope of the projection line.

Problem:

look for the spectrum and eigenvectors of the discrete Laplace operator on the quasiperiodic lattice.

$$\frac{\phi_{n+1} - \phi_n}{\lambda_n} - \frac{\phi_n - \phi_{n-1}}{\lambda_{n-1}} + z\phi_n = 0$$

This problem can be transformed to the following on a regular lattice with quasiperiodic potential.

$$Q_{n+1} + Q_{n-1} - 2Q_n + z\lambda_n Q_n = 0$$

Results: [2,3]

a. For a class of quasiperiodic tilings (i.e. for a class of irrational slopes) there is no localized eigenstate.
b. For tilings obtained by projecting on a line with slope equal to the golden mean, self similarity properties of the spectrum are used to show that the states are neither extended nor localized.

References:

1. M. Duneau, A. Katz, Quasiperiodic tilings, CPT Ecole Polytechnique, Palaiseau preprint
2. F. Delyon, D. Petritis, Absence of localization in a class of Schroedinger operators with quasiperiodic potential, CPT Ecole Polytechnique, Palaiseau preprint
3. J.M. Luck, D. Petritis, Phonon spectra in one dimensional quasicrystals, SPhT, CEN Saclay preprint

RANDOM WALKS ON RANDOM LATTICES

W.Th.F. den Hollander
TH Delft
Lorentzweg 1
2628 CJ Delft
The Netherlands

P.W. Kasteleyn
Instituut Lorentz
Nieuwsteeg 18
2311 SB Leiden
The Netherlands

In the past few years we have studied a model consisting of a random walk on a lattice with each point of which associated a random colour, black or white. The random walk is supposed to take place *independently* of the colouring. We have investigated stochastic properties of the sequence of consecutive colours encountered by the walker while moving through the lattice. Whereas, on the one hand, our model is simpler than that of a random walk in a random environment (where the walk *does* depend on the medium), we are, on the other hand, allowing types of colour distributions and types of random walks that are more general than those usually dealt with in the literature. Thus, for instance, we consider lattices of arbitrary dimensionality, very broad classes of Gibbs states for the colour distribution, and random walks with steps of arbitrary range. Our results concern in particular the distribution of the *interarrival times* between the successive hits of black points, and they come in the form of rigorous bounds on the moments of these interarrival times, limiting behaviour for long times, and a few remarkable identities between various moments. In spite of the assumption of independence of walk and colouring, our model can be used to describe random walks in the presence of (imperfect) traps.

references:
1. Physica 117A (1983) 179
2. J. Stat. Phys. 30 (1983) 363
3. J. Stat. Phys. 39 (1985) 15
4. Further papers are to be published

THE ROLE OF TRANSVERSE FLUCTUATIONS IN MULTIDIMENSIONAL TUNNELING

Assa Auerbach and Pierre van Baal
Institute for Theoretical Physics
State University of New York
Stony Brook, NY 11794
USA

We summarize the Path Decomposition Expansion (PDX) for the situation of multi-dimensional tunneling. The method is applied to the asymptotic form for the energy difference (ΔE) between the even and odd groundstates in a double-well geometry with non-quadratic minima. For a double-cone potential in 3 dimensions ($V(x,y,z) = [(|x|-1)^2 + y^2 + z^2]^{\frac{1}{2}}$) we find:

$$\Delta E = A\,\hbar^{5/3}\exp(-\hbar^{-1}S + \hbar^{-1/3}E_oT).(1 + O(\hbar^{1/3}))$$

with A, S, E_o and T well defined constants. For a potential with a vacuum-valley in 2 dimensions $(V(x,y) = \frac{1}{2}\hbar^{-2}(x^2-1)^2y^2)$ we find:

$$\Delta E = A\,\hbar^{2/3}\exp(-\hbar^{-1}S + \hbar^{-1/3}E_oT).$$

In both cases the quadratic approximation to the potential fails near the minima $x = \pm 1$, $y=(z=)0$ and a better understanding of transverse fluctuations is required. Instanton techniques fail dramatically in these situations.
Finally an application in field theory is discussed.

References:

1. A. Auerbach, S. Kivelson, The Path Decomposition Expansion and multidimensional tunneling, Nucl. Phys. B, to appear.
2. A. Auerbach, P. van Baal, Transverse fluctuations in multidimensional tunneling, Stony Brook preprint ITP-SB-85-28.
3. P. van Baal, On the ratio of the string tension and the glueball mass squared in the continuum, Stony Brook preprint ITP-SB-85-26.

Acknowledgement: This work was supported by MSF Grant # PHY-81-09110A-03.

SELF-SIMILAR TEMPORAL BEHAVIOR OF RANDOM WALKS IN RANDOM MEDIA

J. Bernasconi and W.R. Schneider
Brown Boveri Research Center
CH-5405 Baden
Switzerland

One-dimensional random walks with random transition probabilities are analyzed within the framework of a new real-space renormalization procedure. The renormalized random walks are shown to approach a non-random, directed walk with a limiting distribution $\rho(\tau)$, of transition times. For different types of disorder, $\rho(\tau)$ and its Laplace transform, $\tilde{\rho}(z)$, are evaluated with Monte Carlo methods. Under certain conditions the set of transition times exhibits self-similar clustering with an average fractal dimension $\nu < 1$, and the mean transition time becomes infinite. The corresponding small z behavior of $\tilde{\rho}(z)$ is of the form $\tilde{\rho}(z) \sim 1-z^{\nu}G(\ell nz)$, with G(x) periodic. It follows that the long time asymptotic behavior of the mean displacement, $\langle x(t)\rangle$, is oscillatory, i.e. $t^{-\nu}\langle x(t)\rangle$ approaches a periodic function of ℓnt.

LOOP EXPANSIONS IN THE PRESENCE OF GRIBOV COPIES

Florian Nill
Max-Planck Institut für Physik und Astrophysik
München
B.R.D.

In lattice gauge theories unnormalized expectations of gauge invariant observables in gauge fixed functional integrals are known to differ from the corresponding non-gauge-fixed quantities by a constant factor due to Gribov copies. This factor is called the gauge degree.

It is shown that performing a conventional loop expension of the gauge fixed integral by expanding around the trivial minimum of the action and forgetting about the Gribov copies gives the correct asymptotics of the non-gauge-fixed integral.

This means that this prescription automatically compensates for the gauge degree. It also means that this prescription does not give the correct asymptotics of the gauge fixed integral itself.

However as soon as external sources J are coupled to the gauge fixed systems, this prescription yields the correct asymptotics of the gauge fixed integral, since now Gribov copies are exponentially suppressed.

This implies that in the limit $J \to 0$ loop expansions of gauge fixed integrals become arbitrarily bad approximations. They fail to be asymptotic to the true gauge fixed integral as $J \to 0$. In particular they don't provide conclusive arguments concerning spontaneous symmetry breaking (see also poster contribution F. Nill, C. Borgs to these proceedings).

Reference:

F. Nill, preprint MPI-PAI/PTh 39/85, submitted to Nucl.Phys. B.

NON-EXISTENCE OF LONG-RANGE-ORDER FOR A CERTAIN 1D MODEL AND THE SOLITON PICTURE

H. Grosse
Institute for Theoretical Physics
University of Vienna
A-1090 Vienna
Austria

The Schrieffer model for polyacetylene is studied as a model of quantum statistical mechanics. The system is equivalent to a 1D quantum XY model interacting with a boson field. It is shown that the XY spin correlation functions as well as the fermion correlation functions exhibit exponential clustering for all $\beta < \infty$. This is also the case for the phonon correlation functions for β small.

Besides the standard soliton picture, solitons of MKdV type nonlinear equations are discussed. Finally recent results on the quantization in presence of external soliton fields are stated.

QUANTUM FIELDS OUT OF THERMAL EQUILIBRIUM

Wim Schoenmaker and Roger Horsley
Fachbereich Physik
Universität Kaiserslautern
Kaiserslautern
West Germany

A formulation of the non-equilibrium statistical mechanics for quantum fields is given, by following the equations of motion of finite time intervals. The step length is interpreted in various models, such as the free and weakly interacting boson field. A detailed calculation of the thermal conductivity is presented in these models and moreover an estimate is given for the thermal conductivity in quarkless L-QCD using strong coupling methods with the result:

$$K = 10^{-2}\, T_o e^{-1/T_o}, \quad T_o = \frac{T}{M_g} \lesssim 0.3$$

STATISTICAL MECHANICAL APPROACH TO THE CONSTRUCTION OF FOUR DIMENSIONAL BOSON FIELD THEORIES*

George A. Baker, J.R. and J.D. Johnson
Theoretical Division
Los Alamos National Laboratory
University of California
Los Alamos, NM 87545
USA

We prove the existence of a family of four dimensional, continuum, quantum Boson field theories which we had previously reported to be non-trivial. These theories are proven to satisfy all the usual requirements for such a field theory, except perhaps rotational invariance where no proof is given, but such invariance is consistent with currently known information. For a sufficiently strong ultra-violet cut-off the bare coupling constant expansion is proven summable. These theories are probably not asymptotically free.

We investigate some of these theories numerically by high temperature series methods, and find them to be non-trivial, e.g. non-zero two-body scattering and accessible to such an approach.

*Work performed under the auspices of the US D.O.E.

CORRELATION LENGTHS AT ZERO TEMPERATURE IN ONE DIMENSION

Jean-Jacques Loeffel
Institut de Physique Théorique
Université, Dorigny
CH-1015 Lausanne
Switzerland

Based on the notion of "stochastic dependence degree between two σ-algebras", a precise definition of the correlation length ξ for a stationary Markov chain is given. It is then seen that $exp(-1/\xi)$ is equal to the spectral radius of the difference between the stochastic matrix for the Markov chain and its "dominant projector". Finally, this result is applied to discuss the zero temperature limit of ξ for one-dimensional examples in classical statistical mechanics of equilibrium.

HORIZONTAL WILSON LOOPS IN FINITE TEMPERATURE LATTICE GAUGE THEORIES

C. Borgs
Max-Planck-Institut für Physik und Astrophysik
München
B.R.D.

It has been pointed out that the behavior of space-like Wilson loops in non-zero temperature lattice gauge theories cannot be considered as a confinement criterion. In particular they may show area law behavior in the deconfinement phase. We rigorously establish this for sufficiently high temperatures by means of a cluster expansion. For abelian theories correlation inequalities imply a stronger result: spacelike Wilson loops in the d + 1 dimensional theory (at coupling g^2 and temperature T) show area law whenever the corresponding d dimensional 0 temperature theory (with coupling Tg^2) is confining. Our results remain valid in the τ-continuum limit.

Reference:

C. Borgs: Area Law for Spatial Wilson Loops in High Temperature Lattice Gauge Theories, to appear in Nucl. Phys. B.

THE RANDOM BOND POTTS AND ASHKIN-TELLER MODELS

P.L. Christiano and S. Goulart Rosa Jr.
Instituto de Fisica e
Quimica de Sao Carlos
Caixa Postal 369
13.560 Sao Carlos S.P. Brasil

A unified treatment to study classical spin systems on a Cayley Tree is presented. It is based on the relationship between the properties of the system defined on the regular tree with those of the system put on a related hierarchical tree, the so-called closed asymmetric tree. Exploring the hierarchical nature of the c.a.t. we implement a renormalization group procedure to obtain the thermodynamic properties of the regular as well as random bond Ising, Potts and Askin-Teller models. The study of these systems is reduced to a mapping problem.

SPECTRAL LINE SHAPES IN QUANTUM MARKOV SCATTERING

H. Maassen
University of Delft
Delft
Netherlands

It is shown how in many circumstances a scattering operator can be associated to a quantum Markov process in a natural way. As an example the two-level atom at positive temperature is treated. The transition probabilities between its levels and the damping of its coherence define a Markovian time evolution on it. The associated scattering operator in a Bose field turns out to display a temperature broadened spectral line.
This is a typically non-quasifree feature: it does not occur either in the harmonic oscillator coupled to a Bose field or in the two-level atom coupled to a Fermi field.

EXACT BOGOLIUBOV LIMITS FOR THE BASSICHIS-FOLDY MODEL AND CONTINUED FRACTIONS

D. Masson
Mathematics Department
University of Toronto
Toronto, M5S 1A1

Revolvent convergence and continued fractions are used to derive competing Bogoliubov type limits for the Bassichis-Foldy model. The critical values of coupling are shown to be associated with either transitions from a discrete to a continuous spectrum or a transition from one Bogoliubov type limit to the other.